Differentiated Lymphocyte Functions

Progress in Allergy

Vol. 28

Series Editors

Kimishige Ishizaka, Baltimore, Md.; *Paul Kallós,* Helsingborg;
Byron H. Waksman, New York, N.Y., and *Alain L. de Weck,* Bern

S. Karger · Basel · München · Paris · London · New York · Sydney

Differentiated Lymphocyte Functions

Volume Editor
Alain L. de Weck, Bern

Contributors
J.R. Battisto, Cleveland, Ohio; *J.T. Epplen,* Duarte, Calif.; *U. Hämmerling,* New York, N.Y.; *T. Hozumi,* Duarte, Calif.; *K. Karre,* Kingston, Ont.; *R. Kiessling,* Stockholm; *T. Matsunaga,* Duarte, Calif.; *S. Ohno,* Duarte, Calif.; *D.M.V. Parrott,* Glasgow; *N.M. Ponzio,* Chicago, Ill.; *J.C. Roder,* Kingston, Ont.; *P.C. Wilkinson,* Glasgow

47 figures and 26 tables, 1981

S. Karger · Basel · München · Paris · London · New York · Sydney

Progress in Allergy

Vol. 21: XIV +408 p., 48 fig., 33 tab., 1976. ISBN 3-8055-2342-4
Vol. 22: XII +220 p., 17 fig., 18 tab., 1977. ISBN 3-8055-2419-6
Vol. 23: XIV +366 p., 32 fig., 18 tab., 1977. ISBN 3-8055-2665-2
Vol. 24: XIV +366 p., 29 fig., 5 tab., 1978. ISBN 3-8055-2781-0
Vol. 25: XIV +242 p., 32 fig., 16 tab., 1978. ISBN 3-8055-2849-3
Vol. 26: XVIII +238 p., 17 fig., 40 tab., 1979. ISBN 3-8055-2934-1
Vol. 27: XII +276 p., 26 fig., 19 tab., 1980. ISBN 3-8055-3053-6

National Library of Medicine, Cataloging in Publication
Differentiated lymphocyte functions
Volume editor, Alain L. de Weck; contributors, J.R. Battisto ... [et al.]. – Basel, New York, Karger, 1981.
(Progress in allergy; v. 28)
1. Lymphocytes – immunology 2. Lymphocytes – physiology I. Battisto, J.R. II. Weck, Alain L. de
W1 PR666F v. 28/WH 200 D569
ISBN 3-8055-1834-X

Drug Dosage

The author and publisher have exerted every effort to ensure that drug selection and dosage set forth in this text are in accord with current recommendations and practice at the time of publication. However, in view of ongoing research, changes in government regulations, and the constant flow of information relating to drug therapy and drug reactions, the reader is urged to check the package insert for each drug for any change in indications and dosage and for added warnings and precautions. This is particularly important when the recommended agent is a new and/or infrequently employed drug.

Printed in Switzerland by Werner Druck AG, Basel
ISBN 3-8055-1834-X

Contents

Natural Killer Cells

Prog. Allergy, vol. 28, pp. 1–7 (Karger, Basel 1981)

Introduction

The explosive increase in knowledge and the vitality which have characterized modern immunology during the past two decades appear to continue, unabated, during the beginning of the 1980s. At basic and applied levels, new analytical techniques and procedures of immunological manipulation enable immunologists to trespass former classical boundaries and make the impact of immunology increasingly felt not only in human medicine or general biology but also in veterinary medicine, botanics, zoology, paleontology or agriculture. Not even outer space has remained safe from immunologists! This explosion of activities is unfortunately accompanied by a corresponding inflation in the printed word. It becomes increasingly difficult to keep on the current and to remain abreast with new developments. At the present time, it appears that only economic pressure will ultimately set some limits both to the printing of information and to the means of generating it.

Under such circumstances, the need for competent reviews in specialized fields of immunology has become more pressing than ever.

Progress in Allergy has celebrated its 40th birthday in 1979 and is therefore almost an ancestor by current immunological criteria. Despite the increasing concurrence of some younger and dynamic series, it will continue to present authoritative reviews on the most significant developments in the field, as it did in the past 27 volumes. However, in order to meet our readers requirements for systematic information, a special effort will be made in the future to devote individual volumes to closely related topics.

In the present volume, subtitled *Differentiated Lymphocyte Functions,* the authors address themselves to some of the most fascinating problems of modern immunology. Indeed, the polymorphism and diversity of the elements constituting the immune system and insuring its regulation make the immune apparatus one of the most intricate biological systems known to man. The amazing diversity of immunoglobulin molecules as well as the increasing heterogeneity of cellular subsets involved in immune reactions and of their regulating factors become slowly unraveled.

Whether the generation of antibody diversity is the fruit of some evolution due to natural selection or whether the immune system, like Prometheus, has been gifted with foresight and has anticipated future needs is the topic of the first review by *Ohno, Epplen, Matsunaga and Hozumi.* In an almost philosophical exposé on the evolution of the immune system, *Ohno,* shows us how the ability to generate antibodies, even against new man-made molecules, cannot have been conferred to higher vertebrates by natural selection only. Indeed, the immune system has anticipated future needs and its apparent foresight is due to random generation of a large variety of antigen-binding sites, coupled to the possibility of clonal expansion as needs arise. However, 'in a Darwinian world of evolution by natural selection, foresight appears to be a forbidden fruit'. The curse of Prometheus laid upon the immune system is that V genes are also directed against self-antigens and are thereby beyond the reach of natural selection. Therefore the need arose of differentiating T cells into helper (against pathogenic external antigens) and suppressor (against self-antigens) T cells. The interaction of the immune system with pathogenic agents in its environment may sometimes evolve in a hide-and-seek game. Viruses and bacteria obviously attempted to escape from immunological 'persecution' in disguising their antigens as components of the host. However, a foreseeing immune system can keep up with the rapid adaptive changes of viruses and bacteria by countering such attempts through the polymorphism of its histocompatibility antigens.

The mechanisms by which V and C gene coding sequences may fuse appear to be the central core of Promethean foresight in the immune system. The pretranscriptional fusion mechanism which unites V and J coding sequences seems to have evolved from a ubiquitous splicing mechanism. Equally important to explain the generation of diversity is an understanding of the posttranscriptional splicing mechanisms by which intervening sequences from the nuclear transcripts of genes are eliminated and mature cytoplasmic messenger RNAs generated. *Ohno* et al. also point out in an almost teleological fashion how the same V_H gene may be used for building up immunoglobulin heavy chains on the one hand or T cell receptors on the other. The interactions of classical immunology with modern molecular biology and the new possibilities to achieve V gene splicing will predictably open not only theoretical but also practical new avenues during the next decade. No wonder that some of the greatest minds in molecular biology become increasingly busy with immunological models!

In the second chapter of this volume, *U. Haemmerling* addresses himself to the differentiation of B lymphocytes in lineage development per se as well as during the immune response. Several features of the lymphoid system make it an attractive subject in differentiation research. Among these are the accessibility of the system, the predominance of immature cells in some lymphoid organs, the increasing availability of immunogenetic markers, the fact that several mutant animal strains or congenital diseases offer examples of blocked differentiation, the possibility to raise monoclonal lymphoid lines and finally the fact that a number of biochemical and biological functions of lymphoid cells, present only at a certain stage of differentiation can be measured. Among the currently recognized inducible markers of B lymphocytes are various membrane immunoglobulins, Ia antigens, Lyb 2 and Lyb 5 antigens, plasma cell membrane antigens and C3 or Fc receptors. Such markers have permitted to distinguish at least five sets of inducible B cells which can be characterized by their phenotypes with regard to at least seven cell membrane markers. The identification of such B lymphocyte subsets implies that maturation of the B cell requires at least five separate differentiation steps. In several instances, it has been shown that the mechanism of differentiation induction is directly related to the cyclic nucleotide metabolism. An increase in cyclic AMP appears responsible for induction while a raise in cyclic GMP is inhibitory. This, however, is not true for the final differentiation in plasma cells and the induction of the PC 1 antigen. The differentiation of B cells may be considered, therefore, at three levels: (1) in generic terms, i.e. as differentiation of any tissular cell; (2) as diversification in terms of antibody specificity, and (3) in terms of changes in immunoglobulin classes of the same specificity (IgM/IgG switch).

Also concerned with the differentiation and diversity of lymphoid cells, but this time of T lymphocytes, is the chapter by *Battisto and Ponzio* on autologous and syngeneic mixed lymphocyte reactions and their immunological significance. Autoimmune reactivity, following the lead of *Ehrlich* and *Burnet,* was for a long time considered as a very abnormal phenomenon and as an aberration of the immune system. In recent years, autoimmune recognition has been shown to be a normal event and even required for intercellular cooperation. The capacity of neonatal thymic cells to proliferate when co-cultured with syngeneic adult spleen cells is described as 'isogeneic lymphocyte interaction of type I'. A second type of isogeneic lymphocyte interaction may be observed when both responder and stimulator cells are derived from mice donors older than 10 weeks. In this case, the responsive cells appear to be restricted to lymph nodes and to have developed either following inherent stimuli or as a consequence of exposure to environmental antigens cross-reactive with self-determinants. In both instances, the responder cells are T cells and the stimulating cells appear to be B lymphocytes carrying antigens considered as murine differentiation antigens. Such differentiation antigens appear to be closely associated with Ia antigens and they may possibly function as receptors for T cell helper factors, such as the so-called T cell replacing factor (TRF).

The murine isogeneic lymphocyte interactions have a counterpart in man, referred to as the 'autologous mixed lymphocyte reaction'. Several groups have identified T cells which are capable of proliferating when co-cultured with mitomycin-treated or X-irradiated autologous lymphoid blastoid cell-lines, mitogen-induced blast cells or normal peripheral blood mononuclear cells. The proliferation of autologous responder cells may also represent recognition of and response to altered cell membrane determinants. Such cell recognition at the T cell level appears to be an integral component of normal immune responses. The data published suggest that monocytes as well as B lymphocytes and 'null' cells can serve as stimulators in such autologous mixed lymphocyte reactions. Identification of a common stimulatory determinant on these various cells would explain their ability to activate autologous T cells. The consequences of an autologous mixed lymphocyte reaction may be the generation of cytotoxic and suppressor cell activities, as well as that of helper cells. The recent progress made in characterizing and separating various subsets of human T cells possessing different functional activities is bound to bring more information on this aspect in the near future. The study of autologous mixed lymphocyte reactions is only beginning; the use of assays such as proliferation, induction of cytotoxicity or of suppressor activity will probably be superseded soon by more rapid tests in the evaluation of lymphocyte functions.

Another category of lymphoid cells which has received increasing attention in recent years are the so-called natural killer (NK) cells, reviewed in this volume by *Roder, Karre and Kiessling.* For a long time, NK cells were characterized solely in terms of overall cytotoxic activity of a given lymphoid cell population on some sensitive target cells such as the murine YAK cell line. Recent developments, such as a technique to assess the binding of NK cells to their target (permitting thereby their enumeration at a single cell level) or the description of specific markers such as an asialo-gangliotetraosyl ceramide has slowly brought the NK cell into a category which can be identified positively. NK cells are therefore no longer only 'negatively characterized' by the fact that they possess no membrane immunoglobulin, no C3 receptors, no Ia antigens, and probably also no Fc receptors (or only Fc receptors of a very low avidity). Although natural killer cells bear the Ly 5 marker like thymocytes and mature T cells, several arguments reviewed by the authors speak against a T cell lineage. In the view of *Roder* et al., there are also differences in the genetic make-up of monocytes and of NK cells. Accordingly, the current opinion frequently expressed that NK cells are in fact promonocytes should not be considered as definitive; it is still not excluded that NK cells might represent a heterogeneous population. In terms of genetic control, the NK activity appears to be only weakly linked to the H-2 complex. Two different H-2-linked loci and probably several other non-H-2-linked genes seem to be involved in the control of NK activity. This control is not specific and appears to be an overall phenomenon unrelated to the target cell type. NK cells appear to function in two ways: (1) to lyse antibody-coated tumor targets via an Fc receptor and (2) to effect a selective destruction of tumor cells through some still unknown recognition system. The well-recognized fact that interferon is influencing and moderating NK cell activity has raised much interest in these cells recently. There appears to be some correlation between in vitro NK assays and in vivo resistance to tumors, suggesting that NK cells are involved in surveillance against neoplasms. It is also to be noticed that NK cell activity declines with age. A potentially very useful model to assess the biological role of NK cells is the description by the authors of a special mutant (beige mutant of C57 Bl mice). Such mice are deficient in NK activity, while their T cells, promonocytes, macrophages and B cells appear to be normal.

The last chapter of this volume by *Parrot and Wilkinson* is devoted to one of the essential functions of lymphocytes whithin the lymphoid system, namely their capacity to move within the organism and to translocate from one lymphoid organ to another, throughout lymph, blood and tissues. Due to their capacity to constantly move and translocate, the lymphocytes ensure the function of the lymphoid system as a single organ, although formally separated into thymus, bone marrow, spleen, lymph nodes, gut-associated tissue and other localized lymphoid patches. The amount of time which lymphoid cells spend as free-floating particles in the blood or in the lymph is very short in comparison to the time spent in moving through the tissues. Assessment of lymphocyte locomotion is crucial in understanding their migration and recirculation, their constant repositioning within lymphoid tissues and their selective accumulation at particular sites, following the impact of inflammatory agents and/or the elicitation of immune reactions.

Lymphocyte locomotion, as studied in vitro, reveals the role of morphological elements such as microfilaments and the uropode in lymphocyte locomotion, the changes in membrane and surface properties of various types of lymphocytes following differentiation and/or activation as well as other phenomena such as chemokinesis and chemotaxis. However, studies of lymphocyte locomotion in vitro may have only limited relevance to events occurring in vivo. For example, lymphoblasts which are vigorously mobile in vitro appear for the most part to be noncirculating cells. Through the injection of labeled lymphoid cells, the traffic of lymphocytes from blood to lymph and their migration through various organs such as lungs, liver or spleen have been studied. It has become obvious that the distribution of lymphocytes through lymphoid organs and also in nonlymphoid tissues is a consequence of highly selective processes which are not yet fully understood. Lymphoid cells appear to recognize 'home' when they arrive there. This is most likely due to an interaction with underlying types of reticulum cells. Among the factors which influence lymphocyte extravasation and distribution in tissues and extravascular spaces are: vascular permeability, blood flow, antigen-directed locomotion and possibly chemoattraction. Another factor possibly responsible for selective accumulation of lymphocytes in some sites, for example of mucosal lymphocytes, is antigen recognition. Stimulation and activation of lymphocytes profoundly affect their locomotor properties, in particular their adherence to reticulum cells and macrophages; this may explain selective accumulation. The present review has the obvious merit to point to the important problem of lymphocyte distribution throughout the tissues, which has received comparatively less attention than phenomena leading to activation and function of lymphoid cells analyzed as isolated populations in vitro. What is the use of fostering activation of specific isolated lymphocytes to attack intruding bacteria, viruses or parasites unless these lymphocytes are able to get at the intruder?

Some of the introductions to previous volumes of *Progress in Allergy* represented in themselves a review of the field at a particular time, pointing out the major highlights having occurred in allergy or immunology between the publication of two volumes. Both the pace at which immunological research is currently proceeding and the increase in the number of scientists working in the field preclude such ambition. However, in relation to the general topic of this volume, it should be emphasized that two relatively recent technical developments will profoundly affect our understanding of lymphoid cell differentiation and the characterization of lymphoid cell subsets. The possibility to grow lymphoid cell clones in vitro on a long-term basis, thanks to the addition in culture of various cell growth factors makes now available in a homogeneous form various cell populations which may be analyzed at the morphological, immunochemical, and functional levels [1]. In addition, the possibility to raise monoclonal antibodies by the hybridoma technique following immunization with homogeneous populations of lymphoid cell subsets provides new and extremely valuable tools. In this way, it becomes possible to recognize membrane markers which may be expressed on some subset only at a given stage of differentiation or which may be the permanent marker of a given subset at various stages of differentiation and/or activation [2]. There is little doubt that progress along these lines will considerably enhance our understanding of the complex cellular interactions regulating the immune system.

References

1 Schreier, M.H.; Iscove, N.N.; Tees, R.; Aarden, L.; Boehmer, H. von: Clones of killer and helper T cells: growth requirements, specificity and retention of function in long term culture. Immunol. Rev. *51:* 315 (1980).

2 Reinherz, E.L.; Kung, P.C.; Goldstein, G.; Schlossman, S.F.: Separation of functional subsets of human T cells by a monoclonal antibody. Proc. natn. Acad. Sci. USA *76:* 4061 (1979).

Prof. Dr. A.L. de Weck, Institute for Clinical Immunology, Inselspital, CH-3010 Bern (Switzerland)

Prog. Allergy, vol. 28, pp. 8–39 (Karger, Basel 1981)

The Curse of Prometheus Is Laid upon the Immune System

Susumu Ohno, Joerg T. Epplen, Takeshi Matsunaga and Toyoharu Hozumi

Division of Biology, City of Hope Research Institute, Duarte, Calif., USA

Contents

Promethean Foresight of the Immune System in an Epimethean World of Evolution by Natural Selection

Zeus assigned the task of creating living creatures to the two Titan brothers, Epimetheus and Prometheus. Upon the completion of their task, Zeus punished not Epimetheus for his hindsight, but Prometheus for his foresight. The moral of the above Greek mythology is very profound, for in this Darwinian world of evolution by natural selection, foresight is indeed the forbidden fruit. Consider the following example: Almost any colony of domesticated rabbits contains a good number of individuals whose occular pupils do not dilate in response to locally administered atropine. The above is due to the polymorphism of a certain esterase locus that confers an atropine esterase allele to certain individual rabbits as already known in 1943 [29]. A high gene frequency of this atropine esterase allele in nearly all of the domesticated rabbit colonies apparently reflects the species' past, when the accidental consumption of belladonas and other nightshades in the field must have been rather common. The possesser of atropine esterase no doubt enjoyed a selective advantage under the circumstances. In this manner, an encounter with atropine-containing plants caused one distinctive evolutionary change in the rabbit species. What if the hand of divine guidance foresaw an eventual encounter by the rabbit species of habitats rich in belladonas and other nightshades and endowed every individual rabbit with an atropine esterase locus rather than an allele? This foresight would have precluded the rabbit species from undergoing one evolutionary change noted above. Thus, we come to realize that foresight and adaptive evolutionary changes are incompatible.

All of the organisms of this earth created together a biological world dominated by six carbon *d*-sugars and 20 *l*-amino acids. Needless to say, natural selection made certain that all the metabolic and catabolic pathways were specifically tailored to deal with the above world. For this very reason, if we were to be divorced from our evolutionary kin and placed in another biological world of seven carbon *l*-sugars and *d*-amino acids, our highly refined metabolic, as well as catabolic enzymes would become totally useless, thus revealing an inherent limitation in Darwinian evolution by natural selection.However, we can be reasonably certain that the immune system of higher vertebrates would continue to perform well in that alien world, for this system has demonstrated ample capabilities to deal with all sorts of odd, man-made molecules that are the products of post-World War II chemistry. One can be reasonably certain that this biological world did not experience most of these man-made molecules until the present. Thus, the ability to generate antibodies directed against them was not conferred to higher vertebrates by natural selection. The immune system has apparently risen above the Epimethean world of evolution by natural selection, as it gives us the impression of having evolved to anticipate future needs [6, 19]. As a difference of several orders of magnitude exists between the generation time of parasitic viruses and bacteria and that of vertebrate hosts, in an evolutionary race for adaptive changes, one expects these parasites to invariably gain the upper hand. An anti-viral antibody selected on the basis of its past merit would be useless against a new mutant strain of that virus. The need for apparent foresight is unquestionably clear.

A polymorphic population that is large enough is bound to contain varying numbers of mutant individuals that can cope with various altered environments. It is an attendant selection of a minority as the foundation stock in a new environment that gives us an illusion of purposeful foresight in the evolutionary process. In the immune system, the ability to randomly generate a large variety of antigen-binding sites (variable region sequences) has been transferred from a population to each individual, and the expression of each antigen-binding site by a particular cell type has been made clonal. Accordingly, almost all individuals came to possess a variety of minority clones that can cope with diverse antigens not yet in existence; the encounter with each new antigen causing rapid expansion of a clone or clones that can cope with it. It was in this manner that the immune system was conferred with apparent foresight.

Prometheus was punished by Zeus for his foresight, so has the immune system been penalized for its audacity in acquiring Promethean foresight. In the past, the black plague and smallpox periodically ravaged the Eurasian continent. Yet these great epidemics should have exerted no selective pressure upon variable region genes of the human genome for the simple reason that the ability to generate antigen-binding sites directed against *Pasteurella pestis* and *Variola major* was made inherent in nearly all human individuals long before their encounter with these virulent pathogens. While it is true that during the first encounter with a new virulent pathogen, individuals who happen to be endowed with the greatest number of relevant clones enjoy a decisive selective advantage, such cannot be inherited as a stable trait. Unequal crossing-over constantly occurring among numerous variable region genes would equalize the above kind of advantages and disadvantages in relatively short order [23]. Since V (variable region) genes of the mammalian immune system, placed upon the Promethean pedestal, are beyond the reach of natural selection, autoimmune diseases, no matter how severe, should not cause the elimination of anti-self V genes from the genome for the simple reason that random mutations which affect numerous V genes would constantly be recreating anti-self-V genes in the genome.

Such then is the curse of Prometheus laid upon the immune system for the impertinence of its foresight. Were it not in the nature of V genes to endow nearly every individual with relevant clones directed against any pathogen as well as against any self-antigen, a subsequent development of *helper* and *suppressor* T cells would have been superfluous, for *helper* T cells regulate the speed of expansion of preexisting clones against pathogens, whereas *suppressor* T cells prevent the expansion of anti-self clones already in existence.

A Viral Attempt to Disguise Their Antigens as the Host's Selves is Countered by Polymorphism of MHC Antigens

The development of the Promethean immune system that can keep up with rapid adaptive changes of viruses and bacteria must have forced these parasites to exercise the one remaining option in escaping from immunological persecution by the host. The sound strategy devised was to disguise their antigens as components of the host. Although smallpox epidemics should have exerted no appreciable influence upon the V genes of the human genome, as already noted, those unfortunate individuals whose blood group and/or other plasma membrane antigens too closely resembled relevant antigens of *Variola major* must have been selectively eliminated by such epidemics. The above tells us that natural selection by favoring genetic polymorphism of host plasma membrane antigens succeeded in placing the Promethean immune system under its surveillance, albeit indirectly. Since the list of pathogens is endless, the best that each species can hope for by generating genetic polymorphism of plasma membrane antigens is the survival of a majority in a population after periodical encounters with various virulent pathogens. Extraordinarily extensive genetic polymorphism of MHC antigens, observed in all mammalian species studied, as well as the invariable presence of duplicated gene loci (e.g. H-2D, H-2K, Ia-1 to Ia-5 of the mouse) in the MHC region of the mammalian genome then indicates that these MHC antigen are the principal tool with which the host immune system attempts to distinguish disguised selves of pathogens from true selves.

After every conceivable experiment has been done and all the possible alternatives have been explored, one often finds that the original article on the subject said it all. We found this to be true of the X-inactivation mechanism that operates in mammalian females [16]. Our feeling is that the same is also going to be the case with MHC restrictions in the cell-mediated immune system [34]. The original interpretation was that cytotoxic T cells do not recognize viral antigens per se on the infected target plasma membrane, but specific perturbations of self-MHC antigens caused by their association with viral antigens – 'altered self' phenomenon. It has been proposed that much of the germ line V genes of the immune system are directed against all the allelic MHC antigens of the species [13]. If by viral perturbations self-MHC antigens come to resemble allo-MHC antigens specified by other MHC alleles, infected individuals can readily mobilize anti-allo-MHC V genes in their genome for the persecution of a virus. Thus, the above-noted view on the germ line V genes and the original 'altered self' concept are quite complementary. Viewed in this light, violent graft rejections caused by MHC incompatibilities are but somewhat inadvertent consequences of the immune system's choice of MHC antigens as the principal tool of discrimination between viral disguised pseudoselves and true selves.

In order to significantly alter the configuration of self-MHC antigens, viral antigens as well as H-Y and other minor histocompatibility antigens must engage in an intimate association with each other at one time or another in the life cycle of cells. We can think of only two reasons for such an invariable association. It should be noted that the two are not mutually exclusive; (1) β_2-MHC antigen dimers serve as the plasma membrane anchorage site of testis-organizing H-Y antigen and other organogenesis-directing plasma membrane antigens. Accordingly, viruses found it to their advantage to displace these functionally important host antigens with their own [20]. The fact that in the absence of its proposed anchorage sites, Daudi β_2m(-), HLA(-) human male Burkitt lymphoma cells failed to maintain H-Y antigen on their plasma membrane supports the above notion on the role of β_2m-HLA antigen dimers [1, 10]. Instead Daudi cells excreted H-Y antigen to the culture medium [18]. (2) β_2m-MHC antigen dimers function as the vehicle of transport toward the plasma membrane for those antigens which would otherwise remain in the cellular interior because their *messenger* RNAs are translated by free ribosomes. This would apply to those viral antigens that are found as much in the cellular interior as on the plasma membrane. Conversely, some of the antigens equipped with HyL, therefore, whose *messenger* RNAs are translated by membrane-bound ribosomes, may be kept on the plasma membrane by β_2m-MHC antigen dimers instead of being excreted. With regard to H-Y antigen, the findings on Daudi cells can also be used to support the above notion. If self-MHC antigens can receive lasting imprints of alien antigens during their association in the cellular interior, a continued association after their emergence on the plasma membrane appears irrelevant. Yet it appears that if given a choice, T cell receptors invariably recognize that MHC antigen which remains in closest association with an alien antigen on the target plasma membrane. It may be that the persistent association is a reflection of a profound alteration inflicted on that MHC antigen.

Indeed, on the plasma membrane of F_1 H-2^b/H-2^d mouse cells, Friend MuLV antigen preferentially remains in association with H-$2D^b$ as shown by *Bubbers and Lilly* [4], and it is this (MuLV+ altered H-$2D^b$) antigen complex which is recognized by cytotoxic T cell receptors [2]. Along a similar vein, it has been shown that in the presence of testosterone, H-Y antigen on the plasma membrane of male H-2^b mice associate with H-2^b but not with H-$2K^b$ [11]. Not surprisingly, in the presence of H-$2D^b$ antigen anti-H-Y T cell receptors ignore all other H-2 antigens and recognize (H-Y+ altered H-$2D^b$) antigen complex [31].

Inasmuch as MHC antigens as one class of molecules must be endowed with a finite number of conformational alternatives and most of these alternatives must be manifesting themselves in the form of allelic MHC antigens in a polymorphic population, it is really no surprise if one allelic MHC antigen altered by an alien antigen comes to resemble another allelic MHC antigen in the unaltered state. Indeed, when anti-H-Y cytotoxic T cells derived from female H-2^b mice were cloned and then propagated, aside from lysing male H-2^b cells that were their intended targets, they also indiscriminately lysed male and female cells alike that carried H-$2D^d$ antigen [3]. It would appear that the (H-Y+ altered H-$2D^b$) antigen complex comes to resemble unaltered H-$2D^d$ antigen as schematically illustrated in figure 1. Inasmuch as nature does not practice the transplantation of allografts, the above confusion is of no consequence.

The fact that anti-(H-Y+ altered H-$2D^b$) cytotoxic T cells made no apparent distinction between male and female H-$2D^d$ cells offered as their targets indicated that either H-Y antigen does not associate with H-$2D^d$, or by association, H-$2D^d$ antigen alters but a very little. Indeed, female mice of the H-2^d haplotype have long been known as poor responders against H-Y antigen. As shown in figure 1, when anti-H-Y female cytotoxic T cells that preferentially recognized the H-Y+ H-$2K^d$) antigen complex were raised in vitro, they lysed nearly half as much of the self (H-2^d female cells) as their intended H-2^d male target [17]. It would appear that by association, H-Y antigen altered H-$2K^d$ antigen just enough to provoke the anti-H-Y cell-mediated immune response, but the resulting anti-H-Y T cell receptor confused unaltered H-$2K^d$ and/or H-$2D^d$ antigens with the (H-Y+ H-$2K^d$) antigen complex. In vivo, the role of *suppressor* T cells might be to prevent such a self-destructive, cell-mediated immune response.

What happens to those unfortunate individuals whose MHC antigens are not altered in the slightest by the viral antigen, for they should have no apparent cell-mediated immune defense against that virus. The presence of multiple gene loci for ubiquitously expressed MHC antigens (H-2D, H-2K, H-$2Q_a$ loci in the mouse; HLA-A, -B, and -C loci in man) greatly

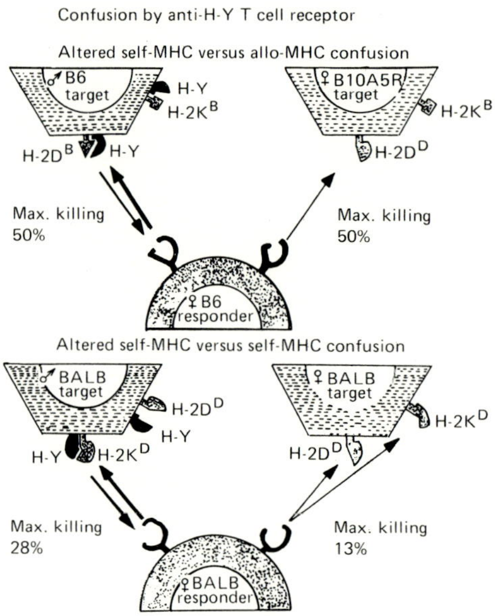

Fig. 1. Anti-H-Y cytotoxic T cells of the mouse confuse altered self-H-2 antigens with unaltered allo- or self-H-2 antigens; the cause of autoimmune response? *Top:* Anti-H-Y cytotoxic female H-2^b T cells confuse (H-Y+ altered H-$2D^b$). Antigen complexes with unaltered allo-H-$2D^d$, thus, indiscriminately lysing male and female H-$2D^d$ cells alike presented as the targets. *Bottom:* In vitro raised anti-H-Y cytotoxic T cells of H-2^d female mice have difficulty in distinguishing (H-Y+altered H-$2K^d$) antigen complexes from unaltered H-$2D^d$ and/or H-$2K^d$ antigens of the self, thus lysing nearly half as much female H-2^d cells (self) as their proper male target.

reduces the probability of the above happening. Furthermore, a saving grace for those individuals is found in the fact that the regulation of humoral antibody response by B cells via *helper* and *suppressor* T cells involves quite another class of MHC antigens in its MHC restriction; I-A to I-J antigens specified by Ia-1 to Ia-5 genes, in the case of mice, and HLA-DW antigens in the case of man. The above reduces the probability of total immune response failure by an individual against any pathogen to practically nil.

What if the immune system was more prudent and dared not to pluck the forbidden fruit of Promethean foresight? The host genome would have remained free of anti-self V genes because of efficient natural selection.

Consequently, everlasting contests between viral and bacterial pathogens and the host immune system would have been fought under an entirely different set of strategies and counterstrategies. It is then likely that we would have seen neither extensive polymorphism of MHC antigens nor the immune system's preoccupation with the differential recognition of these polymorphic MHC antigens. Although the immune system under the yoke of natural selection cannot possibly keep up with the dazzling arrays of new antigenic determinants that parasitic bacteria and viruses are capable of presenting, it is well to remember that it is not in the parasite's interest to be totally victorious, for by exterminating the host, they exterminate themselves.

The V + C Coding Sequence Fusion Mechanism as the Central Core of Promethean Foresight

Inasmuch as the immune system owes its Promethean characteristics to the capacity of each individual to generate a great variety of antibodies and T cell receptors, it was earlier realized that it is necessary for the mammalian genome to contain a rather larger number of V genes (coding sequences) specifying antigen-binding portions of antibodies; the total number being determined by the extent to which somatic mutations contribute to the generation of antibody diversity. In the mammalian genome, three unlinked clusters of genes specify Ig (immunoglobulin) κ-class light chains, λ-class light chains and several classes of heavy chains. Whether or not there exists yet a fourth cluster of genes for T cell receptors shall be discussed later. Since within each gene cluster noted above, a large number of V genes are separated by a million or more base pairs of void from the small number of C (constant region) genes, it was quite clear for sometime that in order to specify each Ig (immunoglobulin) transcript containing one each of the V and C coding sequences, one of the numerous Vs must somehow be brought to the proximity of the C. In the case of κ-class light (L) chains, the above essential has recently been shown to be accomplished by a pretranscriptional coding sequence fusion between one of the Vs and one of the several short J coding sequences; each specifying 14 or so amino acid residues [27, 30].

Since these five or so short Js are likely to be tandem duplicates of HyL of the C when the C was specifying an independent β_2-microglobulin-like peptide [22], the cluster of them accompany the C.

Accordingly, a single V + J coding sequence fusion event causes that V + J, remaining unattached Js and the C coding sequences to be included in the same transcript; subsequent removal of intervening sequences producing mature Ig κL *messenger* RNA. When one each of the Vs and Js are brought together by a pretranscriptional fusion enzyme of enzymes, the cardinal fusion signal sequence $\substack{\text{CCACAGTG}\\ \text{GGTGTCAC}}$, immediately downstream of the V coding sequence, and its complementary fusion signal $\substack{\text{CACTGTGG}\\ \text{GTGACACC}}$, immediately upstream of the J coding sequence initiate extensive side-wise base pairing between them to form a cross-configuration at the V and J coding sequence junction on DNA; the resulting V + J coding sequence fusion presumably discarding a million or more base pairs of DNA (fig. 2).

Rudolph Virchow's dictum *omnis cellula e cellule* applies to the origin of genes and of a biological process as well. The pretranscriptional fusion mechanism that unites V and J coding sequences of IgL chains could not have evolved out of the clear blue sky. We all know that the ancestry of all immunoglobulin subunits can be traced to a β_2-microglobulin-like polypeptide that is 110 amino acid residues long folded upon itself by a single interchain disulfide bridge. Indeed, in the C region of IgH (heavy) chain transcripts, three or four coding sequences, each specifying about 110 amino acid residues, are still separated by intervening sequences as shown by two independent groups [12, 26]. Our view is that there was a time in the remote past of the mammalian ancestors when the number of Vs (β_2-microglobulin-like units) was still small enough to approximate the number of Js in each Ig gene cluster, and that the conventional removal of transcribed intervening sequences was the means then employed to achieve V + J coding sequence fusion. Thus, the pretranscriptional coding sequence fusion mechanism evolved from the ubiquitous splicing mechanism of ancient origin operating on nearly all the transcripts of eukaryotes. Yet, this evolutionary step was the key to the aquisition of Promethean foresight by the immune system, for without this pretranscriptional mechanism, the number of V genes could not increase disproportionally, because the number of C genes which should remain essentially invariant should be kept to a minimum.

It follows then that in order to understand the pretranscriptional V + J coding sequence fusion mechanism which is unique to the immune system, we must first understand the post-transcriptional splicing mechanism that, by eliminating intervening sequences from the nuclear transcripts of genes, produce mature cytoplasmic *messenger* RNAs. From

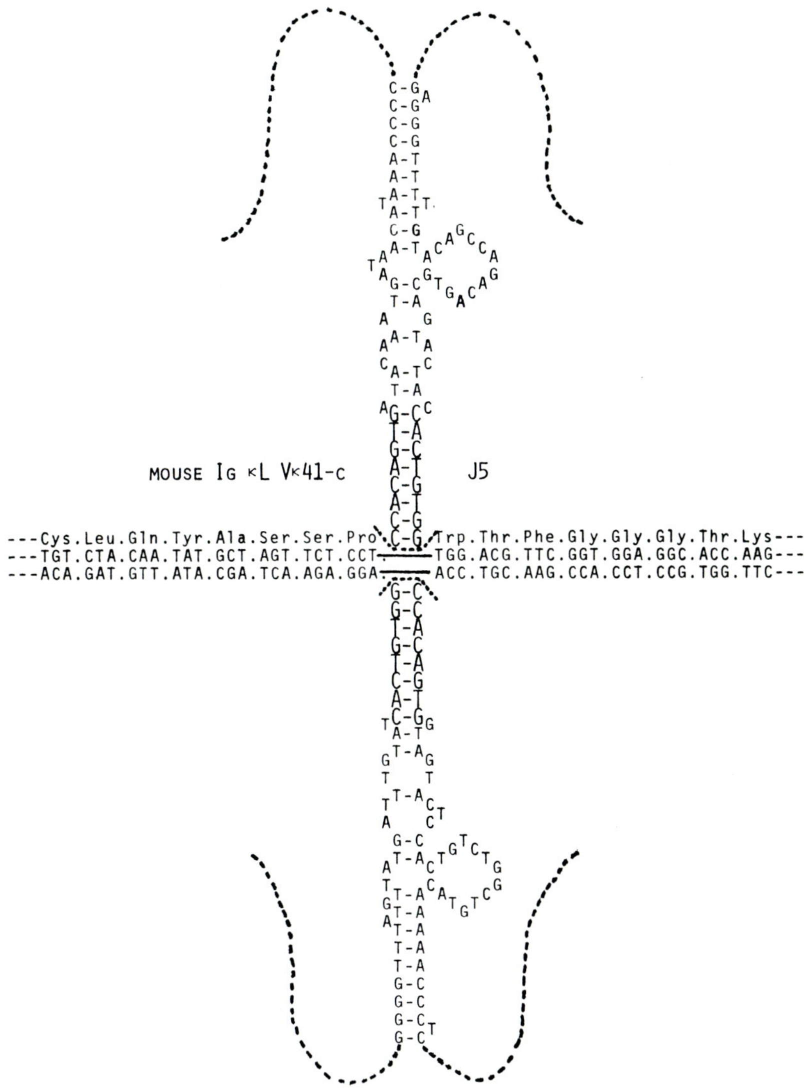

Fig. 2. The cardinal fusion signal sequence $\substack{\text{CCACAGTG}\\ \text{GTGACACC}}$ attached to Vκ41-c coding sequence of the mouse and its complementary fusion partner signal $\substack{\text{CACTGTGG}\\ \text{GTGACACC}}$ adorning Jκ 5 in the coupled, ready-to-be-fused configuration. In the actuel fusion of Vκ41-c and J5, the complementarity between the above noted 8-base-long signal sequences is reinforced by additional complementary sequences. Among them, $\substack{\text{CATAAACC}\\ \text{GTATTTGG}}$ on the V side and $\substack{\text{GTTTTTGG}\\ \text{CAAAAACC}}$ on the J side seemed to have been conserved reasonably well.

various considerations. One of us has concluded that all of the upstream as well as the downstream splicing signals originated from a common 14-base-long prototype sequence CCC ACA.G/GU GAG AG which is an imperfect palindrome[1] [22]. When the two are in a coupled, ready-to-be-spliced configuration, the splice sites represented by two *G/G* base doublets are brought face-to-face and flanked on each side by one *U-A* (A-U) and three *G-C*(C-G) base pairs (fig. 3, top), the act of splicing taking place within the glycine codon GGU [22]. As soon as a single transcript began to contain two or more intervening sequences, the danger of inadvertent splicing necessitated immediate diversification from the prototype of splicing signals. As exemplified in the splice pair for the HyL + Vκ41-c coding sequence union of mouse myeloma MOPC41 κL transcripts shown by *Seidman* et al. [30], sequence diversification affected more of the coding

[1] In molecular biology, perfect palindromes refer to those double-stranded DNA base sequences that are identical in both strands, if each is read in the 5′ to 3′ direction; e.g. $\begin{matrix}^{5'}\mathrm{AGAGCTCT}^{3'}\\ _{3'}\mathrm{TCTCGAGA}_{5'}\end{matrix}$. Consequently, each strand of the palindrome sequence embodies internal complementarity being endowed with the capacity to assume a hairpin loop configuration. For example, the palindrome sequence noted above can assume the following configuration:

```
          C
-AGAGCT : T
 : : : : : :  :
-TCTCGA : 
          G A
```

A given palindrome may become imperfect by insertions of odd bases here and there; e.g. $\begin{matrix}^{5'}\mathrm{AGAAGCTGCT}^{3'}\\ _{3'}\mathrm{TCTTCGACGA}_{5'}\end{matrix}$ is now an imperfect palindrome.

Fig. 3. Each pair of 14-base-long splicing signals are shown in two ways. At the left, upstream and downstream partners of each pair are aligned on the same 5′ to 3′ direction. At the right two partners are presented in the base paired configuration. Bases of intervening sequences are presented in italics, with regard to the left half, bases homologous to the prototype splicing signal sequence are underlined, and each splice site is indicated by a slash. *1st row:* The 14-base-long prototype splicing signal base sequence both as the upstream and downstream signals. *2nd and 3rd rows:* The upstream splicing signal of the hydrophobic leader sequence that accompanies mouse Ig Vκ41-c coding sequence is shown as an example of the diversified signals. In the 2nd row, it is shown with its proper downstream partner of Vκ41-c coding sequence. In the 3rd row, the same upstream signal is shown with its illicit downstream partner of the Cκ coding sequence. Subtle differences between the proper and an illicit pair are seen in the number and positions of paired bases at the right. In the absence of competing Vκ41-c, however, Cκ and HyL can readily engage in splicing. This is known to occur in variant myeloma lines. *4th and 5th rows:* Upstream splicing signals of five mouse Ig Jκs are paired with their common downstream splicing partner of the Cκ. Intervening halves of the upstream signals of all but J3 are identical. In the case of J4, J3, and J2, even their coding halves are identical.

PRIMORDIAL SPLICING SIGNAL BASE SEQUENCE

14 BASE-LONG

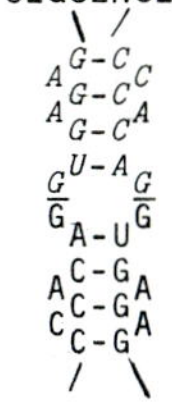

UPSTREAM Pro.Thr.G-
CCC.ACA.G/*GU GAG AG*

DOWNSTREAM *CCC ACA G/*GU.GAG.AG
-ly.Glu.Ar

DIVERSIFIED SPLICING SIGNAL BASE SEQUENCES
MOUSE IG κL

PROGRAMED SPLICING: HγL + Vκ41-C

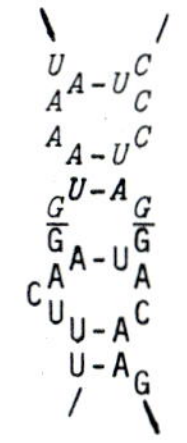

UPSTREAM Phe.Gln.G-
UUU.CAA.G/*GU* AAA AU

DOWNSTREAM *CUC CUA G/*GU.ACC.AG
-ly.Thr.Ar

INADVERTENT SPLICING: HγL + Cκ

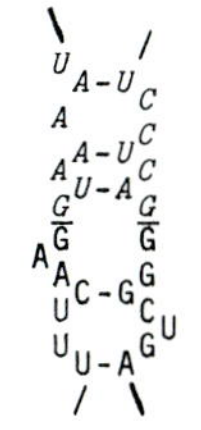

UPSTREAM Phe.Gln.G-
UUU.CAA.G/*GU* AAA AU

DOWNSTREAM *UCC UCA G/*GG.GCU.GA
-Ty.Ala.As

SPLICING OPPORTUNITIES BETWEEN 5 J'S
OF MOUSE IG κL AND Cκ

J5 + Cκ

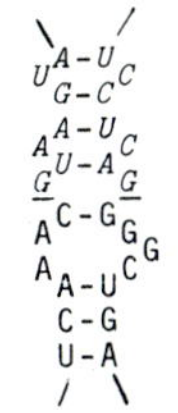

UPSTREAM Ile.Lys.A-
AUC.AAA.C/*GU AAG UA*

DOWNSTREAM *UCC UCA G/*GG.GCU.GA
-rg.Ala.As

J4 + Cκ
J2 + Cκ

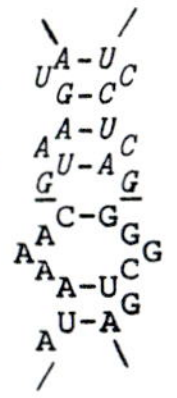

Ile.Lys.A-
AUA.AAA.C/*GU AAG UA*

*UCC UCA G/*GG.GCU.GA
-rg.Ala.As

J1 + Cκ

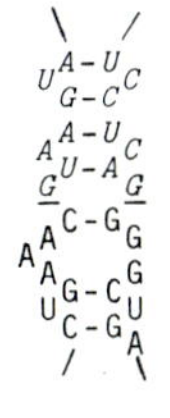

UPSTREAM Leu.Lys.A-
CUG.AAA.C/*GU AAG UA*

DOWNSTREAM *UCC UCA G/*GG.GCU.GA
-rg.Ala.As

J3 (DEFUNCT?) + Cκ

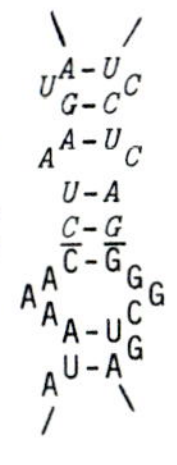

Ile.Lys.A-
AUA.AAA.C/*CU AAG UA*

*UCC UCA G/*GG.GCU.GA
-rg.Ala.As

than the intervening halves of each pair; the above pair differing from a coupled prototype by 8 of the 14 bases of the coding halves and only 6 of the 14 bases of the intervening halves (fig. 3, 2nd row). Because diversification has to be accomplished within a rigid set of rules, deciphered elsewhere [22], often very subtle differences separate the proper splicing partners from illicit ones on the same transcript. In MOPC41 κL transcript, Vκ41-c takes precedence over Cκ as a splicing partner of HyL merely because in coupled HyL + Vκ41-c signals, the pair of G/*G* splice sites are flanked on both sides by paired *U-A* and A-U, whereas in an illicit HyL + Cκ coupling, the above A-U is replaced by unpaired A and G (fig. 3, 2nd and 3rd rows). Accordingly, in the mutational absence of a more competitive Vκ splicing signal, HyL and Cκ of the mutant myeloma κL transcript readily unite to yield a so-called Cκ fragment peptide. Also in the case of IgH chain transcripts, similarly subtle differences separate 5 pairs of proper splicing signals from each other as shown by three groups [8, 12, 26]. Not surprisingly, a mutant mouse myeloma γ1H chain lacking C1 peptide has been reported. Inasmuch as subtle differences among upstream as well as downstream splicing signals mainly reside in their coding halves, the possession by the splicing enzyme of short RNAs complementary to intervening halves of splicing pairs would not increase the enzyme's power of discrimination, although such complementary RNAs have recently been reported [15].

It is of interest to note here that of the 5 Js located in front of the Cκ of the mouse, J3 appears to represent a now defunct relic, for the sequence specified by J3 has not been found in the Ig κL of numerous mouse myelomas analyzed thus far [30]. Indeed, the upstream splicing signal of J3 might have become defunct by a single base substitution; the intervening half of J3's upstream splicing signal begins with *C* instead of *G*. As shown at the top three rows of figure 3, the pair of splice sites in the coupled configuration of most splice partners are brought face to face due to base pairing of their immediate neighbors, while they themselves being G/*G* and *G*/G remain unpaired. In the case of all 5 Jκs, however, the coding half of their upstream splicing signals ends in the codon-straddling A.C instead of the customary codon-straddling A.G (fig. 3, 4th and 5th rows). As a result, in the coupled configuration the coding halves of a pair of splice sites are already engaged in C-G base pairing. An additional substitution of *G* with *C* changes the existing pair of splice sites *G G/*C-G to fully paired *C-G/*C-G. It is likely that the splicing enzyme is incapable of recognizing fully paired C-G base quadruplets as the splice sites (fig. 3, bottom left).

Thus, the finding on an apparently defunct Jκ3 coding sequence was very instructive in knowing what exactly the splicing enzyme sees in the ready-to-be spliced coupled configuration of upstream and downstream splicing signals.

The evolutionary relationship between the ubiquitous and more ancient post-transcriptional splicing mechanism and the apparently immune system-specific pretranscriptional coding sequence fusion mechanism is evident in the fact that the cardinal 8-base-long pretranscriptional fusion signal sequence, CCACAGTG, directly attached Vκ41-c of figure 2 represents the core of the already noted prototype 14-base-long splicing signal sequence CCC ACA.G/GU GAG AG from which all the upstream as well as downstream splicing signals have sprung. The finding represented in figure 2 has not revealed the all critical V+J pretranscriptional fusion mechanism in its entirety, for there remain two unresolved questions begging for answers. First of all, unless a V+J coding sequence fusion happened to involve the last J nearest to the C, a transcript would have to contain a number of unattached J coding sequences situated between the fused (V+J) and the C. Since each J is equipped with its own upstream splicing signal, splicing with the C during the processing of that nuclear transcript is more likely to involve one of the unattached Js than the fused V+J, thus, resulting in the formation of nonsense *messenger* RNA. The pretranscriptional fusion mechanism has to contain a built-in safeguard against the inadvertent splicing noted above (fig. 4). In each of the three chromosomes carrying Ig genes, a large subcluster of numerous V genes and a small subcluster made of several Js and one or more Cs are separated by a million or more base pairs of void as already noted; roughly a genetic distance of one cross-over unit. When viewed in the above light of immense distance, the fusion signal sequence which is barely 40 or so base pairs long, loses its impressiveness. In order for the fusion enzyme in the B cell nucleus to be able to recognize one V and one J separated by so immense a distance, each V and each J must be endowed with additional, more conspicuous characteristics to make their positions known.

In figure 5, the solutions to both questions are shown using J1, J2 and J5 coding sequences of mouse κL as examples. It should be noted that each J coding sequence is an imperfect palindrome, internal base sequence complementarity extending to adjacent non-coding sequences. Accordingly, even in germ line DNA, each J should stand out like a sore thumb by assuming a small cross-configuration (fig. 6). Needless to say, DNA can not assume such cross-configuration without the help of enzymes. The

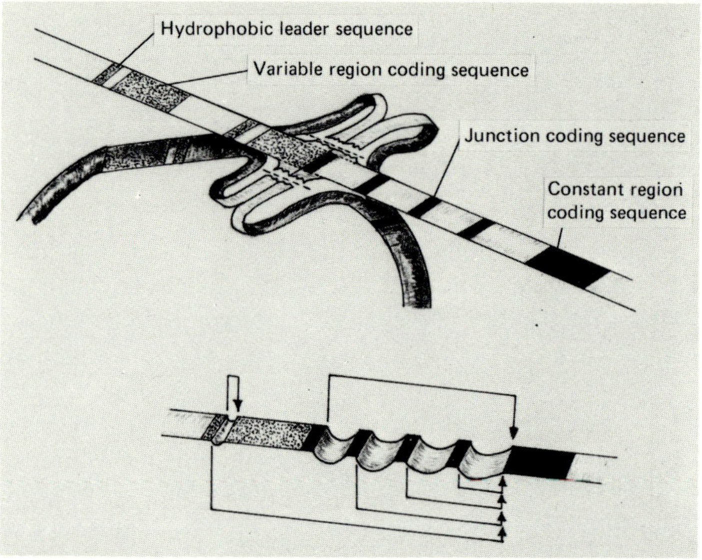

Fig. 4. A pretranscriptional V + J coding sequence fusion event occurring in the gene cluster for mouse Ig κL and the consequent production of transcripts containing one fused V + J, three unattached Js and one C coding sequences are schematically illustrated. *Top:* Because the cardinal fusion signal sequence $\genfrac{}{}{0pt}{}{\text{CCACAGTG}}{\text{GGTGTCAC}}$ resides immediately downstream of each Vκ coding sequence, while its complementary partner signal $\genfrac{}{}{0pt}{}{\text{CACTGTGG}}{\text{GTGACACC}}$ adorns each Jκ, any combination of Vκ and Jκ can be brought together by the enzyme that recognizes the above pair of signals. The side-wise pairing initiated by the pair produces the cross configuration at the Vκ + Jκ coding sequence junction. The resulting V + J fusion eliminates a very long stretch of DNA; longer than one million base pairs containing varying numbers of Vs and Js. *Bottom:* The RNA transcript of the above rearranged DNA segment. For it to become mature cytoplasmic *messenger* RNA for Ig κL, only two proper pairs of splicing signals should be recognized as indicated by two arrows above the transcript. The presence of three unattached Js, however, creates the possibility of four different types of illicit splicing as indicated by arrows below the transcript.

presence of specific enzymes has to be invoked. It should be recalled that the cardinal upstream fusion signal sequence CCACATG directly attached to the Vκ41-c coding sequence of figure 2 represents the core of the prototype upstream splicing signal CCC ACA.G/*GU GAG AG.* It follows then that the 8-base-long downstream fusion signal attached to each J being complementary to the above-noted CCACAGTG is also largely complementary to that Js own upstream splicing signal located at the other end because of the latter's derivation from the prototype CCC ACA.G/*GU*

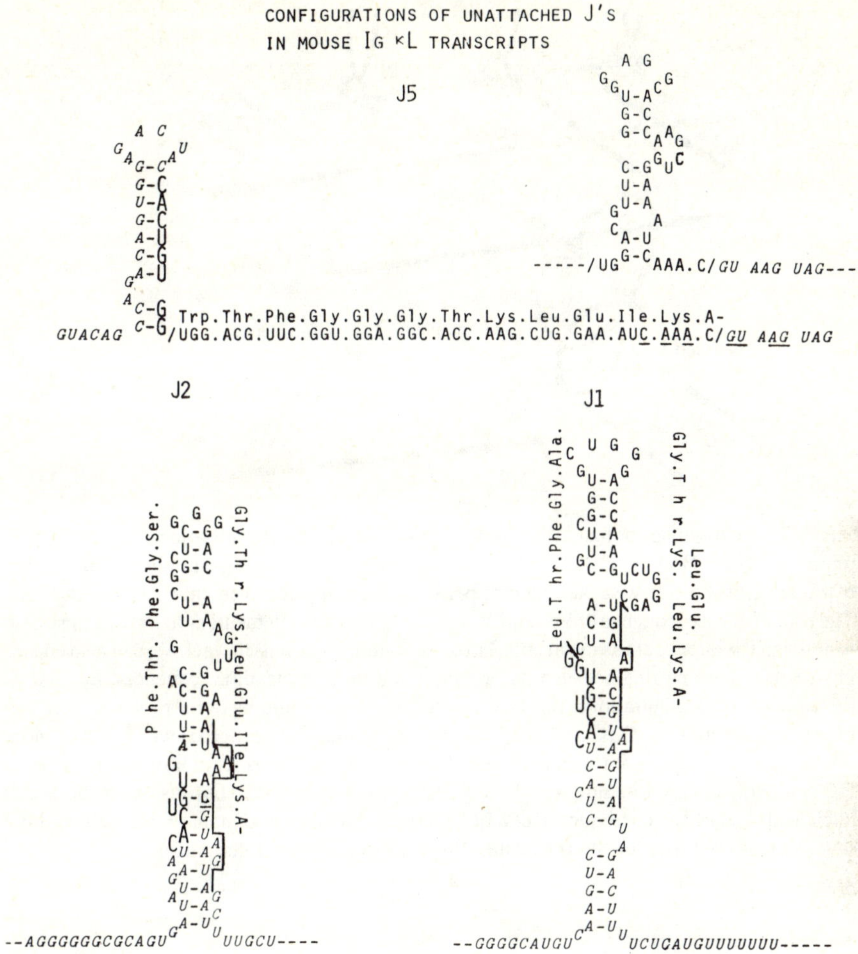

Fig. 5. Hairpin loop configurations of unattached J5, J2, and J1 in mouse Ig κL transcripts. All J coding sequences including J_Hs are imperfect palindromes. Only in J5 does the complementary fusion partner signal CACUGUGG form another hairpin loop with its immediate upstream bases, thus leaving the AA.C/*GU AAA UA* portion the J5's upstream splicing signal sequence accessible to the splicing enzyme. The above apparently reflects the terminal position of J5. Being farthest away from the Cκ , J5 would never be transcribed without the prior fusion with V. In unattached J2 as well as J1, the AAA.C/*GU AAG UA* portion of their upstream splicing signals is made inaccessible to the splicing enzyme, because of base pairing with the complementary fusion signal CACUGUGG.

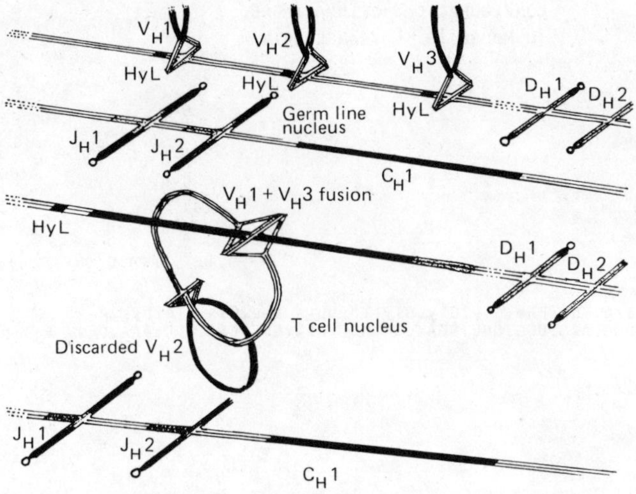

Fig. 6. Cross-like configurations of V_Hs, D_Hs, and J_Hs in a chromosome DNA in the pre-rearranged state. Because D_Hs and J_Hs are palindrome, each can assume the configura tion of a small cross by side-wise base pairing between internally complementary base sequences. The proximal base sequence of V_Hs and V_Ls is also largely complementary to its own upstream fusion signal sequence at the distal end. Thus, by doubling upon itself, each V_H can also engage in side- wise base pairing between its proximal and distal noncoding sequences. Because of these cross-like configurations, the DNA fusion enzyme should have no problem in recognizing individual V_Hs, D_Hs, and J_Hs, although they must be scattered over a vast region. Because the upstream fusion signal of each V_H as well as the downstream fusion signal of each D_H is already engaged in side-wise base pairing with the self, V_H+D_H fusion in the B cell nucleus should be the consequence of a better complementary fit of nonself over self, and the same should hold true for the postulated V_H+V_H fusion in the T cell nucleus.

GAG AG. Accordingly, each unnattached J in Ig transcripts, by assuming a hairpin loop configuration, manages to hide its upstream splicing signal by base pairing; e.g. the upstream splicing signal CUG AAA.C/*GU AAG UA* of unattached J1 in figure 5 is base paired with a sequence centered by the complementary fusion signal CACUGUGG.The above then is a safeguard against inadvertent splicing involving unattached Js and the C. Naturally, the pretranscriptional fusion with V discards the fusion signal sequence from J, and by so doing renders its upstream splicing signal accessible to the splicing enzyme, thus ensuring the representation of one each of the V, J, and C coding sequences in contiguity in mature cytoplasmic *messenger* RNA. As shown at the top of figure 5, the transcribed J5 in the unattached state stands out from the rest of the Jκs by virtue of forming two hairpin loops instead of one. As its complementary fusion signal CACUGUGG is included in the first hairpin loop, J5's upstream splicing signal at the 3′ end remains free and accessible to the splicing enzyme. The above is apparently a reflection of J5's farthest position from the Cκ. Because of this position, J5 would never be transcribed without prior fusion with V. Hence, there exists no need to render its upstream splicing signal inaccessible in transcripts. In the chromosomal DNA, J5 would also be represented by two small crosses instead of one. Such may serve as yet another landmark by which the DNA fusion enzyme locates a cluster of Js in a vast stretch. Figure 5 also shows that the 8-base-long downstream fusion signal CACTGTGG which is attached to J5 and J1 is completely complementary to the cardinal upstream fusion signal sequence CCACAGTG of Vκs, whereas one base mismatch is shown by CACTGTGA of J2 as well as by CACTGTGT of J4 (not shown). Interestingly, defunct J3's downstream fusion signal CACTGTAA reveals two base mismatches. Thus, the importance of 8-base-long fusion signals in successful V+J pretranscriptional coding sequence fusion is proven.

The situation is a little different in the IgH chain gene cluster, for it was found that anti-phosphorylcholine V_Hs used by certain mouse myeloma antibodies first fuse with very short D coding sequences good only for 5 amino acid residues, and only then does (V_H + D_H) fuse with J_H [8, 28], Interestingly, each D_H coding sequence such as $\overset{\text{Tyr.}}{\text{TAC}}.\overset{\text{Tyr.}}{\text{TAC}}.\overset{\text{Gly.}}{\text{GGT}}.\overset{\text{Ser.}}{\text{AGT}}.\overset{\text{T}}{\text{A}}$ is a nearly perfect palindrome. Furthermore, it is equipped with the downstream fusion signal sequence at one end for fusion with V_H, and the upstream fusion signal at the other end for fusion with J_H, and the two partner fusion signals are naturally complementary to each other. It follows then that in the germ line DNA, each D_H stands out as conspicuously as each J by assuming the configuration of a small cross (fig. 6).

Naturally, those J_H genes that are destined to undergo the first fusion with D_H should be endowed with a built-in safeguard to prevent inadvertent direct fusion with H_H, and this calls for diversification of their fusion signals. Indeed, the 8-base-long upstream fusion signal of V_HS107 shows one base deviation from the cardinal sequence being ACACAGTG [8]. Although neither upstream nor downstream fusion signal of any D_H is known, downstream fusion signals of the four J_Hs are indeed diversified from the cardinal sequence CACTGTGG being GACTGTGC, TAGTGTGA, CAATGTGC, and CAATGTGC; 2,3,2, and 3 base deviations [28].

The Differential Use of the Same V_H Gene by IgH Chains and T Cell Receptors

This insertion of an extra step represented by D_H in the pretranscriptional coding sequence fusion mechanism of IgH chains does not appear to be universal, for anti-levan as well as anti-dextran V_Hs appear to undergo direct fusion with J_Hs. It may be that only those V_Hs that are utilized both as IgH chains and as T cell receptors undergo this extra step in the B cell nucleus as an act of irreversible commitment toward Ig heavy chain formation. Indeed, starting with the original findings by *Binz and Wigzell* as well as by *Rajewsky and Eichmann,* increasing evidence indicates that T cell receptors derive their antigen-binding sites from V_H genes, while receiving no contribution from V_L genes. One of the more convincing demonstrations of the above utilized V_HIg-1b genetic marker which is a V_H allotype of the mouse. The anti-NP hapten humoral antibody of $V_{\gamma L}V_HNP^B$ composition exhibits the unique heteroclicity in that this antibody raised against NP demonstrates a far greater binding affinity to a related hapten NIP than to NP itself. Anti-NP T cell receptors of the same mouse also exhibit the same heteroclicity toward NIP, thus suggesting their antigen-binding sites to have been specified by the V_HNP^B allelic gene [7]. The above finding, however, created one unresolved dilemma. In the case of anti-NP humoral antibodies, the heteroclicity towards NIP is the unique property of $V_{\gamma L}V_HNP^b$ and not that of $V_{\kappa L}V_HNP^b$. Thus, the question arises as to how V_HNP^b alone of the T cell receptor without help from $V_{\gamma L}$ can express this heteroclicity. At any rate, the antigen-binding specificity is not often, if ever, demonstrable in $V_H$monomers. Here we show that the pretranscriptional coding sequence fusion mechanism that unites one of the numerous Vs with the C in the B cell nucleus, albeit indirectly, can also unite two or more V_H coding sequences in tandem in the T cell nucleus. We thus suggest that the antigen-binding sites of T cell receptors are furnished by two V_H coding sequences fused in tandem. We acknowledge here that the possibility of V_H+V_H fusion dawned on us during our conversation with *Melvin Cohn.*

What occurs in the nuclei of anti-phosphorylcholine *helper* and *suppressor* T cells? We have noted that the 42 base pair-long sequence made of the last 31 base pairs of the intervening sequence after HyL and the first 11 base pairs of V_HS107 coding sequence, are largely complementary to V_HS107s own upstream fusion signal at its other end. Since the above-noted 42 base pair-long base sequence of V_HS107 is also found unaltered in another anti-phosphorylcholine V_HM603 as well, it would appear that on the germ line DNA, each V_H also stands out by doubling upon itself to assume the configuration of a small cross by side-wise base pairing between its proximal and distal, largely noncoding sequences; the latter representing the upstream fusion signal for fusion with D_H (fig. 6).

Since we found such internal base sequence complementarity in Vκ41-c as well as, we assume the above to be a unique property not only of V_Hs but also of V_Ls. As far as the gene cluster of IgH chains distributed over a few million base pairs is concerned, the pretranscriptional DNA fusion enzyme should have no difficulty in recognizing individual V_Hs, D_Hs, and J_Hs on a chromosomal DNA still in the germ line configuration (fig. 6). Figure 6 immediately reveals the possibility of a V_H+V_H pretranscriptional coding sequence fusion as the means to generate the antigen-binding site of T cell receptors. This possibility is illustrated in figure 7 using V_HS107 and V_HM603 as fusion partners. While the comparison of figure 6 with figure 2 should convince most skeptics as to the feasibility of a V_H+V_H fusion, it should be recalled that at the other end of V_HS107 itself resides the 42 base-pair-long sequence identical with that of V_HM603 engaged in side-wise base pairing with the upstream fusion signal sequence of V_HS107 in figure 7. It follows then that a real downstream fusion partner of V_HS107 should be equipped with a corresponding sequence that shows even greater complementary fit to the upstream fusion signal sequence of V_HS107.

By having two or more V_Hs tandemly fused as their antigen-binding site, the antigen-discriminating power of T cell receptors is elevated to the equivalent at least of IgLH dimers (fig. 8).

Anti-KLH receptors excreted by *suppressor* T cells of H-2^b mice have been shown to contain the KLH-binding site and I-J^b antigenic determinants in unseparable coupling; their combined molecular weight being in the range of 50,000 [32]. The above appears to limit the number of V_Hs involved in tandem fusion to two (fig. 8).

After the V_H+D_H pretranscriptional fusion in the B cell nucleus, V_HS107 contributes the amino acid sequence down to Asp of the 101st position to IgH chains directed against phosphorylcholine. If V_HS107 is involved as a downstream partner in the V_H+V_H fusion taking place in the T cell nucleus, the translation of V_HS107 in processed *messenger* RNA can continue until the chain-terminating codon UGA, thus, adding 8 extra amino acid residues beyond the above-noted Asp (fig. 9, top). Conversely, immediately beyond GAU for the terminal Asp of V_HS107, one finds the upstream splicing signal sequence CAC ACA.G/*U GAG AG* which differs

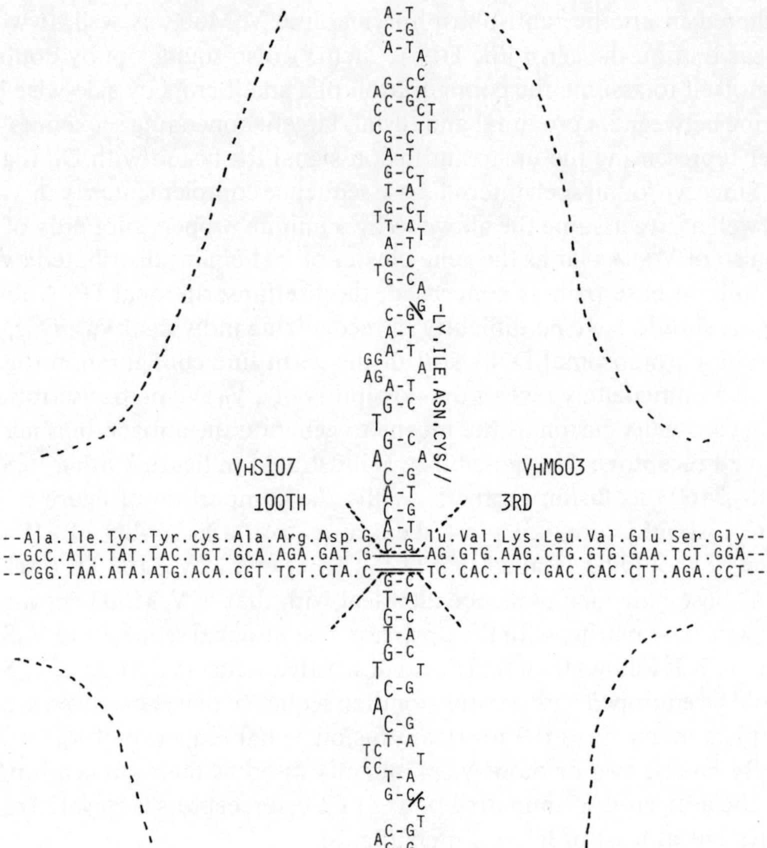

Fig. 7. Utilizing the same set of fusion signal sequences as that utilized for V_L+J_L as well as for V_H+D_H+J_H fusions, the feasibility of V_H+V_H pretranscriptional coding sequence fusion in the T cell nucleus is illustrated choosing anti-phosphorylcholine V_HS107 and V_HM603 as upstream and downstream fusion partners. The upstream fusion signal sequence of V_HS107 and the proximal 42 base-pair-long sequence of V_HM603 can engage in almost as extensive side-wise base pairing as that shown by Vκ 41-c and J5 engaged in bona fide side-wise base

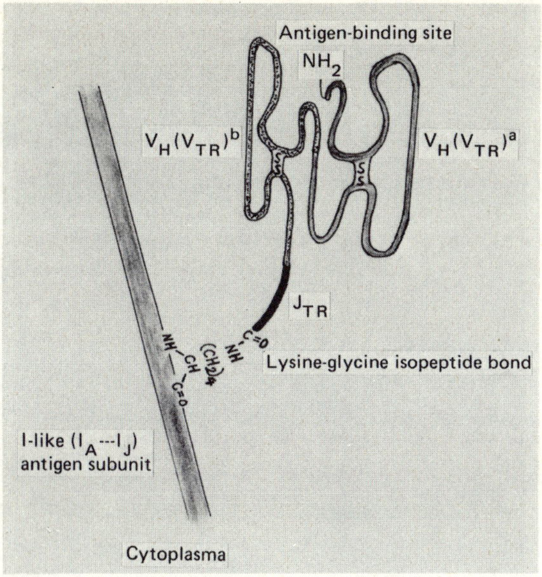

Fig. 8. The schematic illustration of T cell receptors as we conceive them. This scheme is heavily influenced by the finding of *Tada's* group on anti-KLH *suppressor* T cell receptors. Its antigen-binding site is furnished by two V_Hs directed against the same antigen in tandem fusion. The two alternative sources of J_{TR} carboxyl terminal amino acid sequence are shown in figure 9. We believe that J_{TR} and one or the other subunit of I-J and other I-like antigens are linked together by the isopeptide bond, here shown as lysine-glycine.

from the already noted prototype upstream splicing signal sequence CCC ACA.G/*GU GAG AG* only by one base substitution and one base deletion; bases of an intervening sequence in italics [22]. 41 bases further downstream, one finds the equally respectable but incomplete downstream splicing signal *CCA ACA G/*A.GG (fig. 9, bottom). A further extention of the sequence data on the germ line V_HS107 may show that the splicing between the above pair of splicing signals adds a considerable number of amino acid residues beyond the terminal Asp of V_HS107. In the schematic drawings of figure 8, the amino acid sequence beyond that corresponding

---

pairing for the fusion shown in figure 2. However, as explained in the text, the real fusion of V_HS107 should involve another V_H as its downstream partner whose 42 base-pair-long proximal sequence shows even better complementary fit to the upstream fusion signal sequence of V_HS107.

```
1) FURTHER TRANSLATION

              VHM603-LIKE + VHS107

                      101st     102nd

----Tyr.Tyr.Cys.Ala.Arg.Asn.G//lu.Val.Lys.Leu.Val.Glu.Ser----------
----UAU.UAC.UGU.GCA.AGA.AAU.G//AG.GUA.AAG.CUG.GUG.GAA.UCU----------

          202nd
----Cys.Ala.Arg.Asp//ALA.THR.SER.ASP.ARG.THR.SER.LEU.
----UGU.GCA.CGA.GAU//GCA.CAC.AGU.GAG.AGG.ACG.UCA.UUG.[UGA] GCCCAGCACAAA---

               + DH


2) SPLICING

              VHM603-LIKE + VHS107

                      101st     102nd

----Tyr.Tyr.Cys.Ala.Arg.Asn.G//lu.Val.Lys.Leu.Val.Glu.Ser----------
----UAU.UAC.UGU.GCA.AGA.AAU.G//AG.GUA.AAG.CUG.GUG.GAA.UCU----------

          202nd
----Cys.Ala.Arg.Asp//ALA.THR.AR-
----CGU.GCA.CGA.GAU//GCA CAC AG/ U GAG AG GACGUCAUUG UGA GCC CAG CAC

               + DH
                                                 -G.GLY
AAA CCU CCA UUG CAG GGG UGU UCU GGA CCA ACA G/A.GG (END OF THE SEQUENCE
                                                              DATA)
```

Fig. 9. Two alternative sources of J_{TR} when V_HS107 as the downstream partner of V_H+V_H fusion in the T cell nucleus is transcribed and translated. *Top:* The continuous translation of V_HS107 until the first chain terminating codon *UGA* can add 8 extra amino acid residues beyond Asp or the 202nd position (101st position in IgH chains). *Bottom:* Conversely, the base sequence immediately beyond GAU for the 202nd Asp contains a pair of prototype-like splicing signals (intervening sequence bases in italics). Bases in the splicing signals conserved from the prototype CCC ACA.G/GU GAG AG are underlined. This splicing can create $J_{TR}$of an indeterminant length depending upon the position of the first chain terminating codon in the spliced sequence.

to the terminal Asp of V_HS107 is identified as the J_{TR} (T cell receptor junction) sequence in that through J_{TR}, the antigen binding site of the T cell receptor furnished by two tandemly fused V_Hs is joined to one of the two subunits of I-J and other I-like MHC antigens (fig. 8).

In view of the inseparable coupling between the antigen-binding site and I-J antigenic determinants observed in anti-KLH *suppressor* T cell receptors [32], we propose an isopeptide linkage between the antigen-binding site furnished by two tandemly fused V_Hs and a subunit of I-like MHC antigen; e.g. nuclear protein ubiquitin and histone H-2A are linked by the lysine-glycine isopeptide bond (fig. 8). It must be this linkage with the I-like antigen subunit that anchors T cell receptors to the plasma membrane. The third alternative origin of V_{TR} is the constant region coding sequence reserved only for T cell receptors; the pretranscriptional fusion between two V_Hs being followed by a second fusion between (fused V_H+V_H) and one of the D_Hs. According to this view, D_Hs are duplicates of the hydrophobic leader of the T cell-specific C. Thus, each T cell receptor transcript should contain (fused V_H+V_H+D_H), a number of unattached D_Hs and the C. Evidence suggesting the existence of such V_{TR} (C) which is class-specific for receptors of *suppressor* and *killer* T cells has been presented [24].

Along a similar vein, we found the base sequence in the middle of the intervening sequence of mouse Vκ41-c:

GGGATGTCCTCTTTTATCATGCTTATCTATGTGG
CCCTACAGGAGAAAATAGTACGAATAGATACACC

to be largely complementary to its own upstream fusion signal sequence at the other end which is:

CCACAGTG	ATACAAATCATAA	CATAAACC	CC
GGTGTCAC	TATGTTTAGTATT	GTATTTGG	GG;

conserved signal sequences in boxes [30]. Thus, it would appear that the antigen-binding sites of exceptional T cell receptors can also be furnished by tandemly fused V_L+V_L. A certain class of T cells may indeed utilize V_L as their receptor antigen-binding sites [9]. The problem with the above-noted type of pretranscriptional V_H+V_H as well as V_L+V_L tandem fusion is that the use of the same signal sequences implies the dependence upon the same DNA fusion enzyme. Accordingly, V_H+V_H as well as V_L+V_L fusions for T cell receptor are as likely to occur as V_L+J_L as well as V_H+D_H+J_H fusions for B cell immunoglobulins in either the B cell nucleus or the T cell nucleus. Although there has been a report of the IgH C_μ expression by T cells [14], we are inclined to believe in the absence of such wasteful duplicity in the pretranscriptional coding sequence fusion mechanism.

In view of the very early ontogenic separation of the T cell lineage from the B cell lineage [5], we feel that the pretranscriptional DNA fusion enzyme responsible for V_L+J_L as well as V_H+D_H+J_H coding sequence fusion in the B cell nucleus is not likely to be utilized by T cells for their tandem V_H+V_H fusions. It is necessary that the T cell fusion enzyme must recognize a related but different set of fusion signal base sequences. We found such downstream fusion signals residing within the first two thirds of the intervening sequence that follows HyL of V_HM141 [28]. The postulated fusion between B_HM141 and V_HS107 is illustrated in figure 10.

It should be recalled that $V_L + J_L$ (fig. 2) and presumably also $V_H + D_H$ and then $(V_H + D_H) + J_H$ coding sequence fusions in B cell

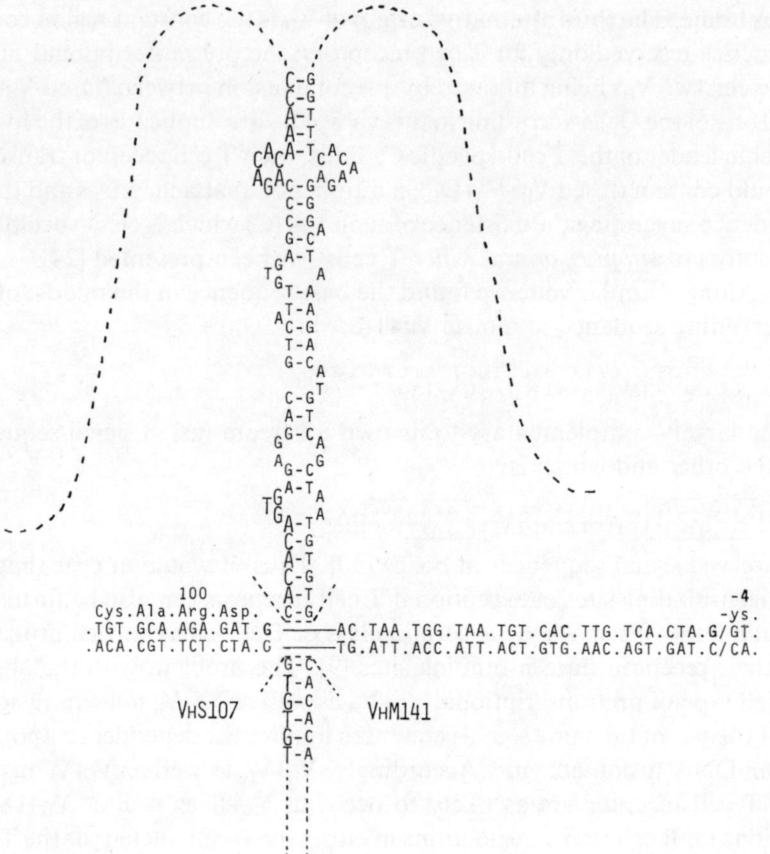

Fig. 10. The feasibility of the V_H+V_H pretranscriptional coding sequence fusion in the T cell nucleus, using unique fusion signal sequences, is illustrated choosing V_HS107 [8] and V_HM141 [28] as upstream and downstream fusion partners. Both of the two conserved upstream fusion signal sequences for V_H+D_H fusion residing immediately adjacent to V_HS107 coding sequence are shown in larger capital letters. When the extensive side-wise base pairing of this figure is compared with that shown in figure 2, a strong measure of support is found for our notion that V_H+D_H+J_H fusions in B cell nuclei and V_H+V_H fusions in T cell nuclei are catalyzed by two different enzymes that recognize pairs of related but slightly different fusion signal base sequences.

nuclei depend upon the complementarity between the two sets of fusion signal sequences; that between the cardinal $^{\text{CCACAGTG}}_{\text{GGTGTCAC}}$ and $^{\text{CACTGTGG}}_{\text{GTGACACC}}$ or their immediate derivative supplemented by that between $^{\text{ACAAAAACC}}_{\text{TGTTTTTGG}}$ and $^{\text{GGTTTTTGT}}_{\text{CCAAAAACA}}$ or their immediate derivatives [8, 27, 28, 30]. Figure 10 suggest that for proposed tandem $V_H + V_H$ coding sequence fusions in T cell nuclei, only specific portions of the above-noted pair of upstream fusion signals are utilized for the side-wise base pairing; the complementarity between $^{\text{CACACA}}_{\text{GTGTGT}}$ and $^{\text{TGTGTG}}_{\text{ACACAC}}$ supplemented by that between $^{\text{AAACC}}_{\text{TTTGG}}$ and $^{\text{GGTTT}}_{\text{CCAAA}}$. The differential use of fusion signal base sequences would insure that $V_H + D_H$ and then $(V_H + D_H) + J_H$ coding sequence fusions can occur only in B cell nuclei, whereas $V_H + V_H$ fusions can take place only in T cell nuclei. Anti-KLH receptors excreted by mouse *suppressor* T cells have been shown to contain the KLH-binding sites and I-J antigenic determinants in unseparable coupling; their combined molecular weight being in the range of 50,000 [32]. The above appears to limit the number of V_Hs involved in tandem fusion to two (fig. 7).

After the $V_H + D_H$ pretranscriptional fusion in the B cell nucleus, V_HS107 contributes the amino acid sequence down in Asp of the 101st position to IgH chains. In the transcript of the fused V_HS107 + V_HM141 of figure 11, one finds the upstream splicing signal sequence $\overset{\text{T-yr.\ T-yr.\ C-y-}}{\text{U.AU U.AC U.G/}\textit{U GCA AG}}$ within the V_HS107 coding sequence; the first of the two Tyrs of the above occupying the 93rd position (bases of an intervening sequence in italics). Since the absence of *G* at the head of the intervening half of the above upstream signal has to be compensated by an addition of an extra *G* at the intervening sequence end of its partner downstream splicing signal [22], the regular downstream splicing signal $\textit{UCA CUA G}\overset{\text{-ys.\ Ile.\ Le}}{\text{/GU. AUC. CU}}$ of V_HM141 for the removal of its intervening sequence should now be read as $\textit{UCA CUA GG}\overset{\text{-s.\ Ile.\ Le}}{\text{/U.AUC.CU}}$. Accordingly, the processing of the (V_HS107 + V_HM141) transcript by T cells should unite the two V_H coding sequences at the junction $\overset{\text{Tyr.\ Try.\ Cy-s.\ Ile.\ Le}}{\text{UAU.UAC.CG/U.AUC.CU}}$ (fig. 11). Since V_HM141 is involved as a downstream partner of this $V_H + V_H$ fusion, the translation of V_HM141 coding sequence can continue beyond Ser of the 97th position until the chain terminating codon is encountered which is not yet evident in the sequence date presented by *Sakano* et al. [28], thus gaining at least 30 additional amino acid residues (fig. 11). Alternatively, the second fusion between (fused V_HM141) and one of the D_Hs may bring the T cell-specific constant region coding sequence into the same nuclear transcript as already noted.

```
          SPLICING OF (VHS107 + VHM141) TRANSCRIPT

 97      VHS107
--T-yr.T-yr.C-y -s.Ala.Asp.
--U.AU U.AC U.G/ U GCA AGA [UGA] CUA [AUG] GUA AUG UCA CUU
            ---

  G UC A CU A G G/U.AUC.CUU.UCC.CAG.------------------
                -s.Ile.Leu.Ser.Gln.------------------
                  -3          1

                VHM141


                JTR OF (VHS107 + VHM141)
          T CELL RECEPTOR ANTIGEN BINDING SITE


          97
--TYR.CYS.ALA.SER.Asp.Thr.Val.Arg.Glu.Val.Gln.Cys.Glu.Pro.Ala.Gln.Ilc.
--UAC.UGU.GCC.AGA.GAC.ACA.GUG.AGG.GAA.GUC.CAA.UGU.GAG.CCU.GCA.CAA.AUA.

  Pro.Leu.Cys.Arg.Asp.His.Asn.Gln.Gln.Gly.Ala.Val.Arg.Thr.Gln.Gly.Leu.
  CCU.CUC.UGC.AGG.GAU.CAC.AAC.CAG.CAG.GGG.GCG.GUG.AGG.ACC.CAA.GGA.CUU

(the sequence data end)
```

Fig. 11. A possible manner of processing the fused $V_HS107+V_HM141$ transcript and subsequent translation of its processed cytotoplasmic *messenger* RNA to yield T cell receptor antigen binding sites are illustrated. *Top:* The removal of a newly created intervening sequence from the fused $V_HS107+V_HM141$ transcript. Bases of an intervening sequence are shown in italics and the precise positions of a pair of intervening-coding boundaries are indicated by slashes. The rules that enable us to identify each pair of splice sites in all transcripts have been described in detail elsewhere [22]. *Bottom:* The continuous translation of the processed *messenger* RNA adds 30 extra amino acid residues to the carboxyl terminus of the fused $V_HS107+V_HM141$ polypeptide beyond Ser of the 97th position (199th in the fused antigen-binding site) of V_HM141. The presence of a chain terminating codon is not yet in evidence in the sequence data [28].

The Curse of Prometheus and Numerous Defunct V Genes

There was a time when germ-line theorists as to the source of antibody diversity postulated the presence of nearly a million V genes (coding sequences) per genome, whereas the somatic mutation school insisted on a few V genes per class; i.e. κL, γL, and H. It seems that nearly everyone is now content with the number of V genes per class of the order of 10s or 100s. The fact is that there is no way of measuring the number of V genes per genome or per class. When the base sequence complementary to the V_L or V_H portion of the *messenger* RNA derived from a monoclonal myeloma is used as a probe for nucleic acid hybridization, absolutely no allowance for infidelity in base pairing should yield data consistent with one V gene per class or at the most a few V genes per class, if that V gene happens to exist in exact duplicate or triplicate. Greater allowances for base pairing infidelity would progressively increase the estimated number of V genes per class or per genome.

We should point out that the curse of Prometheus extends to this realm of the immune system as well. The immune system owes its Promethean foresight to the random generation of numerous antigen-binding sites by nearly every individual of a population. Yet, the functional V regions that we see as the antigen binding sites of humoral antibodies as well as T cell receptors are those in which amino acid substitutions are confined to a few hypervariable segments, the amino acid sequences of longer framework segments showing remarkable evolutionary conservation. Since random mutations apparently know no bounds, we can only assume that extensive amino acid substitutions affecting the framework regions are incompatible with the antigen-binding capacity of V regions. Since V genes have been placed beyond the reach of natural selection as the price that had to be paid for the acquisition of Promethean foresight, the elimination of defunct V genes have to depend upon periodical clean up jobs by unequal crossing-overs that eliminate segments containing functional as well as defunct V genes [23]. The above ineffectiveness in eliminating constantly arising defunct V genes may be the very reason that within each Ig gene cluster a group of V genes is apparently separated by a million base pairs or longer void from a group made of several Js and a few Cs. We have already noted that of the 5 Js accompanying Cκ of the mouse, J3 is apparently defunct [27, 30]. If the above indicates a proportion of defunct genes among the V genes, a minimum of 25% degeneracy is expected.

If, on the other hand, there is to be a big surprise stored for the future in that the thorough analysis of germ-line V gene clusters shall reveal no great number of defunct V genes, one would have to assume either natural selection managed to place these randomly mutating V genes under surveillance, or mutations sustained by individual V genes are not random in their distribution. As far as we can see, there exists only one means by which natural selection can accomplish this apparently impossible feat. The possession of two totally independent antigen-binding sites had to be made inherent in the amino acid sequence specified by each V gene. We

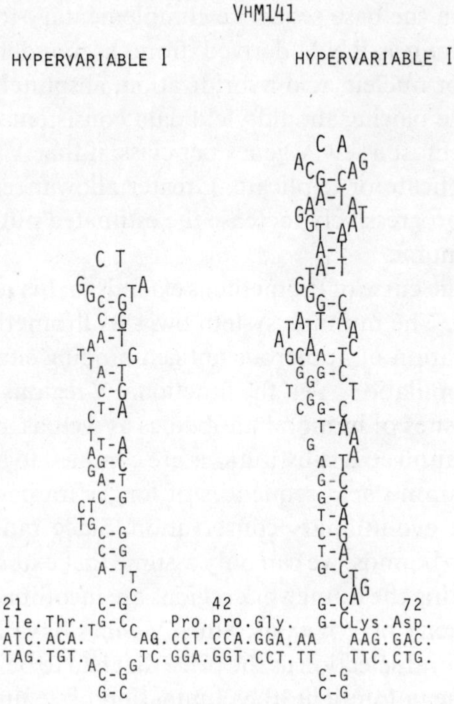

Fig. 12. Contrary to certain previous claims, the first and second hypervariable regions residing within each V_L and V_H coding sequence do not show sufficient internal base sequence complementary within themselves. But each is invariably included in a longer palindrome sequence. This point is illustrated on V_HM141. Accordingly, the enzyme responsible for pre-transcriptional fusion shall have no difficulty in recognizing individual Vs, Ds, and Js on the germ-line DNA. Bases of first and second hypervariable regions are shown in larger capital letters. While the first specifies only five amino acid residues, the second specifies 19. Naturally, their inclusions in palindrome hairpin loops might account for selectively high mutational base substitutions, deletions, and insertions affecting these regions; not only first and second hypervariable regions within each V, but also individual Ds and Js (fig. 5) that constitute the third hypervariable region of antigen-binding sites.

might mention here that at least certain monoclonal antibodies have indeed been shown to be endowed with two independent antigen-bindings sites [25]. If what holds true for the L_2H_2 antibody molecules also holds true for individual V_Ls and V_Hs, those V genes whose one of the two binding sites are directed against the most common pathogens of the immediate past as well as the present can be placed under the customary surveillance by natural selection. Promethean foresight then resides exclusively in the second binding site which is free to accumulate randomly sustained mutations in the germ line as well as in somatic cells [21]. The second means by which to avoid the accumulation of defunct V genes by the genome is to confine mutations to hypervariable regions by giving these regions special characteristics. We have already pointed out that each J_L, J_H as well as D_H is a pseudopalindrome, thus demonstrating the tendency to form a small cross-configuration in the germ-line DNA (fig. 5, 6). It should be recalled that J_L as well as J_H+D_H constitutes a part of the third hypervariable regions of V_L as well as V_H. As suggested by others [33], we also found that the first and second hypervariable regions are invariably included in independent pseudo-palindrome sequences. This point is illustrated in the first and second hypervariable regions of mouse V_HM141 [28] in figure 12.

Unless we are dealing with extraordinary coincidences, there apparently exists the cause-and-effect relationship between its inclusion in a pseudo-palindrome hairpin loop and hypermutability of each specific region in individual V_L as well as V_H.

References

1 Beutler, B.; Nagai, Y.; Ohno, S.; Klein, G.; Shapiro, I.: The HLA dependent expression of testis-organizing H-Y antigen by human male cells. Cell *13:* 509 (1978).

2 Blank, K.J.; Lilly, F.: Evidence for an H-2/viral protein complex on the cell surface as the basis of H-2 restriction of cytotoxicity. Nature, Lond. *269:* 808 (1977).

3 Boehmer, H. von; Hengartner, H.; Nabholz, M.; Lernhardt, W.; Schreier, M.H.; Haas, W.: Fine specificity of a continously growing killer cell clone specific for H-Y antigen. Eur. J. Immunol. *9:* 592 (1979).

4 Bubbers, J.E.; Lilly, F.: Selective incorporation of H-2 antigenic determinants into Friend virus particles. Nature Lond. *266:* 458 (1977).

5 Cantor, H.; Boyse, E.A.: Lymphocytes as models for the study of mammalian celluar differentiation. Immunol. Rev. *33:* 105 (1977).

6 Cohn, M.: The molecular biology of expectation; in Plescia, Braun, Nucleic acids in immunology (Springer, Berlin 1968).

7 Cramer, M.; Krawinkel, U.; Melchers, I.; Imanishi-Kari, T.; Ben-Neriah, Y.; Givol, D.; Rajewsky, K.: Isolated hapten-binding receptors of sensitized lymphocytes. IV. Expression of immunoglobulin variable regions in (4-hydroxy-3-nitrophenyl) acetyl (NP)-specific receptors isolated from murine B and T lymphocytes. Eur. J. Immunol. *9:* 332 (1979).

8 Early, P.; Huang, H.; Davis, M.; Calame, K.; Hood, L.: An immunoglobulin heavy chain variable region gene is generated from three segments of DNA: V_H, D_H, and J_H. Cell *19:* 981 (1980).

9 Eichmann, K.; Ben-Neriah, Y.; Hezelberger, D.; Polke, C.; Givol, D.; Lonai, P.: Correlated expression of V_H framework and V_H idiotypic determinants on T helper cells and on functionally undefined T cells binding group A streptococcal carbohydrate. Eur. J. Immunol. *10:* 105 (1980).

10 Fellous, M.; Günther, E.; Kemler, R.; Wiels, J.; Berger, R.; Guenet, J.L.; Jakob, H.; Jacob, F.: Association of the H-Y male antigen with β_2 microglobulin on human lymphoid and differentiated mouse teratocarcinoma cells. J. exp. Med. *148:* 58 (1978).

11 Flaherty, L.; Zimmerman, B.; Wachtel, S.S.: H-Y antigen: cell surface mapping and testosterone-induced supramolecular repatterning. J. exp. Med. *:* 1020 (1979).

12 Honjo, T.; Obata, M.; Yamawaki-Kataoka, Y.; Kataoka, T.; Kawakami, T.; Takahashi, N.; Mano, Y.: Cloning and complete nucleotide sequence of mouse immunoglobulin γ1 chain gene. Cell *18:* 559 (1979).

13 Jerne, N.K.: Somatic generation of immune recognition. Eur. J. Immunol. *1:* 1 (1971).

14 Kemp, D.J.; Harris, A.W.; Cory, S.; Adams, J.M.: Expression of the immunoglobulin $C_\mu$gene in mouse T and B lymphoid and myeloid cell lines. Proc. natn. Acad. Sci. USA *77:* 2876(1980).

15 Lerner, M.R.; Boyle, J.A.; Mount, S.M. Wolin, S.L.; Steitz, J.A.: Are snRNPs involved in splicing? Nature, Lond. *283:* 220 (1980).

16 Lyon, M.F.: Gene action in the X-chromosome of the mouse (*Mus musculus* L.). Nature, Lond. *190:* 372 (1961).

17 Matsunaga, T.; Ohno, S.: Cellular immune response to non-MHC plasma membrane antigens. Transplant. Proc. *12:* 135 (1980).

18 Nagai, Y.; Ciccarese, S.; Ohno, S.: The identification of human H-Y antigen and testicular transformation induced by its interaction with the receptor site of bovine fetal ovarian cells. Differentiation *13:* 155 (1979).

19 Ohno, S.: Promethean evolution as the biological basis of human freedom and equality. Perspect. Biol. Med. *19:* 527 (1976).

20 Ohno, S.: The original function of MHC antigens as the general plasma membrane anchorage site of organogenesis-directing proteins. Immunol. Rev. *33:* 59 (1977).

21 Ohno, S.: The significance of gene duplication in immunoglobulin evolution (Epimethean natural selection and Promethean evolution); in Litman, Good, Comprehensive immunology, vol. 5, (Plenum Publishing, New York 1978). p. 197.

22 Ohno, S.: Origin of intervening sequences within mammalian genes and the universal signal for their removal. Differentiation *17:* 1 (1980).

23 Ohta, T.: Evolution and variation of multigene families. In Lecture notes in biomathematics, vol. 37 (Springer, Berlin 1980).

24 Owen, F.L.; Finnegan, A.; Gates, E.R.; Gottlieb, P.D.: A mature T lymphocyte subpopulation marker closely linked to the Ig-1 allotype C_H locus. Eur. J. Immunol. *9:* 948 (1979).

25 Richards, F.; Konigsberg, W.; Rosenstein, R.; Varga, J.: On the specificity of antibodies. Science *187:* 130 (1975).

26 Sakano, H.; Rogers, J.H.; Hüppi, K.; Brack, C.; Traunecker, A.; Maki, R.; Wall, R.; Tonegawa, S.: Domains and the hinge region of an immunoglobulin heavy-chain are encoded in separate DNA segments. Nature Lond. *277:* 627 (1979).

27 Sakano, H.; Hüppi, K.; Heinrich, G.; Tonegawa, S.: Sequences at the somatic recombination sites of immunoglobulin light-chain genes. Nature, Lond. *280:* 288 (1979).

28 Sakano, H.; Maki, R.; Kurosawa, Y.; Roeder, W.; Tonegawa, S.: The two types of somatic recombination necessary for generation of complete immunoglobulin heavy chains. Nature, Lond. *286:* 676 (1980).
29 Sawin, P.B.; Glick, D.: Inheritance of 'atropinesterase', a blood enzyme of the rabbit. Genetics, N.Y. *28:* 88 (1943).
30 Seidman, J.G.; Max, E.E.; Leder, P.: Aκ-immunoglobulin gene is formed by site specific recombination without further somatic mutation. Nature, Lond. *280:* 370 (1979).
31 Simpson, E.; Gordon, R.D.: Responsiveness to H-Y antigen, Ir gene complementation and target cell specificity. Immunol. Rev. *35:* 59-75 (1977).
32 Taniguchi, M.; Saito, T.; Tada, T.; Antigen-specific suppressive factor produced by a transplantable I-J bearing T-cell hybridomas. Nature, Lond. *278:* 255 (1979).
33 Wuilmart, C.; Urbain, J.; Givol, D.: On the location of palindromes in immunoglobulin genes. Proc. natn. Acad. Sci. USA *74:* 2526 (1977).
34 Zinkernagel, R.M.; Doherty, P.D.: Immunological surveillance against altered self components by sensitized T-lymphocytes in lymphocytic choriomeningitis. Nature, Lond. *251:* 547 (1974).

Susumu Ohno, PhD, Division of Biology, City of Hope Research Institute,
Duarte, CA 91010 (USA)

Prog. Allergy, vol. 28, pp. 40–65 (Karger, Basel 1981)

Differentiation of B Lymphocytes in Lineage Development and in the Immune Response

Considerations Derived from Differentiation Events Induced in vitro

Ulrich Hämmerling

Memorial Sloan-Kettering Cancer Center, New York, N.Y., USA

Contents

Introduction

To understand the whole is to know its parts. This old axiom has guided generations of experimentalists in their quest to unravel complex biological phenomena, but in no other field than differentiation research has there been such a resistance to crack the black box, labeled 'differentiation', and to reveal its inner workings. This is not to say that the phenomenon, the process of differentiation, is not definable as a research goal, although some confusion exists as to its meaning. Commonly, we understand as differentiation those processes that govern the formation of different strains of cells from the zygote, which in their specialized form interact to produce tissues, organs and lastly a whole individuum. In a narrower sense, differentiation applies to that process that leads to irreversible change and restriction of the potential of a cell. In *Jacob and Monod's* [37] words: 'Differentiation is present when cells with the same genome synthesize different proteins.' *Weiss* [84] has postulated, however, that this process be 'unidirectional and irreversible' engendering 'true constitutional differentials among the emerging daughter generations'. In light of experiments involving nuclear transplantation [14, 15, 17, 80] and somatic hybridization [25, 26] which show that the nuclear clock can be rewound by exposure to a different cytoplasmic environment, the definition of irreversibility can apply only to the intact cell. On this count, alteration of the genetic program by mutation, or by introduction of added genetic information by, for example, viral infection, has no part in differentiation; although, when investing a state of transformation in the cell, it may produce an effect indistinguishable from differentiation.

Differentiation then means an irreversible advancement of cells in time during which they attain a new phenotype, but in this course become progressively restricted in overall potential. Why then, have decades of differentiation research yielded such a dearth of understanding? The answer lies in the immense complexity of organization and regulation of tissues or organs in the context of which differentiation of individual cells occurs.

In most differentiating systems, we know neither the script, the actors, the director, the chorus line, nor the stage set. And inevitably, as late-coming observers we are ignorant of the happenings of the preceding scenes of the play, which may present itself to our eyes for only a short period of time.

Lymphopoiesis

The development of special strains or lineages of cells (T and B) during lymphopoiesis is governed in principle by similar laws of differentiation that apply to the development of other cellular lineages. As a graphic picture, one can view the development of an embryo as a succession of steps to bring about the specialization of millenia of cells which compose the complete animal. Since all of this begins with the zygote a vertical continuum of lines can be envisioned originating with the zygote and branching out to form an elaborate fan. In the adult individuum, the lower trunk is then severed leaving only the distal ends. Because of the axiom of irreversibility, and because primary embryonal tissues are not maintained in the grown animal, a problem exists in those tissues which are continuously regenerating (e.g. epidermis) or which can do so on demand (e.g. wound healing). Obviously, only distal parts of cell lineages can be regenerated in adult animals, necessitating the existence of specialized stem cells which continuously replenish themselves and furnish daughter cells to replenish the lineage. It follows that during and on completion of embryonic development cell lineages are formed which attain a limited autonomy for further development of each lineage. In due time, when diversification within a lineage has occurred, some degree of autonomy may again be delegated to separate branches of the lineages. There are, therefore, stem cells of stem cells, which in turn are derived from earlier stem cells.

The picture of several tiers of branching cell lineages can be used to symbolize the hemopoietic system. It is well established that a pluripotent stem cell exists which seeds all four major strains: erythroid, myeloid, lymphoid, and megakaryocytic cell lines [45, 78]. Pluripotentiality is ascribed to these stem cells in the sense that they can give rise eventually to all blood cells, but whether the stem cell has an immediate choice of becoming any one of the four cell types (or their committed progenitor cells), or whether an elaborate branching system gradually furnishes the major blood cell lineages is under dispute. It is, furthermore, unclear, whether or not lymphoid stem cells exist as autonomous progeny of the pluripotent stem cell. However, the distinctive sites where lymphocyte maturation occurs (e.g. thymus for T cells and bursa, or its equivalent, for B cells), and which do not seem to support maturation of other blood cells, suggest some degree of autonomy of lymphocyte lineages themselves [41]. Recent experiments using chromosomal markers have demonstrated the existence of a stem cell restricted to the T lineage [2].

Several features of the hemopoietic system in general and of lymphocytes in particular invite their use as subjects in differentiation research: lymphocytes are readily accessible; immature or mature lymphocytes predominate in different organs (e.g. bone marrow versus lymph nodes); there is a good deal of information on population kinetics; immunogenetics has provided numerous markers; several strains of mutant animals or congenital diseases in man provide convenient examples of blocked differentiation; numerous lymphoid leukemia cell lines have been established; lymphocytes display a number of measurable biochemical and biological functions.

Despite such detailed knowledge, the bewildering complexity of the lymphoid system has defied a precise description of the differentiation events leading to the establishment of lymphocyte strains, and the regulatory control imposed on the system. In this paper, some of the parts constituting the 'working parts' of one lymphocyte strain, the B cells, and their ontogenetic connections, will be described. The new information has been gathered by the systematic application of an experimental technique that allows the induction of surface antigenic changes in vitro [40]. In the course of these experiments, regulatory phenomena came to light which bear on the concept of differentiation itself. It turns out that induction of surface antigens may constitute a decisive part of the differentiation process, and as such may provide one means for analytical dissection of this enigmatic process.

Tissue Induction

In the 1920s experimentation with amphibian embryos by *Speman and Mangold* [71] and others revealed that the development of tissues in the ectoderm depended on influences exerted by the underlying endoderm. The term 'induction' was adopted to describe the action of one tissue, or group of cells, upon another tissue as an obligatory condition of progression in a pre-ordained developmental program (for instance, the formation of the nervous system from competent ectoderm). The concept of induction contains the element of determination, as revealed by the experimental observations that in all systems the time and the site of origin in the embryo were decisive in the type of tissue which was newly formed. On the contrary, a tight specificity of the inducing tissue for the competent responding tissue was not always apparent. Indeed, to this day, the nature of the inducing principle remains unknown, whether or not it is mediated by chemically definable and diffusible substances, or by contact between surfaces of the inducing and responding cells, where matching grids of orientated molecules provide the essential signals [6]. Because of the outspoken predetermination of the reacting tissue and the frequent lack of specificity, the inductor is assumed to have no informative role but to act solely as a regulator. This is not unlike the role attributed to antigen in the initiation of the immune response (and to the role of polyclonal activators which, in short-circuiting normal regulation, set essentially similar events in motion).

Induction of Surface Antigens

Induction of surface antigens in lymphocytes was first described in 'in vivo' experiments by *Komuro* et al. [38], when they noted that mice responded to injection of spleen cells by antibody formation to the thymocyte antigen TL, even though the transferred cells did not possess demonstrable amounts of the antigen. But these authors reasoned that the spleen population contained progenitor cells which mature in the recipient animal to become thymoyctes and as such, on expression of the TL antigen, might provide the antigenic stimulus. This assumption was corroborated by the finding that an intact thymus was essential. Subsequently, *Komuro and Boyse* [40] gained positive proof of the existence of pre-thymic T cells in bone marrow, and spleen populations when they showed that the appearance of TL and Thy-1, as well as of other thymocyte markers, can be induced in vitro. The experimental procedure required the enrichment of precursor cells by density gradient centrifugation and short-term tissue culture of cells harvested from the low density portion of the gradient. The decisive finding was that the conversion of TL^- to TL^+ cells occurring in a period of only 2 h. was dependent on the presence in the culture of a cell free saline extract of thymus [39]. Extracts of other organs (e.g. spleen, kidney) did not elicit a similar inductive effect. Hence, it was concluded that thymus extracts contained one or more activities which could mimic similar inductive events in vitro and which characteristically occurred in the thymus in vivo.

The observation that competent precursor cells can readily change their phenotype has not been unique to prothymocytes (a term adopted to describe the immediate precursor of cortical thymocytes, comprising most likely those cells which populate the thymus anlage in the embryo, and continuously replenish the thymocyte pool in adult life). Analogous changes in phenotype can also be induced experimentally in precursors of B lymphocytes, suggesting the generality of the phenomenon as a crucial event in the differentiation process [18, 19, 22, 61, 69]. Among the markers which can be induced on B lymphocytes are: surface-associated immunoglobulin M (IgM and IgD); the I region-associated antigens Ia-AB and Ia-EC; the B cell-specific surface antigens Lyb 2 and Lyb 5; the plasma cell antigen PC.1; the receptors for the third complement component CR, and for the Fc fragment of IgG (FcR). In all cases, the frequency of precursor cells is around 1% or less, and enrichment by serum albumin or Ficoll gradient centrifugation is required to allow evaluation of induction in serological assay procedures (mostly cytotoxicity assays). The proportions of precursor cells inducible for the above B cell markers vary significantly within different lymphoid organs. For instance, adult bone marrow contains appreciably more precursor cells inducible for IgM than cells inducible for CR or IgD. Conversely, CR-inducible cells are more frequent in spleen than are IgM-inducible cells. These differences in tissue distribution suggest that several distinct populations of inducible cells exist and, as will be discussed below, this assumption is well supported by serveral pieces of evidence.

The induction assay measures a change of serologically unreactive cells to responsive ones. The serological procedures commonly used are cytotoxicity (for surface antigens) and rosette assays (for Fc and C receptors). Both assays do not measure quantitatively increases in antigen expression at the cell surface. Rather, they provide a means to enumerate the phenotypically positive cells in the population. That the increase in sensitivity to cytotoxic antibody is related to increased antigen expression is an accepted standard. (For instance, the lower sensitivity of thymocytes to cytotoxic anti-H-2 antibody as compared with peripheral lymphocytes correlates well with a sixfold difference in H-2 sites. Conversely, thymic T cells with 5×10^5 Thy.1 molecules per cell are more sensitive to anti-Thy 1 antibody than postthymic cells with 1×10^5 Thy.1 sites [24].) However, given the indirectness of cytotoxicity assays, one can propose that the inducing agent confers to the cell an increased sensitivity to cytolytic antibody. That this trivial explanation for induction is unlikely follows from the facts that: (a) only a selected group of cells migrating to the upper layer of the BSA gradient responds in the induction experiment and that cells from lower layers do not show a similar antigenic change; (b) induction can be inhibited by seemingly innocuous hormones (see below), and (c) induction can also be visualized by serological binding techniques (rosette formation, immunofluorescence) which directly measure antigen quantity. Moreover, by the method of quantitative absorption it has been possible to demonstrate that induction is associated with an increase in surface antigen expression [22].

The biochemical and molecular parameters of induction are not well understood. In the absence of methods to purify inducible precursor cells to homogeneity no direct biochemical measurements have been possible. In particular, the question is unresolved whether or not the surface antigen under study is synthesized during culturing, as attempts to biosynthetically label surface antigen in the induction experiment have been unsuccessful.

However, experiments with biosynthetic inhibitors inform us that while protein and RNA synthesis are required for antigen changes to occur, DNA synthesis is not necessary [7, 74].

Sequence of B Lymphocyte Differentiation

Our current interpretation of induction of surface antigens is one of a cell which has made a commitment to a new genomic program, as a result of which new biochemical potential emerges, inclusive of the potential to synthesize new surface antigens. Such new potential may be at the expense of other, existing potential which may be lost. How such selection and implementation may be controlled is not known, but the new concept emerging from lymphocyte induction experiments stipulates that the final expression of the altered potential, the change in phenotype, is a tightly and independently regulated end phase of the whole differentiation process. One can visualize inducible cells as cells which have undergone genomic reprogramming and have implemented the new potential, but which depend for expression on extracellular control.

In a graphic picture, the inducible cell would be symbolized by one that has moved along a differentiation pathway and has reached the border of its present differentiation compartment. Induction could then be likened to ushering the cell into the next compartment. In practical terms, the possibility of monitoring the change in phenotype during the passage of a cell from one compartment to the next offers the opportunity to identify two adjacent differentiation compartments. Given enough phenotypic traits to be monitored, and given that these traits are not switched on simultaneously but are expressed at different time points, one can identify a likely sequence of differentiation steps prevailing in the lineage by using the induction assay. These conditions are indeed fulfilled for B lymphocytes [20]. There are at present seven markers (table I) which can be induced on the surface of B lymphocytes under the conditions of the *Komuro and Boyse* [40] assay.

The seven markers identify five different sets of inducible cells. This conclusion is based on the analysis of the phenotypes of inducible cells by the method of negative immunoselection. This is a method commonly used in many studies in which the functional properties of a group of cells within a mixed population are to be determined. The most convenient technique is selctive cytolysis with antibody and complement. Where

Table I. Effect of negative selection on the inducibility of various B cell markers

Target of negative selection	Induction observed for surface marker						
	Lyb 2	IgM	Ia	IgD	CR	Lyb 5	PC. 1
None	yes	yes	yes	yes	yes	yes	yes
Lyb 2	yes	no	no	no	no	no	n.d.
IgM	yes	yes	no	no	no	no	no
Ia	yes	yes	yes	no	no	no	no
IgD	n.d.	n.d.	yes	yes	yes	n.d.	n.d.
CR	yes	yes	yes	yes	yes	yes	no
Lyb 5	n.d.	yes	yes	yes	yes	yes	n.d.

surface receptors, rather than antigens, are involved specific rosetting techniques in conjunction with differential centrifugation have been developed. The rationale of experiments of this kind has been to selectively eliminate a subset of lymphocytes demarcated by a particular surface antigen (for instance the Ia antigen) and test in the induction assay whether or not other surface markers (e.g. surface IgM or complement receptor) can be induced among the surviving cells. Depending on the outcome one can reach a definitive conclusion whether or not a given inducible precursor cell carries other surface markers. But more significantly, one can deduce from such data the probable sequence in which various markers arise in ontogeny of the lineage. As an illustrating example, we have found that elimination of Ia$^+$ cells affects the induction of CR whereas it does not interfere with the induction of IgM [31]. The obvious conclusions are that CR- and IgM-inducible cells comprise phenotypically different cell sets, and that Ia expression must precede CR expression in ontogeny (because the CR-inducible cells is already Ia$^+$). By these criteria alone, Ia- and CR-carrying cells are not identified as B lymphocytes and they could belong to monocytes which also express these markers. Crucial evidence is rendered in experiments in which elimination by cytotoxic anti-μ antibody has abrogated both Ia and CR induction. Thus, these two inducible cells must be of IgM$^+$ phenotype; hence they are B cells. Detailed data concerning the phenotypes of the known inducible B cells are summarized in figure 1. Of the seven markers studied, four (Lyb 2, IgM, Ia and PC.1) are inducible on separate cells distinguished by different phenotypes, whereas three (IgD, CR, and Lyb 5) are inducible simultaneously on cells of a single phenotype. The existence of five sets of inducible cells distin-

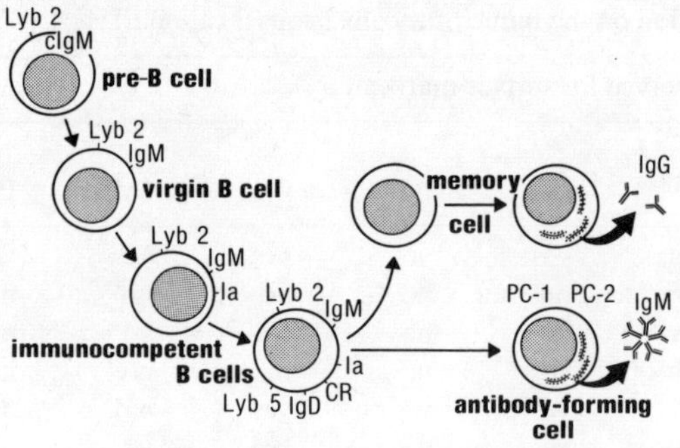

Fig. 1. Sequence of B lymphocyte differentiation deduced from induction of phenotype changes.

guishable in their phenotypes with regard to the seven B cell markers implies that maturation of the B cell requires at least five separate differentiation steps, each of which is characterized by the additon of a new surface marker; a sixth step is indirectly implied in the formation of the Lyb 2 inducible cells, which are characterized by the absence of all known surface antigens.

Inductive events are not necessarily associated only with gain of phenotypic traits but may also entail loss of such traits. It may only be due to the lack of early lymphocyte markers that we fail to recognize such losses. However, in the case of IgD, induction of conversion from IgD^+ to IgD^- has been documented [69]. This finding implies that the elimination of IgD from the surface demarcates a normal differentiation event. Thus, like TL in thymocytes, IgD may be expressed only during a limited stretch of the B lymphocyte life cycle.

Induction as a Regulatory Process

While the reaction of the cell is preordained, and no deviation from a pre-programmed phenotype is permitted, the mode of induction does not display a narrow specificity. As in many experimental differentiating systems, the induction stimulus for B progenitor cells can be provided by a number of agents in dose-dependent fashion. These include lipopolysaccharide, adrenergic hormones, poly A:U, choleratoxin, and prostaglandins (table II). A common feature of these agents is their ability to raise cellular

Table II. Inducing agents of pro-B cells and prothymocytes

Material	Effective concentration	Induction	
		pro-B cells[a]	prothymocytes[b]
LPS	10 μg/ml	yes	yes
Lipid A	5 μg/ml	yes	yes
Thymopoietin	10^{-8} *M*	no	yes
Ubiquitin	10^{-7} *M*	yes	yes
Isoproterenol	5×10^{-5} *M*	yes	yes
Epinephrine	5×10^{-5} *M*	yes	yes
PGE_1	5×10^{-4} *M*	yes	yes
Dibutyryl cAMP	5×10^{-4} *M*	yes	no
Tumor-necrotizing serum	1:2000 dilution	yes	no

[a] Induction of $Ia^- \rightarrow Ia^+$ and $IgM^- \rightarrow IgM^+$.
[b] Induction of Thy-$1^- \rightarrow$ Thy-1^+.

cyclic AMP levels in some systems. On the basis of these observations, and the persistent finding that exogenously added dibutyryl cyclic AMP can trigger induction, a close association of the mechanism of induction with cyclic nucleotide metabolism has been proposed [62]. Although at present a causative role as a second signal in induction cannot be attributed to cyclic AMP, primarily because purified inducible cells are not available in quantities that would allow direct cyclic nucleotide measurements, there is considerable circumstantial evidence to suggest a close association.

Among the two basic mechanisms of signal transduction in cells, namely activation of membrane-associated nucleotide cyclases, and ion flux through gated channels [55], most features of phenotype induction fit better with the first mechanism: (1) The inducing agents act at the surface, and there appears to be no requirement for them to enter the cell. (2) The presence of the inducing agent is required only for a crucial few minutes, and thereafter the inducing agent can be removed without affecting the magnitude or time course of induction. (3) Once initiated, the process seems to be insensitive to further regulation and is irreversible, provided that favorable biochemical conditions exist. (4) Drugs known to modulate phosphodiesterase activity associated with the adenylate cyclase complex can influence induction positively or negatively, in accordance with their inhibitory or stimulatory action on this enzyme. Methylxanthines commonly block phosphodiesterase and by slowing decay of cyclic AMP indirectly enhance cellular cyclic AMP levels; although the methylxanthines themselves do not act as inducing agents, they act synergistically with β-adrenergic hormones in the prothymocyte induction system [62]. This is revealed by a shift of the minimal effective dose to a significantly lower concentration. Conversely, imidazole, a drug which potentiates phosphodiesterase, has been shown to antagonize induction. In the presence of this drug, higher doses of β-adrenergic hormones are needed to effect induction. (5) Hormones classically associated with cyclic AMP elevations have consistently been found to act as inducing agents. These include isoproterenol, epinephrine, and choleratoxin [22, 61, 63]. In this context, the inducing activity of cyclic AMP and its dibutyryl derivative should be emphasized again, although the unphysiologically high concentrations required may stimulate other cellular reactions. (6) Cholinergic hormones, foremost acetylcholine and carbamylcholine, persistently induce elevations in cellular cyclic GMP levels. These hormones have not been shown to commonly serve as inducing agents of B lymphocytes (but note the exemption of induction of the PC.1 alloantigen) [19]. (7) *Parker and Morse* [51] have isolated a protein from *B. pertussis* which on much circumstantial evidence can directly block adenylate cyclase. This protein has been shown to inhibit the changes of phenotype in B lymphocytes inducible by a number of hormones [*Hämmerling*, unpublished results]. It remains to be seen whether or not the pertussis protein interferes with the cyclase complex in this system. But the fact that the block can be overcome by exogenously added cyclic AMP would point to something more than a non-specific interaction of pertussis protein with cellular metabolism and implicate the cyclic AMP pathway as the primary target.

In addition to the set of inducing agents another group of substances has been compiled which can inhibit induction of phenotype changes [21]. Because many of these substances are well characterized hormones (table III) it is likely that they act by imparting a negative signal to the cell. The principal experimental finding has been that exposure of B or T precursor cells to carbamyl choline induces a refractory state during which the cells cannot be induced to undergo phenotype conversion. This inhibition is not immediately reversible by removal of the inhibitor from the system, but regeneration of inducibility requires culturing for several hours. In view of the diverse nature of inhibitors, all interacting with different sets of receptors, and in view of the fact that a given inhibitory hormone neutralizes the effect of all inducing agents, competition for a common 'induction'

Table III. Inhibitors of induction of pro-B cells[a]

Agent	Effective concentration, *M*
Carbamyl choline	10^{-5}
Human growth hormone	10^{-8}
Insulin	10^{-7}
8-bromo cyclic GMP	10^{-3}
Thymopoietin	10^{-8}

[a] Induction of $Ia^{-} \rightarrow Ia^{+}$.

receptor site seems unlikely. Rather, the inhibitory agent interacts positively with the cell and delivers a negative signal, thereby imposing a refractory state on the cell.

In context with the earlier discussion of the commitment of cells and the presumed irreversibility of differentiation, it is noteworthy that once the induction signal is received and those events are initiated which lead, 2 h later, to the appearance of new surface antigens, the addition of inhibitory hormones is ineffective. This has been convincingly demonstrated in experiments in which inducing and inhibitory agents were added sequentially to cultures of inducible prothymocytes [21]. When the inhibitor was added for a period as short as 5 min before the inducing agent no induction was seen. However, when the order of additions was reversed, and the inducing agent preceded the inhibitor by only 5 min., full expression of the thymocyte phenotype was observed. The action of inhibitory and inducing agent is comparable, and entirely compatible, with the action of a hormone on a tissue. Indeed, many of these agents, notably adrenergic and cholinergic agents, are classically defined as hormones, which exert regulatory function by triggering inherent cell responses.

A molecular basis for negative regulation of inducible cells has not been identified but a future search will have to take into account that many of the inhibitors tested (but not all) show an association with cyclic nucleotide metabolism. Foremost, on the list is carbamyl choline, which regularly causes elevations of cyclic GMP in many tissues [35]. This finding and the observation (which is not without reservations) that 8-bromo cyclic GMP is also inhibitory in the lymphocyte induction system have led to the speculation that elevations in cyclic GMP may exert restraining influences on the differentiating cell. It should be noted that associations with cyclic GMP have not been established for all inhibitory agents.

Opposite effects of hormones causing cyclic AMP elevations vis-à-vis other hormones causing cyclic GMP elevations have been noted in several systems [reviewed in 16]. For instance, initiation of DNA synthesis and cell division is thought to be associated with an augmentation of cyclic GMP, and this cellular event is opposed regularly by agents which elevate cyclic AMP levels [82]. If future work can substantiate the claim that cyclic nucleotides are secondary mediators in the lymphocyte induction then the concept that the two protagonists, cyclic AMP and cyclic GMP, promote mutually opposing effects has been enriched by another example, and this may underscore the generality of this regulatory process. Contrary to the promotion of cell division by cyclic GMP elevation, the induction of phenotype changes, which presumably constitutes the final phase in differentiation processes, seems to be promoted by cyclic AMP elevation, but is restricted by cyclic GMP elevation. A simple formula, in which a secondary signal of induction is ascribed to an elevated cellular cyclic AMP over cyclic GMP concentration is not tenable in view of the finding that in one induction system symbolized by the expression of the PC.1 alloantigen on late B lymphocytes, the secondary signal appears to be mediated by precisely the reverse ratio, requiring a high cyclic GMP over cyclic AMP concentration [19].

Regardless of the nature of secondary mediators and whether or not these may be attributable to cyclic nucleotides, the picture of an inducible lymphocyte is one of a cell which is controlled by two opposing signals transmitted by two different sets of surface receptors. The immediate fate of the cell, whether it can complete the differentiation step on which it has embarked, is regulated by the relative strength of the two signals.

B Cell Development and Antibody Specificity

It is well established that the pathway of B lymphocyte development occurs in mammals and birds within the context of hemopoiesis, but follows a special route distinct from the other hemopoietic lineages. The derivation from the pluripotent stem cells is well established by a variety of experiments involving hemopoietic chimeras and cell marking techniques relying on either inherited chromosomal anomalies, as for instance, in the CBA/T6 strain, or radiation-induced chromosomal abberations [reviewed in 45]. The experiments of *Moore and Metcalf* [46], *Osmond and Nossal* [49], *Moore and Owen* [47], and *LeDouarin and Jotereau* [41] have unequivocally demonstrated the origin of hemopoietic cells from primordial stem cells generated in the blood islands of yolk sac. Although in the mouse no traces of lymphoid development are noticeable during the time of maximum function of this hemopoietic anlage, the essential progenitor cells are clearly present, as evinced by the potential to generate superficial spleen colonies when injected into irradiated mice [76]. The responsible cell, the colony-forming unit in spleen, CFU-S, has been shown to fully reconstitute the hemopoietic system of lethally irradiated recipient mice and hence qualifies as a pluripotent stem cell.

The evidence that CFU-S consist of a single cell type with pluripotentiality comes from chromosome marker studies. Upon transfer of CFU-S in which unique chromosome anomalies were induced, recipient mice displayed the same anomaly in several cell types of a heterogeneous spleen colony [86, 87]. Although a link between thymus cells and CFU-S had been already established in the chimera experiments of *Wu* et al. [86, 87] the common origin of all lymphoid cells and myeloid cells was demonstrated only recently by *Abramson* et al. [2]. These authors also furnished evidence for the existence of two types of restricted stem cells, committed to the T lymphocyte and myelocyte lineages, respectively. These findings suggest that lymphoid and myeloid cells acquire a measure of autonomy by deriving from restricted stem cells, but the lineages may not be entirely truncated from pluripotent stem cells. The yolk sac ceases to function as a major resevoir of stem cells after day 13 of embryonal life, and this function is delegated in succession to fetal liver and beginning at shortly before birth to bone marrow and to a lesser degree, to spleen. Bone marrow becomes the major site of stem cell generation throughout life [reviewed in 45].

As mentioned before, lymphoid development does not occur in the yolk sac. The onset has been traced to fetal liver at day 12 when primitive B cells are first discerned [43, 44, 57]. The potential of fetal liver organ cultures to generate Ig-positive cells has been known for some time, but the direct demonstration of surface-Ig positive cells requires special, highly sensitive techniques, such as radioautography, immunofluorescence, and rosette formation [48, 59, 70]. While these techniques have all yielded convincing evidence that a small proportion of cells in 13-day fetal liver display immunoglobulin, the precise classes of immunoglobulin remain controversial. There is agreement, however, that 7S IgM is synthesized in lymphoid cells of fetal liver.

The knowledge of B cell ontogeny made a leap forward by the

discovery that primitive B cells have cytoplasmic Ig at a time when little or no Ig is seen at the cell surface. *Raff* et al. [57] were the first to notice IgM in the cytoplasm of fetal liver cells and bone marrow cells. Although initially thought that a complete immunoglobulin molecule was synthesized, it is now more likely that these cells contain only μ-chains. Confirmatory evidence comes from the study of tumor cell lines with pre-B cell characteristics, notably the chemically induced tumor cell line 70Z/3 and several Abelson virus-induced cell lines [50, 58]. To the extent that these malignant lines reflect a normal analogue, it is clear that B cells pass through a stage in which the μ heavy chain is synthesized before light chain synthesis commences. The uncoordinate onset of transcription and translation of immunoglobulin genes is peculiar and one might speculate that it is a consequence of the unique genomic rearrangements prerequisite of Ig gene expression [77].

As a general picture emerges, B cell differentiation appears to encompass three tiers: First, there is development of the B cells in generic terms, governed by the laws of tissue differentiation. And second, superimposed on the first, is the diversification of B lymphocytes with respect to antibody specificity [77]. This can be regarded as a special type of differentiation aimed at providing B cells with their immunological individuality. The basic facts of assembly of V, J, and C gene bits to a functional immunoglobulin gene are becoming known in rapid progression, but the actual mechanism of gene splicing and the regulation of the process remain unknown [9, 12, 67, 83]. What is quixotic and at the same time unique to immunocytes is the phenomenon of allelic exclusion, which describes the fact that only one haploid set of immunoglobulin genes is assembled and expressed in a given B cell clone [34]. The third tier refers to the fact that whereas B cells, once committed, maintain their specificity for antigen irreversibly in subsequent daughter generations, they have the option of changing the class of immunoglobulin they synthesize. In molecular terms, this infers the switch from one C_H gene to another, and most experimental evidence is in accord with this hypothesis. DNA hybridization studies with cloned, class-specific cDNA pobes have unequivocally shown rearrangements of $C_{H\gamma}$ genes in IgG producing plasmacytomas associated with deletions of the $C_{H\mu}$ genes from one or both chromosomes [9, 12, 34]. Whether $C_{H\mu}$ deletion is obligatory in C_H switching remains to be seen. In at least one situation, namely during the switch form μ to δ expression, which seems to be an essential step during normal B cell differentiation, the evidence suggests that the B cells can later on revert back to μ expression

[68]. Deletion of the $C_{H\mu}$ gene at this stage of B cell differentiation would generate a paradox.

A fourth condition, peculiar to lymphocytes warrants a comment, namely that many lymphocytes after establishing the immunological identity are eliminated or permanently inactivated because they happen to be directed against self-antigen. In addition, because of failure to encounter their respective foreign antigen, the vast majority of the mature B lymphocytes will never become terminally differentiated. Whereas this vast overproduction of cells is not unrivaled (for instance, it is easily surpassed by germ cell production) it is unique to the lymphoid system because most lymphocytes never have the opportunity of exercising the full potential for differentiation inherent in them. Yet, lymphocytes are capable of responding rapidly and in great numbers, even after many years of quiescence.

Regulation of the Immune Response

In the preceding it has been proposed that with exception of expression of immunoglobulin genes which is a unique event for lymphocytes, all other events can be explained in the context of general tissue differentiation. In the following an attempt will be made to place the induction of the immune response into the same general cell biological framework.

The essence of initiation of antibody formation is that foreign antigen serves the role of selecting among the library of lymphocytes with individual specificity those cells which match the antigen specificity closely. This match of antigen-determinant and surface-associated antibody must be understood in terms of affinity above a certain threshold. In principle, the events set in motion and the end result are clear: the B lymphocyte is activated, begins to proliferate to form a clone and the cells of this clone differentiate to plasma cells. However, the details of this happening, the actual sequence and number of steps, as well as the possibility of divergent pathways to yield memory cells, and plasmacytes devoted to different immunoglobulin classes are yet to be identified. There are numerous indications that the initiation depends on the interaction of the B lymphocyte with several cell types, notably macrophages [81] and T helper cells [4]. However, the concerted action of these cells may subserve primarily the roles of presenting antigen properly to the B cell and to ensure the efficacy of the system. The question arises whether the cellular focusing devices can be replaced by soluble factors.

Before entering a meaningful discussion we must identify the structures of the cell surface utilized in the alignment of cooperating cells. Surface immunoglobulins and products of the major histocompatibility complex are among the molecules regulating cell contact. Their precise role remains to be defined. When the first reports concerning expression of IgM on B cells appeared a decade ago [53, 56] it was widely believed that the formal demonstration of receptor function for antigen-binding and activation of the cell would soon be forthcoming. Instead, the biochemical scrutiny of B cells revealed that two immunoglobulins, IgM and IgD, are expressed on the same immunocompetent cell [52, 79], dashing the hope for a simple explanation. Among much speculation that IgM and IgD have different functions in the induction of either tolerance or immunity, the reasons for dual Ig expression remain obscure.

On the other hand, much progress has been made in the understanding of the role of I region products in cellular cooperation. It seems reasonable that specialized cells of the T cell circuit form coalitions with macrophages [66] to either enhance or subdue the humoral immune response, and that the intricate relationships between these different factions are largely regulated by cell contact. The mechanism of cellular recognition seems to reside in the MHC products. This realization came about by several observations that matching of H-2 or Ia determinants was required for efficient cooperation. Thus, H-2 restrictions were noted in effector-target cell interaction of cytotoxic T cells, but Ia differences were irrelevant [88]. Conversely, the initiation of effector T cells and B cells required I-A region correspondence [36, 70], whereas suppressive cells appear to be focussed to their targets by I-J interactions [11; and *Gershon*, personal communication]. As suppressor cells are thought to communicate primarily with amplifier T cells (preventing expression of the full potential of the latter) the main issue concerning the activation of B cells in the immune response is whether the I-A molecules themselves have signal function, and whether I-A interactions between B and T cells are obligatory. This issue is at present unresolved, but the role of I-A on cells ancillary to antibody induction is well documented. T cells in general recognize an Ia-antigen complex on macrophages [60], and there are good reasons to believe that this dual function, i.e. binding of antigen and recognition of syngeneic MHC determinants, is an intrinsic feature of the immune receptor on T cells. As a result of such interaction T cells and macrophages are activated (i.e. they probably differentiate) and release factors regulating B lymphocytes [10, 29]. In studies employing intact activated T helper cells it became apparent that antigen specific B cells accept help only of helpers of matching Ia. This phenomenon is best explained as an active recognition of the B cell Ia by T cells, and this is demonstrated by the fact that T cells heterozygous for Ia can individually recognize only one of the parental Ia types [42, 73, 75].

In apparent contrast to the MHC restrictions of cellular interactions, there is evidence that soluble factors can provide help irrespective of their MHC type. Macrophage-derived factors can replace the intact macrophage in experimental antibody formation to sheep red blood cells, and no H-2 or Ia restrictions have been noted. The factor responsible for this regulatory function has been termed interleukin 1 (Il-1) [1], and is known to constitute the active principle of a number of biological factors, such as lymphocyte-activating factor [13], tumor-necrotizing factor [32], B cell differentiation factor [28], B cell activation factor [85], or a thymocyte-stimulating factor [8]. One of the measurable properties of Il-1 has been its ability to induce B cell differentiation in the Komuro-Boyse system [31, 32]. These latter findings are revealing, as they pin-point the role of Il-1 as an inducer of differentiation, rather than a mitogen, for B cells. Il-1 has not been shown to have mitogenic properties for B cells, but it can promote proliferation of thymocytes [5]. It may be relevant in this context to reminisce on the dual and apparently opposite effects which regulatory molecules can exert on B cells versus T cells. Thymopoietin has been shown to induce differentiation primarily in prothymocytes and to block B cell differentiation; vice versa, Il-1 acts as an inducer of differentiation of B cells, but inhibits the induction of prothymocytes [21]. Evidently, the response to either of these regulatory molecules is preordained by the cell type and also by the state of differentiation.

The soluble antigen non-specific T helper factor identified in concanavalin A-stimulated or allogeneically stimulated culture fluids neither carries Ia determinants nor does it appear to be subject to I restrictions in its action on B cells [65]. Although not entirely proven, it is likely that the major, antigen-nonspecific factor of activated T cells, variously termed T cell-replacing factor (TRF) [64] or allogeneic effect factor (AEF) [4] (reflecting its mode of preparation), acts on B cells by promoting clonal outgrowth, once these cells have been activated by specific antigen, macrophages, and/or macrophage-derived factors. This mode of function of TRF is suggested by the kinetics of interaction in the in vitro immune response, primarily by the observation that TRF is not needed at initiation of culture but is most effective 24–48 h later [33, 65]. Il-1 and TRF have been shown to act in synergy. While either factor alone can provide a moderate degree of help, the concerted action of both dramatically increases the number of plaque-forming cells which is the common measure of the success of the immune response [33].

Further insight into the mechanism of activation of B cells came from studies of the frequencies of antigen-responsive precursor B cells in the limiting dilution analysis introduced by *Quintans and Lefkovitz* [54]. As shown by *Hoffmann* [27], the addition of Il-1 to cultures of purified B cells resulted in a dramatic increase in the number of cells that can give rise to anti-SRBC plaque-forming cells, and this effect is best explained as a recruitment of immature or dormant B cells into a state where they can enter the immune response. This effect of B cell recruitment is entirely in accord with the demonstrated capacity of Il-1 to induce differentiation in B precursor cells.

TRF does not mediate a similar inducing activity. Neither has it been possible to induce B cell phenotypic conversion [*Hämmerling*, unpublished results], nor is TRF engaged in the recruitment of B cells into the immune response.

Activation of B cells from the dormant state is most likely an event involving the shifting from G_0, or a particular phase within G_0, toward G_1. As experimentation has been impractical because of a general inablility of defining or manipulating G_0, information on the relationship between cell cycle status and B cell responsiveness is scarce. The recruitment of B cells by Il-1 may reflect in part the induced maturation of CR^- cells into the CR^+ pool [30], but the magnitude of the measured effect is unlikely to be accountable by a mere increase in the CR^+ pool of cells most readily engaged in the immune response because the relative proportion of CR^- cells in the peripheral lymphoid organs, lymph node and spleen, are small. Instead, the action of Il-1 in concert with specific antigen may well have the effect of lifting the dormant B cells into the G_1 phase. This transition may then enable T cell factors (notably TRF) to drive the cells further in the cell cycle and thus promote the essential clonal outgrowth.

Recent studies by *Andersson* et al. [3] address this question by aid of cloned helper T cells. They conclude that whereas no MHC restriction exists in the polyclonal activation of B cell blasts, resting B cells require histocompatible T cell help for proliferation and initiation of Ig secretion.

A further parameter influencing the ability to initiate antibody formation is the state of differentiation of the responding B cell. The question is whether a relationship exists between the ontogenetic scheme derived from investigations of the surface phenotype (fig. 1) and the propensity to initiate antibody formation. Information on this issue is scarce and somewhat conflicting, but it appears that the most advanced B cells, i.e. those displaying the complement receptor, and IgD are more readily induced to immunoglobulin secretion than cells of the phenotype IgM^+ Ia^+ IgD^- CR^-. This has been deduced from the markedly different requirements for accessory cells and/or factors in the immune response of isolated CR^- B cells compared to CR^+ cells [30]. Whereas CR^- cells are obligatorily dependent on intact T helper cells and macrophages, the induction of the immune response to sheep cells in CR^+ cells requires only the presence of Il-1 and TRF. Similarly, the CR^+ cells have a greater facility to respond to polyclonal activation by LPS, reaching maximal levels of Ig synthesis 24–36 h earlier than CR^- cells [23].

Coda

The phenomenon of induction of phenotype changes suggests more than any other system of differentiating lymphocytes that the immediate reaction of the cell is predetermined and dictated by its momentary condition. It would be desirable to define this condition in the context of the cell cycle, but other than concluding that inducible cells are somewhere in G_0 (because they do not seem to have the immediate option of division or have recently undergone division, and hence are neither in G_1, S, G_2 or M phases) there is no fitting description. The condition of inducibility may well represent a 'gap phase' akin to G_2, but related to a differentiation cycle (fig. 2). Currently, all differentiation events are assigned to the G_0 phase. The necessity follows that G_0 be divided into measurable phases, and this

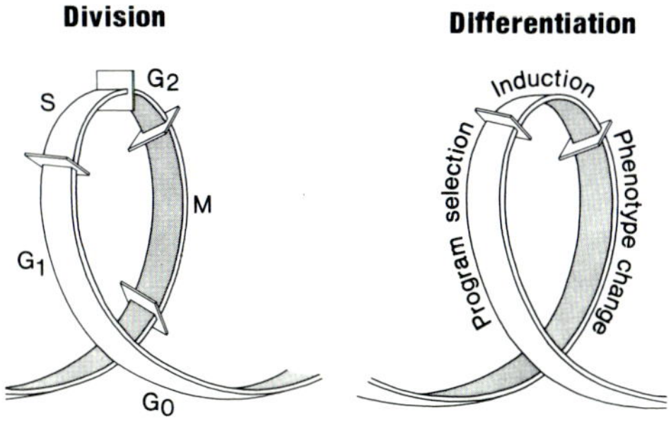

Fig. 2. Division and differentiation.

is a major task for contemporary cell biologists. The condition of inducibility may reflect such a phase and since it is accessible for experimental analysis and definition it may constitute a step toward refinement of the cell cycle.

Acknowledgement

The expert and patient help of *Barbara LoFaso* in the preparation of the manuscript is greatly appreciated.

This work has been supported by the following grants of the National Institutes of Health: CA 07848, AI 14831, and CA 25508; and the National Science Foundation: PCM 77-08018.

References

1 Aardeen, L.; et al.: Revised nomenclature for antigen-nonspecific T cell proliferation and helper factors. J. Immun. *123:* 2928–2929 (1979).

2 Abramson, S.; Miller, R.G.; Phillips, R.A.: The identification in adult bone marrow of pluripotent and restricted stem cells of the myeloid and lymphoid systems. J. exp. Med. *145:* 1567–1579 (1977).

3 Andersson, J.; Schreier, M.H.; Melchers, F.: T-cell dependent B-cell stimulation is H-2 restricted and antigen dependent only at the resting B-cell level. Proc. natn. Acad. Sci. USA *77:* 1612–1616 (1980).

4 Armerding, D.; Katz, D.H.: Activation of T and B lymphocytes in vitro. I. Biological and biochemical properties of an allogeneic effect factor (AEF) active in triggering specific B lymphocytes. J. exp. Med. *140:* 19–37 (1974).

5 Beller, D.I.; Farr, A.J.; Unanue, E.R.: Regulation of lymphocyte proliferation and differentiation by macrophages. Fed. Proc. *1978:* 37–91.

6 Bennett, D.; Boyse, E.A.; Old, L.J.: Cell surface immunogenetics in the study of morphogenesis; in Silvestri, Cell interactions. 3rd Lepetit Colloquium, pp. 247–263 (North-Holland, Amsterdam 1972).

7 Boyse, E.A.; Abbott, J.: Surface reorganization as an initial inductive event in the differentiation of prothymocytes to thymocytes. Fed. Proc. *34:* 24-27 (1975).

8 Calderon, J.; Kiely, J.M.; Lefko, J.L.; Unanue, E.R.: The modulation of lymphocyte functions by molecules secreted by macrophages. I. Description and partial biochemical analysis. J. exp. Med. *142:* 151-164 (1975).

9 Cory, S.; Adams, J.M.: Deletions are associated with somatic rearrangement of immunoglobulin heavy chains. Cell *19:* 37–51 (1980).

10 Dutton, R.W.; Falkoff, R.; Hirst, J.; Hoffmann, M.K.; Kappler, J.; Kettman, J.; Lesley, J.; Vann, D.: Is there evidence for a non-antigen specific diffusible chemical mediator from the thymus derived cells in the initiation of the immune response? in Amos, Progr. Immunol., pp. 355–359 (Academic Press, New York 1972).

11 Eardley, D.D.; Hugenberger, J.; McVay-Boudreau, L.; Shen, F.W.; Gershon, R.K.;

Cantor, H.: Immunoregulatory circuits among T-cell subsets. I. T-helper cell induce other T-cell sets to exert feedback inhibition. J. exp. Med. *147:* 1106–1115 (1978).
12 Early, P.W.; Davis, M.M.; Kaback, D.B.; Davidson, N.; Hood, L.: Immunoglobulin heavy chain gene organization in mice: analysis of a myeloma. Proc. natn. Acad. Sci. USA *76:* 857–861 (1979).
13 Gery, I.; Handschuhmacher, R.E.: Potentiation of the T lymphocyte response to mitogens. III. Properties of the mediators from adherent cells. Cell. Immunol. *11:* 162–169 (1974).
14 Gurdon, R.B.: Nuclear transplantation and the control of gene activity in development. Proc. R. Soc. B *176:* 303–314 (1970).
15 Gurdon, J.B.; DeRobertis, E.M.; Partington, G.: Injected nuclei in frog oocytes provide a living cell system for the study of transcriptional control. Nature, Lond. *260:* 116–120 (1976).
16 Hadden, J.W.: Cyclic nucleotides in lymphocyte proliferation and differentiation; in Hadden, Coffey, Spreafico, Immunopharmacology, pp. 1–28 (Plenum Publishing, New York 1977).
17 Hämmerling, J.: Über formbildende Substanzen bei *Acetabularia mediterranea.* Ihre räumliche und zeitliche Verteilung und ihre Herkunft. Roux Arch. EntwMech. Org. *131:* 1–81 (1934).
18 Hämmerling, U.: B lymphocyte differentiation and regulatory signals; in Siskind, Litwin, Weksler, Developmental immunobiology, pp. 59–88 (Grune & Stratton, New York 1979).
19 Hämmerling, U.; Chin, A.F.: The ontogeny of B lymphocytes. IV. Induction signals mediating phenotypic conversion of $PC.1^-$ to $PC.1^+$ B cells. Eur. J. Immunol. *7:* 533–554 (1977).
20 Hämmerling, U.; Chin, A.F.; Abbott, J.: Ontogeny of B lymphocytes: Sequence of B-cell differentiation from surface immunoglobulin-negative precursors to plasma cells. Proc. natn. Acad. Sci. USA *73:* 2008–2012 (1976).
21 Hämmerling, U.; Chin, A.F.; Abbott, J.A.; Goldstein, G.; Sonenberg, M.; Hoffmann, M.K.: The ontogeny of B lymphocytes. III. Opposite signals transmitted to B lymphocyte precursor cells by inducing agents and hormones. Eur. J. Immunol. *6:* 886–872 (1976).
22 Hämmerling, U.; Chin, A.F.; Abbott, J.; Scheid, M.P.: The ontogeny of murine B lymphocytes. I. Induction of phenotype conversion of Ia^- to Ia^+ lymphocytes. J. Immun. *115:* 1425–1431 (1975).
23 Hämmerling, U.; Chua, R.; Hoffmann, M.K.: Polyclonal activation of CR^+ and CR^- B lymphocytes: the kinetics of initiation of DNA and immunoglobulin synthesis by lipopolysaccharide. J. Immun. *120:* 750–753 (1978).
24 Hämmerling, U.; Eggers, H.J.: Quantitative measurement of uptake of alloantibody on mouse lymphocytes. Eur. J. Biochem. *17:* 95–99 (1970).
25 Harris, H.: The reactivation of the red cell nucleus. J. Cell Sci. *2:* 23–32 (1967).
26 Harris, H.: The Croonian Lecture. Cell fusion and the analysis of malignancy. Proc. R. Soc. Lond. B *179:* 1–20 (1971).
27 Hoffmann, M.K.: Macrophages and T cells control distinct phases of B cell differentiation in the humoral immune response in vitro. J. Immun. (in press, 1980).
28 Hoffmann, M.K.: Control of B cell differentiation by macrophages. Ann. N.Y. Acad. Sci. *332:* 557–563 (1979).

29 Hoffmann, M.K.; Dutton, R.W.: Restoration of the immune response with macrophage culture supernatants. Science *72:* 1047–1051 (1971).

30 Hoffmann, M.K.; Gilbert, K.: Immune functions controlled by an LPS-induced macrophage factor (Il-1). Microbiology. pp. 55–61 (1980).

31 Hoffmann, M.K.; Koenig, S.; Mittler, R.S.; Oettgen, H.F.; Ralph, P.; Galanos, C.; Hämmerling, U.: Macrophage factor controlling differentiation of B cells. J. Immun. *122:* 497–502 (1978).

32 Hoffmann, MK.; Oettgen, H.F.; Old, L.J.; Chin, A.F.; Hämmerling, U.: Endotoxin-induced differentiation of bone-marrow derived lymphocytes. Proc. natn. Acad. Sci. USA *74:* 1200–1204 (1976).

33 Hoffmann, M.K.; Watson, J.: Helper T cell replacing factors secreted by thymus-derived cells and by macrophages: cellular requirements for B cell activation and synergistic properties. J. Immun. *122:* 1371–1376 (1979).

34 Honjo, T.; Kataoka, T.: Organization of immunoglobulin heavy chain genes and allelic deletion model. Proc, natn. Acad. Sci. USA *75*: 2140–2144 (1977).

35 Illiano, G.; Tell, G.P.E.; Siegel, M.I.; Cuatrecasas, P.: Guanosine 3′:5′-cyclic monophosphate and the action of insulin and acetylcholine. Proc. natn. Acad. Sci. USA *70:* 2443–2447 (1973).

36 Katz, D.H.; Hamaoka, T.; Dorf, M.E.; Benacerraf, B.: Cell interactions between histoincompatible T and B lymphocytes. The H-2 gene complex determines successful physiologic lymphocyte interactions. Proc. natn. Acad. Sci. USA *70:* 2624–2628 (1973).

37 Jacob, F.; Monod, J.: Genetic repression, allosteric inhibiton, and cellular differentiation; in Locke, Cytodifferentiation and macromolecular synthesis, pp. 30–42, (Academic Press, New York 1963).

38 Komuro, K.; Boyse, E.A.; Old, L.J.: Production of TL antibody by mice immunized with TL^- cell populations. A possible assay for thymic hormone. J. exp. Med. *137:* 533–536 (1973).

39 Komuro, K.; Boyse, E.A.: Induction of T lymphocytes from precursor cells in vitro by a product of the thymus. J. exp. Med. *138:* 479 (1973).

40 Komuro, K.; Boyse, E.A.: In vitro demonstration of thymic hormone in the mouse by conversion of precursor cells into lymphocytes. Lancet *April 7:* 740–743 (1973).

41 LeDouarin, N.M.; Jotereau, F.V.: Tracing the cells of the avian thymus through embryonic life in interspecific chimeras. J. exp. Med. *142:* 17–40 (1975).

42 McDougal, J.S.; Cort, S.P.: Generation of T helper cells in vitro. IV. F_1 T helper cells primed with antigen-pulsed parental macrophages are genetically restricted in their antigen-specific helper activity. J. Immun. *120:* 445–451 (1978).

43 Melchers, F.; Andersson, J.; Phillips, R.A.: Ontogeny of murine B lymphocytes: development of immunoglobulin synthesis and of reactivities to mitogens and to anti-Ig antibodies. Cold Spring Harb. Symp. quant. Biol. *41:* 147–158 (1976).

44 Melchers, F.; Von Boehmer, H.; Phillips, R.A.: B lymphocyte subpopulations in the mouse. Organ distribution and ontogeny of immunoglobulin-synthesizing and mitogen-sensitive cells. Transplant. Rev. *25:* 26–58 (1975).

45 Metcalf, D.; Moore, M.A.S.: Haemopoietic cells (North-Holland, Amsterdam 1971).

46 Moore, M.A.S.; Metcalf, D. Ontogeny of the haemopoietic system: yolk sac origin of in vivo and in vitro colony forming cells in the developing mouse embryo. Br. J. Haemat. *18:* 279–296 (1970).

47 Moore, M.A.S.; Owen, J.J.T.: Stem cell migration in developing myeloid and lymphoid systems. Lancet *ii:* 658–659 (1967).

48 Nossal, G.J.V.; Pike, B.L.: Studies on the differentiation of B lymphocytes in the mouse. Immunology *25:* 33–45 (1973).

49 Osmond, D.G.; Nossal, G.J.V.: Differentiation of lymphocytes in mouse bone marrow. I. Quantitative radioautographic studies of antiglobulin binding by lymphocytes in bone marrow and lymphoid tissues. Cell. Immunol. *13:* 117–131 (1974).

50 Paige, C.J.; Kincade, P.W.; Ralph, P.: Murine B cell leukemia line with inducible surface immunoglobulin expression. J. Immun. *121:* 641–647 (1978).

51 Parker, C.W.; Morse, S.I.: The effect of *Bordetella pertussis* on lymphocyte cyclic AMP metabolim. J. exp. Med. *137:* 1078–1090 (1973).

52 Parkhouse, R.M.E.; Abney, E.R.; Bourgois, A.; Willcox, H.N.A.: Functional and structural characterization of immunoglobulin on murine B lymphocytes. Cold Spring Harbor Symp. Quant. Biol. *41:* 193–200 (1977).

53 Pernis, B.; Forni, L.; Amante, L.: Immunoglobulin spots on the surface of rabbit lymphocytes. J. exp. Med. *132:* 1001–1018 (1970).

54 Quintans, J.; Lefkovitz, I.: Precursor cells specific to sheep red cells in nude mice. Estimation of frequency in the microculture system. Eur. J. Immunol. *3:* 392–397 (1973).

55 Raff, M.C.: Histamine release from mast cells: an unusually accessible model of a ligand-induced cell response; in Talwar, Regulation of growth and differentiated function in eucaryotic cells, pp. 177–184 (Raven Press, New York 1975).

56 Raff, M.C.: Two distinct populations of peripheral lymphocytes in mice distinguishable by immunofluorescence. Immunology *19:* 637–650 (1970).

57 Raff, M.C.; Megson, M.; Owen, J.J.T.; Cooper, M.D.: Early production of intracellular IgM by B lymphocyte precursors in mouse, Nature, Lond. *259:* 224–226 (1976).

58 Rosenberg, N.; Baltimore, D.; Scher, C.D.: In vitro transformation of lymphoid cells by Abelson murine leukemia virus. Proc, natn. Acad. Sci. USA *72:* 1932–1936 (1975).

59 Rosenberg, Y.J.; Parish, C.R.: Ontogeny of the antibody-forming cell line in mice. IV. Appearance of cells bearing Fc-receptors, complement receptors and surface immunoglobulin. J. Immun. *118:* 613–617 (1977).

60 Rosenthal, A.S.; Shevach, E.M: Function of macrophages in antigen recognition by guinea pig T lymphocytes. I. Requirement for histocompatible macrophages and lymphocytes. J. exp. Med. *138:* 1194–1212 (1973).

61 Scheid, M.P.; Goldstein, G.; Boyse, E.A.: The generation and regulation of lymphocyte populations. Evidence from differentiative induction systems in vitro. J. exp. Med. *147:* 1727–1741 (1978).

62 Scheid, M.P.; Goldstein, G.; Hämmerling, U.; Boyse, E.A.: Lymphocyte differentiation from precursor cells in vitro. Ann. N.Y. Acad. Sci. *249:* 531–538 (1975).

63 Scheid, M.P.; Hoffmann, M.K.; Komuro, K.; Hämmerling, U.; Abbott, J.; Boyse, E.A.; Cohen, G.H.; Hooper, J.A.; Schuloff, R.S.; Goldstein, A.L.: Differentiation of T cells induced by preparations from thymus and by non-thymic agents. J. exp. Med. *138:* 1027–1032 (1973).

64 Schimpl, A.; Wecker, E.: Replacement of T cell function by a T cell product. Nature new Biol. *237:* 15–17 (1972).

65 Schimpl, A.J.; Wecker, E.: A third signal in B cell activation given by TRF. Transplant. Rev. *23:* 176–188 (1975).

66 Schreier, M.H.; Tees, R.: Clonal induction on helper T cells: conversion of specific signals into nonspecific signals. Int. Archs Allergy appl. Immun. *61:* 227–237 (1980).

67 Seidman, J.G.; Leder, P.: The arrangement and rearrangement of antibody genes. Nature, Lond. *276:* 790–795 (1978).

68 Simon, M.; Hämmerling, U.: Functional analysis of B cell subpopulations. Responsiveness of phenotypically characterized CR^+ and CR^- B cell subsets to a T dependent antigen in vitro. J. Immun. *119:* 1700–1705 (1977).

69 Sitia, R.; Abbott, J.; Hämmerling, U.: The ontogeny of B lymphocytes. V. Lipopolysaccharide induced changes of IgD expression on murine B lymphocytes. Eur. J. Immunol. *9:* 859–864 (1979).

70 Spear, P.G.; Wang, A.; Rutishauser, U.; Edelman, G.: Characterization of splenic lymphoid cells in fetal and newborn mice. J. exp. Med. *138:* 557–573 (1973).

71 Speman, H.; Mangold, H.: Arch. mikrosk. Anat. EntwMech. *100:* 599 (1924).

72 Sprent, J.: Role of H-2 gene products in the function of T helper cells from normal and chimeric mice measured in vivo. Immunol. Rev. *42:* 108–148 (1978).

73 Sprent, J.: Restricted helper function of F_1 hybrid T cells positively selected to heterologous erythrocytes in irradiated parental strain mice II. Evidence for restrictions affecting helper cell induction and T-B collaboration both mapping to the K-end of the H-2 complex. J. exp. Med. *147:* 1159–1174 (1978).

74 Storrie, B.; Goldstein, G.; Boyse, E.A.; Hämmerling, U.: Differentiation of thymocytes. Evidence that induction of surface phenotype requires transcription and translation. J. Immun. *116:* 1358 (1976).

75 Swierkosz, J.E.; Rock, K.; Marrack, P.; Kappler, J.W.: The role of H-2 linked genes in helper T cell function. II. Isolation on antigen-pulsed macrophages of two separate populations of F1 helper T cells each specific for antigen and one set of parental H-2 products. J. exp. Med. *147:* 554–570 (1978).

76 Till, J.E.; McCulloch, E.A.: A direct measurement of the radiation sensitivity of normal mouse bone marrow cells. Radiat. Res. *14:* 213–222 (1961).

77 Tonegawa, S.; Hozumi, N.; Mattysses, G.; Schuller, R.: Somatic changes in the content and context of immunoglobulin genes. Cold Spring Harb. Symp. quant. Biol. *41:* 877–899 (1976).

78 Trentin, J.J.; McGarry, M.P.; Zenkins, V.K.; Gallagher, M.T.; Speirs, R.S.; Wolf, N.S.: Role of inductive microenvironments on hemopoietic (and lymphoid?) differentiation and role of thymic cells in the eosinophil granulocyte response to antigen; in Lindahl-Kiessling, Alm, Hanna, Morphological and functional aspects of immunity, pp. 289–298 (Plenum Press, New York 1970).

79 Vitetta, E.S.; Uhr, J.W.: Immunoglobulin receptors revisited. Science *189:* 964–969 (1975).

80 Wabl, M.R.; Brun, R.B.; DuPasquier, L.: Lymphocytes of the toad *Xenopus laevis*, have the gene set for promoting tadpole development. Science *190:* 1310–1312 (1975).

81 Waldman, J.; Munro, A.: T cell dependent mediator in the immune response. Nature, Lond. *243:* 356–357 (1973).

82 Watson, J.: The influence of intracellular levels of cyclic nucleotides on cell proliferation and the induction of antibody synthesis. J. exp. Med. *141:* 97–111 (1975).

83 Weigert, M; Gatmaitan, L.; Loh, E.; Schilling, J.; Hood, L.: Rearrangement of genetic information may produce immunoglobulin diversity. Nature, Lond. *276:* 785–790 (1980).

84 Weiss, P.A. Differentiation and its three facets: facts, terms and meaning. Differentiation *1:* 3–10 (1973).

85 Wood, D.D.; Cameron, P.M.; Poe, M.T.; Morris, C.A.: Resolution of a factor that enhances the antibody response of T cell-depleted murine splenocytes from several other monocyte products. Cell. Immunol. *21:* 88–96 (1976).

86 Wu, A.M.; Till, J.E.; Siminovitch, L.; McCulloch, E.A.: Cytological evidence for a relationship between normal and hemopoietic colony-forming cells and cells of the lymphoid systems. J. exp. Med. *127:* 455–463 (1968).

87 Wu, A.M.; Till, J.E.; Siminovitch, L.; McCulloch, E.A.: A cytological study of the capacity for differentiation of normal hemopoietic colony-forming cells. J. cell. Physiol. *69:* 177–188 (1967).

88 Zinkernagel, R.M.; Doherty, P.C.: The concept that surveillance of self is mediated via the same set of genes that determines recognition of allogenic cells. Cold Spring Harb. Symp. quant. Biol. *41:* 505–510 (1976).

Ulrich Hämmerling, Memorial Sloan-Kettering Cancer Center, New York, NY 10021 (USA)

Prog. Allergy, vol. 28, pp. 66–159 (Karger, Basel 1981)

Natural Killer Cells[1]

J.C. Roder, K. Karre and R. Kiessling

Department of Microbiology and Immunology, Queen's University, Kingston, Ont., Canada, and Department of Tumor Biology, Karolinska Institute, Stockholm, Sweden

Contents

[1] This work was supported by the MRC and NCI of Canada, the Swedish Cancer Society, and National Institutes of Health (USA).

I. Introduction

Cell-mediated immune reactions are known to play an important role in the rejection of tumors or grafts, and the destruction of virus-infected cells and normal tissues during autoimmune reactions. To many investigators the most familiar effector cells mediating these cytotoxic reactions include cytotoxic T lymphocytes [32], K cells participating in antibody-dependent, cell-mediated cytotoxicity [201], and activated macrophages [180]. Recently, however, a new, non-conventional lymphocyte, designated natural killer (NK) cell, has been described by a number of independent laboratories [109, 138, 300]. Although poorly understood, these cells are clearly important as a T cell-independent surveillance mechanism against certain tumors and possibly bone marrow grafts [144, 167] and, in addition, may play a role in regulating subpopulations of autologous thymocytes [101] or virus-infected cells [12]. A number of previous reviews [114, 145] summarize the early developments in the field and NK cells in the rat and human are dealt with at more length elsewhere [178, 207]. The purpose of the present review is to critically summarize recent developments in this rapidly expanding field and emphasize those areas of research that we believe are most important.

II. Assays for NK Cells

A. Cytolysis

By general consensus 'natural cytotoxicity' is defined as cytotoxicity in vitro exerted by lymphocytes from apparently normal unimmunized donors without addition of external antibodies or deliberate in vitro activation of the effector. In one well-described in vitro system YAC cells, a Moloney virus-induced T cell lymphoma of A/Sn origin, are labelled passively with sodium chromate and used as targets in a ^{51}Cr release assay [138, 139]. Target cell lysis is a unidirectional, contact, and cell dose-dependent reaction which kinetically resembles the parabolic curve of an antibody-dependent cell-mediated cytotoxicity (ADCC) reaction rather than the sigmoidal curve of a T cell-mediated reaction on a linear × linear plot of time vs percent lysis. A variety of compounds including ^{125}I-IUdR [38], ^{75}Se-selenomethionine [163], ATP, ^{86}Rb and ^{3}H-TdR [107] can also be used to isotopically label target cells as convenient indicators of cell death. The most widely used isotope, ^{51}Cr [$t^{1/2} = 27.8$ days] binds non-specifically to large molecular weight, intracellular proteins [25, 107], and is released upon damage of the plasma membrane after a lag period of 23 ± 5 min [271]. Released ^{51}Cr is not reutilized. Under standard assay conditions the fraction of ^{51}Cr released is nearly proportional to the number of cells damaged [271]. A more detailed account of the ^{51}Cr release assay [23] and other in vitro assays for ADCC or T cell-mediated cytolysis can be found in the excellent review of *Perlman and Cerottini* [201]. One of the more recent and useful in vivo assays involves the i.v. injection of tumor cells prelabelled with ^{125}IUdR [115, 214]. 4 h later the isotope is measured in various organs such as lungs, spleen and liver, and the amount of isotope remaining is inversely correlated with NK activity. A number of additional in vivo NK assays are described in section XI.

B. Measuring NK Cell Frequency

Isotope release assays measure events at the entire population level and, without the application of Poisson statistics and a limiting dilution analysis, yield little information regarding the actual frequency of NK cells in a heterogeneous population. Relative comparisons of cytotoxicity between experimental groups are usually made using lytic unit (LU) calculations. LU are calculated from effector dose response curves and 1 LU is defined as the absolute number of effector cells required to yield a given percent lysis. However, non-quantitative data can be obtained and even relative comparisons between groups are questionable since real data seldom meet the mathematical restraints required. A slope analysis is often possible only over a very limited range of the effector titration curve. If the times of the cytolytic reaction are varied then standard enzyme kinetic analysis can be applied to derive useful parameters for comparison such as K_m and V_{max} which measure the affinity of interaction and rate of lysis. A few of the difficulties in relating percent lysis to NK frequency are: (1) the reaction is influenced by the geometry of the well and the volumes employed; (2) most assays are not performed in target excess and therefore each NK cell is not 'saturated'; consequently all measures will be underestimates; (3) some cells in the effector population may in fact release 'growth' factors and reduce the spontaneous isotope release by the target; (4) there is no completely satisfactory way to determine how much isotope would be released by the targets even if they were all killed, and (5) each effector can recycle to lyse more than one target in a variable fashion. What is clearly required therefore is an assay to visualize NK cells at the single cell level. Plating effectors on a 'lawn' of targets and measuring plaques is only practical with non-nucleated erythrocyte targets [reviewed in 201]. A simple and reliable technique has recently been reported for detecting NK cells by virtue of their requirement to first bind to the target which they subsequently lyse [216, 218]. First, adherent cells are removed on nylon wool columns and then the nylon non-adherent cells are labelled with FITC or FDA and mixed with an excess of target cells. The percent of FITC-labelled lymphocytes binding to targets (fig. 1) can be used as an estimate of the NK frequency. What is the evidence that target-binding cells (TBC) represent NK cells?

First, the strain distribution of nylon-non-adherent, YAC-binding cells is strongly correlated ($r = 0.93$) with the strain distribution of NK cells which lyse the same target, as previously reported. In addition, both target cell binding and lysis are controlled by dominant genes in F_1 crosses between low and high-reactive strains, and in backcross mice TBC frequently was linked to H-2^d. Second, the age-related rise and fall of NK-mediated lysis is closely paralled to the frequency of TBC. Third, when comparing a panel of different tumors there is a correlation ($r = 0.90$) between the frequency of TBC and the sensitivity for lysis. Fourth, the organ distribution of TBC in nude mice closely parallels ($r = 0.98$) the organ distribution of NK cell activity. Finally, there is a good correlation ($r = 0.86$) between binding and lysis in kinetics experiments [218] and, if binding is inhibited by EDTA, lysis does not occur. All of these studies on the covariance of target cell binding and lysis provide strong indirect evidence that the TBC and the NK cell are closely related.

It was important to purify TBC in order to provide direct proof that the TBC represents the NK cell. TBC sedimented at 7–14 mm/h in a peak separeted from unbound spleen cells (2–7 mm/h), using the velocity sedimentation methods of *Miller and Phillips* [176] at unit gravity. In further

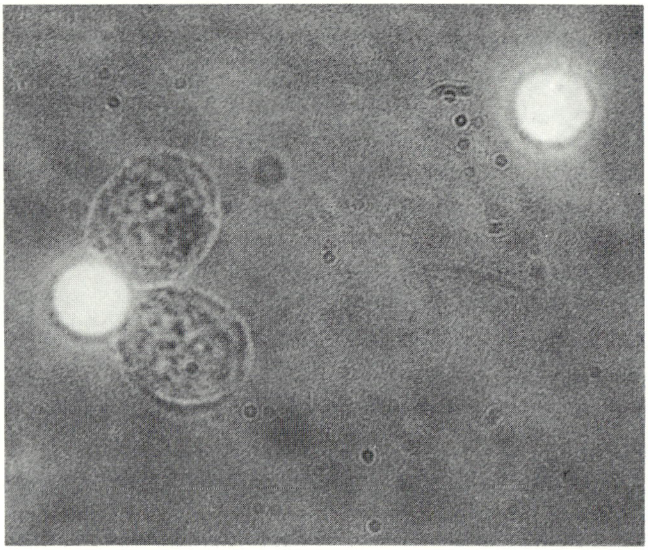

Fig. 1. A NK cell (center) from the mouse attached to Moloney lymphoma (YAC) target cells. Cells from the spleen, lymph nodes, or peripheral blood of normal, non-immunized mice were treated with carbonyl iron and a magnet followed by passage through a nylon wool column in order to remove non-specifically adherent cells. The remaining lymphoid cells were passively labelled with fluorescein isothiocyanate or fluorescein diacetate, to allow positive identification, and were mixed with a 10-fold excess of YAC cells. Micromanipulation experiments with single conjugates revealed that the targets (large cells) were lysed by the very small, attached lymphocytes. The frequency of binding cells measured in this assay ($\sim$ 1–2% of the total lymphocytes) correlated ($r > 0.9$) with variations in the levels of lysis dependent on the effects of age (fig. 5), organ distribution, genotype distribution, target selectivity, H-2 linkage, and kinetics. Enrichment or depletion of conjugates by velocity sedimentation resulted in a corresponding increase or decrease in the number of lytic units recovered. It can be concluded therefore that this target-binding cell assay allows visualization of the NK cell and its quantitation. Adapted from *Roder* et al. [218] and *Roder and Kiessling* [216].

experiments using a variety of sedimentation techniques it was shown that fractions enriched in TBC were approximately 4-fold greater in lytic capacity on a per-cell basis and, conversely, fractions depleted of TBC were up to 10-fold less efficient in a cytolytic assay. Care was taken in these experiments to control for unlabelled targets remaining from the separation procedures since these contaminants could compete with ^{51}Cr-labelled targets in a cytolytic assay, thereby giving spurious depletion values; or induce interferon production, which might yield artifactually high enrichment values.

In a final series of experiments TBC from nylon-passed spleen cells were enriched by velocity sedimentation and then greatly diluted in agar or dispensed as single conjugate droplets under oil. The majority (>80%) of TBC lysed the attached target under these conditions, as demonstrated by trypan blue uptake or complete disintegration.

Finally TBC formation was inhibited >50% by preincubation of lymphocytes with glycoproteins isolated from surface of tumor cells which are lysed by NK cells [220]. Glycoproteins isolated from the surface of tumors not lysed by NK cells did not inhibit TBC formation and specificity patterns were evident in TBC cross-inhibition studies using NK-TS from several different NK-sensitive tumor cell lines. These results show that the majority of TBC are indeed NK cells. The frequency of NK cells detected using these methods varies from 0.6 to 2.4% between various inbred strains. These methods are currently being developed in the human NK system. We believe that the TBC assay is important since target cell binding is a necessary but not a sufficient condition for target cell lysis. For example in studying the site of action of the bg gene [219] or the NK-suppressive effect of ^{89}Sr [157], the level of TBC was normal which allowed these investigators to conclude that the lesions affected a post-recognition event. On the other hand H-2-linked genes may control recognition events [216]. The TBC assay will also be useful in the visualization of surface markers on NK cells. In summary this simple TBC assay should greatly expand the methodological armamentarium of the NKologist.

C. In vitro Culture

The NK cell in the mouse was initially found to be very labile to inactivation at 37°C (24 h) and therefore investigators were discouraged from pursuing longer term culture. One recent study however [133] has shown an early (day 3) peak of cytotoxicity in murine mixed lymphocyte cultures (MLC). This early killing exhibited the same selectivity pattern as fresh NK cells but not mature cytolytic T lymphocytes (CTL). Unlike fresh NK cells however the early in vitro cytolytic activity was mediated by large, thy 1.2-positive cells which were termed 'anomalous killers' (AK) since it remains to be clarified whether NK and in vitro AK cells lie on the same lineage.

A number of recent investigations in human mixed lymphocyte systems have also shown that co-cultivation of normal lymphocytes with various autologous cell lines or allogeneic cells leads to an early (day 3) peak of NK-'like' cytotoxicity to autologous lines or third-party lymphocytes and a later peak of classical T cell-mediated killing [128, 250]. Unlike NK cells which are strongly FcR^+ the early AK cells lack Fc receptors and can be generated from peripheral blood lymphocytes depleted of FcR^+ cells [251]. In addition T cell activity but not NK cells require cell division during the MLC [28]. Therefore, these results suggest that in vitro techniques do not allow the clonal expansion of NK cells as occurs in T cell and B cell populations stimulated by antigen. Consequently, the activity of each NK cell in culture must be enhanced and the only agent described thus far which has this effect at the single cell level is interferon [218]. Since interferon is also released during co-cultivation [276, 280] this remains the most likely explanation of enhanced NK activity in MLC cultures. These conclusions, if valid, limit the usefulness of MLC techniques for generating NK cells or elucidating the sequence of differentiation steps from precursors. In contrast, in vitro culture in the absence of stimulator cells could be a useful system for studying the interaction of NK cell precursors with other cell types [195]. For example culture of peripheral blood lymphocytes alone, in the presence of FCS, leads to the generation of some NK-like activity from radiosensitive FcR^-, E^- precursors in synergy with radioresistant FcR^-, E^+ helper cells [196].

One of the exciting recent developments in the in vitro culture of NK cells centers around attempts to clone the NK cell. *Claesson and Olsson* [35] were successful in obtaining colonies of NK-like cells from murine bone marrow grown in agar. *Dennert* [46] and others [*Cantor,* personal communication] have obtained NK clones by limiting dilution of long-term cultures grown in Con A-derived growth factor. The results of further analysis on these clones is eagerly awaited.

III. Origin and Characteristics of NK Cells

A. Origin and Development

Experiments with radiation chimeras repopulated with small numbers of bone marrow or fetal liver cells reveal that NK cells ultimately derive from stem cells in the bone marrow [90]. In addition NK function is markedly suppressed in mice treated with (1) the bone-seeking isotope, ^{89}Sr [91, 144], (2) chronic β-estradiol administration [248] or (3) in congenitally osteopetrotic mice [249]. In ^{89}Sr-treated mice however the actual frequency of NK cells is normal as judged by target-effector binding [157]. The absence of suppressor cells and the failure to restore NK cytolytic function in ^{89}Sr-treated mice with bone marrow infusions indicates that NK cells are not only bone marrow-derived but also marrow-dependent, as suggested by *Kumar* et al. [157].

In contrast to their marrow dependence NK cells do not require an intact thymus for development and may in fact be somewhat inhibited by the presence of a thymus and/or thymic hormones as discussed previously [114]. Hence, congenitally athymic nude mice [109, 138] or thymectomized mice [92, 114] develop normal or elevated levels of NK activity. Splenectomized [92] and congenitally asplenic mice [114] were also within normal range. The failure to inhibit NK function by chronic B cell suppression in newborn mice by anti-IgM administration further suggests that normal B cell differentiation pathways are not involved in NK development [17, 63]. These findings are supported in the human where relatively normal NK activities are observed in x-linked Bruton type agammaglobulinemia and common variable hypogammaglobulinemia [155, 208].

B. Surface Markers

Until recently NK cells were classified as null cells, lacking the surface characteristics of mature B cells, T cells or macrophages. However, new and exciting work by *Kasai* et al. [135] has revealed an unusual glycolipid on the surface of NK cells, namely gangliotetraosyl ceramide or asialo-GMI (Gal β1→ 3 Gal NAcβ1→ 4 Gal β1→ 4Glc β→ 1 Ceramide). This reactivity was initially discovered by absorption studies using an antiserum against brain-associated T cell antigen which was cytotoxic for both NK and T cells [85]. Absorption with asialo-GMl removed the NK cell activity but not the T cell activity. Subsequently a rabbit antisera was generated against purified asialo-GMl antigen and this reagent was highly reactive with NK cells as confirmed by *Young* et al. [298] and others [*Roder; Wigzell,* unpublished observations]. MLC-generated CTL, thymocytes, bone marrow cells, carrier-specific helper T cells, and Con A-reactive T cells were not susceptible to anti-asialo-GM plus complement, which indicates that this antigen may be restricted to NK cells [135]. Recent results in our laboratory [*Roder and Beaumont,* unpublished observations] using fluorescent antibody techniques support this conclusion in that splenic lymphocytes bound to target (NK) cells preferentially stained with anti-asialo-GMl compared to non-bound lymphocytes (non-NK cells). It cannot be concluded that asialo-GM is specific for NK cells only however since it has been reported that some peripheral T cells react weakly [135,

261]. The results suggest rather that asialo-GMl, or a cross-reacting determinant, is expressed in high concentration on NK cells relative to other cell types, and functionally this difference is large enough that together with complement asiolo-GMl can be used as an NK-specific marker.

There is universal agreement that NK cells do not express surface immunoglobulin [110, 139, 218] as indicated by lack of depletion on anti-Ig columns or after anti-Ig plus complement treatment. Furthermore, NK cells do not stain with fluoresceinated anti-Ig at the single cell level [*Roder,* unpublished observation] as detected in the TBC assay [218], nor do they express Ia antigens as indicated in cytotoxic depletion experiments using anti-H-2 sera and complement [*Kiessling, Hogg,* unpublished observations]. NK cells do however express the serologically detectable antigens of the H-2 K and H-2 D end. In addition NK cells do not posses C3 receptors, as shown in rosette depletion assays using C3-coated erythrocytes [142]. Finally NK function is not depressed in mice chronically depleted of B cells by anti-IgM treatment since birth [17, 63]. These results taken together strongly suggest that NK cells are not B cells and may not lie in the B cell lineage. The one possible exception is the observation that cell fractionations with the B cell-selective lectin [213] soybean agglutinin removes about 50% of the splenic NK activity [*Reisner and Karre,* unpublished observation]. Therefore, it cannot be ruled out at this time that NK cells are very early, immature B cells which lack surface cytoplasmic Ig but express terminal *N*-acetyl-*D*-galactosamine or *D*-galactose residues.

Initial attempts to remove NK cells on Ig-anti-IgG columns [139] or to block NK activity with free immune complexes were unsuccessful [142]. Later, however, low activity FcR were detected by NK depletion on IgG-coated monolayers, a process which could be blocked by the Fc-binding protein A [112]. These findings, together with the observations that NK cells can kill antibody-coated tumor cell targets [187, 219, 244], suggest that murine NK cells like human NK cells possess FcR, albeit of lower avidity.

Endogenous NK cells resemble small lymphocytes morphologically [218], and it is generally agreed that they are non-adherent and non-phagocytic since they pass through columns of nylon wool, Sephadex G-10 and glass beads and are not removed by treatment with carbonyl iron and a magnet [110, 134, 252, 300]. In contrast LCMV-activated NK cells (cf. table VI) or NC cells [199] are partially adherent to nylon wool. The non-adherence of endogenous NK cells suggests that they are not mature macrophages. In support of this view a detailed comparative study revealed that NK cells and activated macrophages had a distinct range of target selectivities and a different genotype pattern of high and low responders [224]. In addition NK cells have recently been shown to be peroxidase-negative [218], whereas promonocytes and monocytes in the mouse are peroxidase-positive [279]. Moreover, mouse NK cells possess receptors for the A hemagglutinin of the snail *Helix pomatia (HP)* [93], while cells of the monocyte-macrophage lineage lack this receptor [97, 106]. Recent evidence also indicates that macrophage killing differs from NK cytotoxicity also with respect to organ distribution and age dependence [136]. NK cells, therefore, are not mature macrophages, but a modifying role for macrophages in the development of NK cells in vivo cannot be excluded. For example, in vivo treatment with antimacrophage agents, such as silica or carrageenan, depress NK activity in vitro [144]. In addition, certain agents (poly I:C) may stimulate macrophages to produce interferon [185, 186], which has been shown to greatly augment NK activity [62]. It is possible, therefore, that further investigations will show that NK activity, like cytotoxic T lymphocyte activity [129], is modulated by macrophages.

It is somewhat more controversial whether NK cells represent an earlier stage of monocyte differentiation, namely the promonocyte. Both NK cells and promonocytes are non-adherent, and non-phagocytic cells with a nucleus/cytoplasm ratio >1 and stain negatively or very slightly positive for esterase [49, 218]. Functionally, both cell types exhibit spontaneous killing of YAC but not P815 cells and are also capable of ADCC [49, 115, 165, 187, 219]. Furthermore, cytolytic activities are enhanced by interferon in both promonocytes and NK cells. In spite of these striking parallels we believe NK cells and promonocytes belong to separate lineages for a variety of reasons. First, each cell type responds differently to macrophage factor (MCF) which induces promonocytes [175] but not NK cells [224] to become adherent phagocytic, activated macrophages with a change in the selectivity pattern of targets sensitive to cytolysis. Both YAC and P815 are lysed by these activated macrophages [224] whereas promonocytes lyse only YAC [165]. Secondly, promonocytes [165] but not NK cells [189] express a macrophage alloantigen [4], although these experiments should be repeated in parallel, preferably with the new monoclonal reagents available. Thirdly, detailed analysis of NK cells and a new promonocyte permanent cell line reveal distinct target selectivity patterns and kinetics of lysis [137]. Fourthly, promonocyte effectors do not follow the same genetic control as NK cells. For example C57Bl/10 and A/J mice are both equally high in promonocyte cytolytic activity whereas C57Bl/10 is relatively high and A/J is low in NK activity. It is however possible that NK activity from A/J mice is activated by interferon-like agents during in vitro culture of the bone marrow-derived promonocytes so that A/J would then appear equal to C57Bl/10 in promonocyte killing, as suggested by *Lohmann-Mathes* et al. [165]. Finally, the most convincing argument against NK-promonocyte identity comes from the observation that the beige mutation in the mouse causes a selective impairment of NK cytolysis with no apparent effect on macrophage or promonocyte effector functions [219, 222]. Therefore, one must conclude at present that promonocytes represent a class of cells distinct from NK cells although a partial parallelism of effector mechanisms has occurred.

Perhaps the most controversial area in the characterization of NK cells lies in their phenotypic properties which are shared with T cells. At least two independent groups have now confirmed that NK bear the Ly5 alloantigen [30, 205] which is also present on thymocytes and mature T cells but not on B cells or macrophages [154]. Both the Ly5.1 and Ly5.2 alleles were expressed in the appropriate congenic strains [30]. It is interesting to note that pretreatment of NK cells with anti-Ly5 antisera in the absence of complement also blocks NK cytolysis, which suggests that Ly5 or some closely related structure may be directly involved in target cell recognition or lysis [30, 205]. Another presumptive alloantigen, NK-1, has also been found on NK cells of some strains [30, 66, 205], but the data on the tissue distribution of this marker is too limited at present to warrant the conclusion that NK-1 is a unique NK marker. NK-1 was originally discovered as an unrecognized antibody present in an anti-Lyl.2 serum [66]. It is known that NK-1 is not expressed on alloimmune CTL [205] and therefore this marker may serve to distinguish NK and CTL in mixed populations. Most groups have been unable to detect thy 1 on NK cells from common murine strains using conventional antibody + complement cytotoxicity techniques [66, 69, 76, 110, 139, 142, 205, 252] or monoclonal anti-thy 1 [133]. However, one group using thy 1 congenic mice or nude mice in particular were able to demonstrate a subpopulation of thy 1-positive cells which exhibited 'NK-like' activity [113]. In contrast to the negative cytotoxicity results reported by others [37, 133], *Mattes* et al. [173] could successfully separate some but not all NK cells on a fluorescence-activated cell sorter using the same monoclonal reagent. The results reemphasize that NK cells may be a heterogeneous population of effectors.

In addition, the expression of some T cell antigens on NK cells does not in itself warrant the conclusion that NK cells belong to the T cell lineage. Thy 1 for example is found not only on all T cells and some NK cells, but also on cells far removed from the T cell lineage, such as brain cells, epithelial cells, and fibroblasts [292]. One final argument against the notion that NK cells lie on the T lineage is the observation that cytotoxic T cell precursors activated by alloantigens express a unique glycoprotein T145 which is not found on NK cells [*Kimura and Wigzell,* personal communication].

Lectins, with diverse carbohydrate specificities, have also been used to analyse cell subpopulations. The snail lectin, HP, is a 79,000-u protein with binding specificity for *N*-acetyl-*D*-galactosamine [96]. Receptors for HP, analogous to human blood group A, are uncovered on neuraminidase-treated T cells and NK cells [93] but not B cells or macrophages [106]. Since NK cells possess low affinity HP receptors compared to T cells, affinity chromatography on sepharose-coupled HP has allowed a relative degree of NK cell enrichment. It has been claimed however [unpublished observation in ref. 173] that not all mouse NK cells are HP-positive, in particular that subpopulation which is thy 1-positive. If true, these two reagents could provide a useful tool for dissecting the heterogeneous NK pool into HP^{+}thy 1^{-}, HP^{-}thy 1^{+}cells. Preliminary studies have also been completed with additional lectins. *Vicia villosa* lectin has been found to bind strongly to CTL but not at all to NK cells as judged by selective retention on lectin-sepharose columns [*Kimura* et al., to be published]. Peanut agglutinin (PNA) is specific for non-reducing terminal *D*-galactosyl residues [D-galactosyl-B (1→3)-*N*-acetyl-*D*-galactosamine] on immature, hydrocortisone-sensitive, murine thymocytes, fetal liver lymphocytes and stem cells in the spleen and bone marrow, whereas these residues are masked by sialic acid on mature lymphocyte populations [211]. NK cells were not agglutinated by PNA and therefore it is unlikely that they belong to the majority of early immature T cells [*Reisner and Karre*, unpublished observation]. Soybean agglutinin on the other hand binds to terminal *N*-acetyl-*D*-galactosamine on splenic B cells but does not agglutinate splenic T cells [212]. Using agglutination techniques coupled with gravity sedimentation, it was found that about 50% of splenic NK cells had receptors for SBA [*Reisner and Karre*, unpublished observation]. This suggests that NK cells are also heterogeneous regarding expression of the SBA receptor and imply that splenic NK cells are more similar to B cells than T cells regarding the expression of surface glycoproteins.

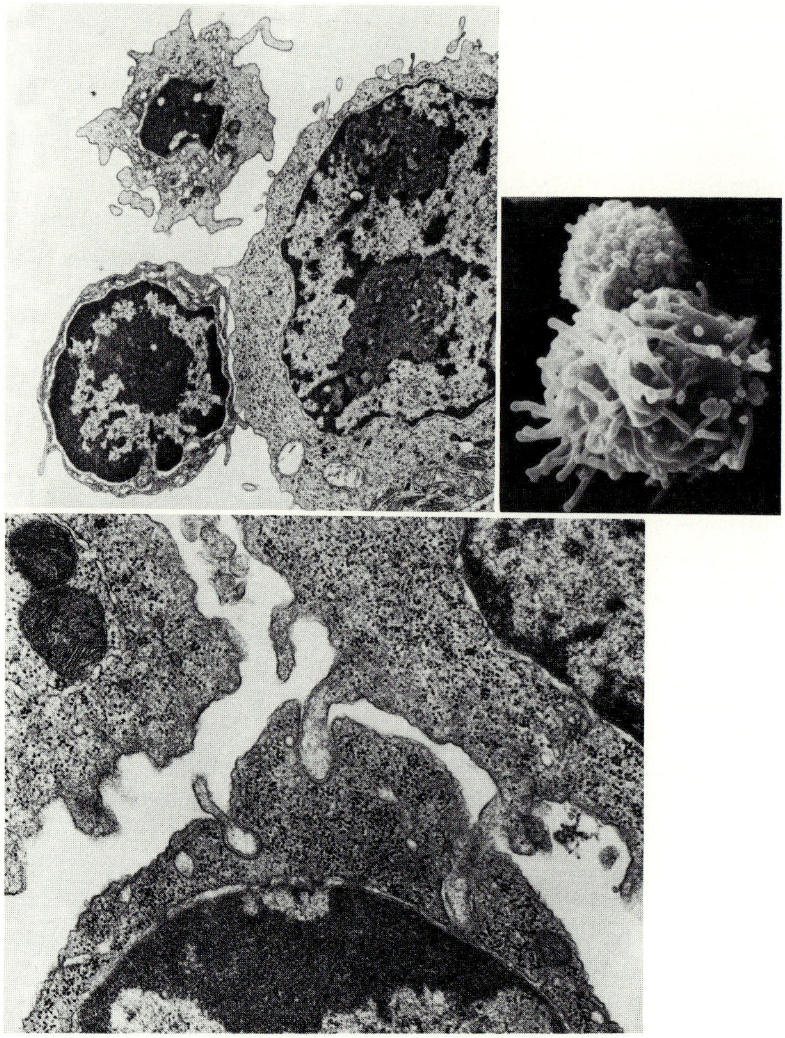

a b c

Fig. 2. a Transmission electron micrograph of a conjugate formed at 0°C. *b, c* Conjugates formed at higher temperatures (20 or 37°C) had a more dynamic interaction with the target. The NK cells shown have the typical morphology of a small, 'resting' lymphocyte, as confirmed by cytochemical techniques [218]. ×. 16,000. *b* Scanning electron micrograph showing a large YAC cell attached to a small NK cell by numerous microvilli. X. 7,000. *c* The dynamic nature of the target-effector interaction. An NK cell (bottom) engulfs a microvillous process of two YAC targets. ×. 54,000. All photos courtesy of Dr. *P. Biberfeld.*

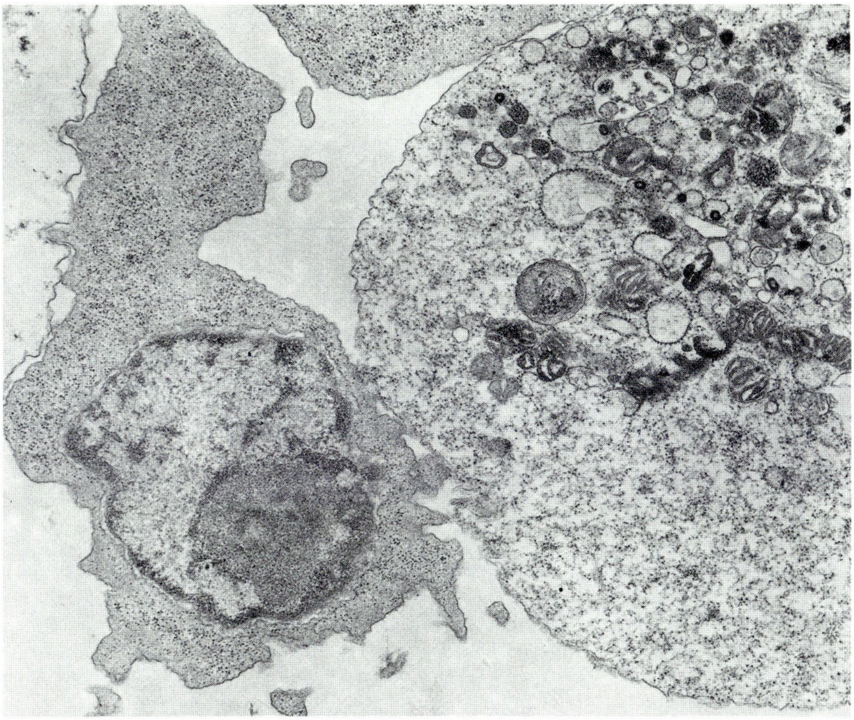

Fig. 3. Target cells lysis. The YAC cell on the right is in the process of osmatic swelling and the target cell on the right is almost completely devoid of cytoplasmic constituents and presumably lysed. the effector cell (center) is somewhat atypical since it contains more cytoplasm than the effector cell type usually encountered (fig. 2). ×. 27,000. Courtesy of Dr. *P. Biberfeld.*

C. Morphology, Ultrastructure, and Cytochemistry

Morphologically endogeneous NK cells in contrast to LCMV-activated NK cells (cf. table VI) resemble small, metabolically inactive lymphocytes as revealed by electron micrographs of target-effector conjugates (fig.2) [218]. The majority of effector cells comprising such conjugates have been shown to consist of functionallly active NK cells [218]. The NK ultrastructure consisted of a high N:C ratio, scanty cytoplasm, relatively numerous large mitochondria, inconspicuous Golgi zones, a lack of endoplasmic reticulum and polyribomes and a predominately heterochromatic nucleus with a 'crescent' nucleolus. In scanning EM (fig.2) the surface was villous but less so than the YAC target used to form the TBC. Point contacts were established between target and effector villi, and NK cells would often be seen actively engulfing processes on the target (fig.2). Although most attached targets were eventually lysed (fig.3) no regions of specialized membrane structures or local cytopathogenic changes were noted and morphologically there was nothing unique about NK-TBC when compared to target effector conjugates in the CTL system. Cytochemical staining of TBC revealed that NK cells were negative for α-naphthyl acid esterase (ANAE) which was previously shown to stain T cells but not B cells in the mouse [179]. NK cells were negative for peroxidase and only weakly stained by acid phosphatase, two agents which normally stain in cells of the monocyte-macrophage lineage [73]. Although it would be preferable to have data from parallel studies we can tentatively conclude on the basis of cytochemical staining that NK cells are lymphoid in nature and distinct from B cells, T cells or cells of the monocyte-macrophage lineage.

D. Physical Properties

Some of the physical properties of NK cells have also been examined. Free flow electrophoresis indicated that NK cells and cells mediating ADCC had a unique electrophoretic mobility midway between that of splenic CTL (high net negative surface charge) and B cells (low net negative surface charge) [132]. At unit gravity NK cells sedimented with a velocity (4–5 mm/h) comparable to that of small to medium-sized lymphocytes. The sedimentation profile was broad however which suggests some size heterogeneity in cells capable of mediating spontaneous killing of tumor cell targets. Fractionation on linear iso-osmotic density gradients revealed that NK cells were of somewhat higher density (1.070 – 1.082 g/mm^2) than the median value for the whole spleen cell population. Similar results were obtained in the rat using the more refined countercurrent distribution in an aqueous two-phase system [182]. Three peaks of cells were partitioned and the peak of NK and ADCC activity was distinct from immunized cells mediating specific cytolysis, cells responding to Con A or PHA, and cells bearing surface Ig [182]. In summary then, physical separation techniques also reveal that NK cells are distinct from the majority of classical B and T cells.

Sensitivity to pharmacological agents can also be used to distinguish NK cell activity. For example NK cells but not cells mediating hybrid resistance to parental marrow grafts are inhibited by in vivo administration of hydrocortisone [121]. Silica for example also inhibits macrophages and some NK activity whereas NK cells are much more sensitive to ^{89}Sr than other cell functions.

E. Conclusions

As shown in table I, NK cells in the mouse possess their own unique spectrum of surface antigens and physiological properties. The available evidence does not firmly place the NK lymphocyte on any known developmental pathway within the lymphoreticular system but rather supports the existence of a separate and unique lineage. The point at which NK cells diverge from T and B cells during development remains an avenue for future research. It is encouraging that several surface markers, some unique, are finally available to allow the unequivocal identification of NK cells. Further work in this area will undoubtedly involve further attempts to subdivide NK cells into subpopulations.

IV. Genetic Control of NK Cells in vitro and in vivo

Genes on several different chromosomes are known to control various aspects of NK function. Hence H-2-linked genes on chromosome 17 control population size and/or surface recognition structures, the bg gene on chromosome 13, controls a post-recognition step in the NK cytolytic pathway, the nude gene on chromosome 11 exerts a positive influence and it can be speculated that genes coding for interferon production may be involved in positive feedback of certain post-recognition events.

A. In vitro NK Reactivity

The starting point for the genetic analysis of the NK system was the finding that mouse strains could be grouped according to the levels of in vitro lytic activity of their spleen cells [109, 138, 300]. This observation has subsequently led to a more extensive genetic analysis aimed at understanding the mode of inheritance of the genes confering high NK reactivity, as well as mapping their position and determining their numbers within the mouse genome.

The least reactive group includes the A strains A/J, A/Sn, and A

Table I. Phenotypic characteristics of murine NK cells

Marker	NK	B	T	Promonocyte	Macrophage
sIg	–	+	–	–	–
C_3R	–	+	–	+	++
FcR	–/+wk	+	–/+	+	+
Ly 1	–	–	+	nt	–
Ly 2	–	–	+	nt	–
Ly 3	–	–	+	nt	–
Ly 5	+	–	+	nt	–
Ly 6	?	+act.	+	nt	nt
Thy 1	–/+wk	–	+	nt	–
NK-1	+	–	–	nt	nt
mph-1	–	–	–	+	+
H-2K, D	+	+	+	+	+
Ia	–	+	–	nt	+/–
asaiol-GM1	++	–	/+wk	nt	nt
GM1	–/+wk	+	++	nt	nt
HP lectin	+	–	++	nt	–
PNA	–	–	+pre-T	nt	nt
SBA	–/+	+	–	nt	nt
V. villosa	–	–	+	nt	nt
ANAE	–	–	+	–	–
Esterase	–	–	–	–/+wk	+
Acid phos.	–/+wk	–	–	+	+
Peroxidase	–	–	–	+	+
Adherence	–	–/+wk	–	–	++
Phagocytosis	–	–	–	–	++
Thymus-dep.	–	–	+	–	–
BM-dep.	++	+	+	–	–
Spleen-dep.	–	–	–	–	–
Cortisone-sens.	+	–	+	–	–
89 Sr-sens.	++	–	–	–	–
Silica-sens.	–/+	–	–	nt	++

Symbols denote the presence (+) or absence (–) of a given marker or property. Subpopulations which are both positive and negative are designated –/+. ? = Uncertain data; wk = the marker is weak and diffucult to detect; act. = present on activated cells only; nt = not tested to the best of our knowledge. References are given in text.

congenic resistant strains A.CA, A.SW, and A.BY, as well as the AKR, 129, and SJL strains. Highest reactivity is invariably seen using effector cells from the CBA or C3H/Strong. The C57Bl, C57Leaden, DBA/2, and C3H/HeJ strains, classified earlier as high reactive, should more appropriately be termed intermediate, as they in fact fall below the CBA or C3H strain in reactivity. In our hands the genetic classification of mouse strains as to NK levels does not depend on the tumor target used for the NK assay, and since the original study with YAC-1 several other tumor targets (C57Bl lymphomas, Balb/c plasmacytoma and human leukemia) have been shown to follow the above-described pattern.Some investigators, however, have concluded that the genetic reactivity of a particular mouse strain depended on what target cell was used in the assay [67]. For example *Herberman* et al. [111] showed that A/J mice were equally NK-reactive compared to C57Bl/6 mice when RL♂ lymphoma cells were used as targets, and *Stutman* et al. [264] made a similar observation using a ^{3}H-proline assay and Meth A targets. Part of the reason for these discrepancies may lie in the poor control that most laboratories exert over mycoplasma infection of their tissue culture cell lines; an event which can markedly alter target sensitivity to NK lysis and possibly obscure genetic differences [20].

In the YAC system the first analysis was to determine if high NK activity was inherited as a dominant trait. When the low reactive A strain was crossed to various other strains, and cells from these F_1 hybrids tested against the semisyngeneic YAC lymphoma of A origin for NK activity, reactivity resembled the high reactive parent [202]. Thus, in these crosses high reactivity was dominant. Lately, however, in other F_1 crosses we found evidence of an inheritance pattern which is not a simple dominance of high reactivity, but rather suggests some type of genetic complementation. Thus, in the C57Bl × DBA/2F_1 hybrid NK reactivity exceeds that observed with either parental strain [148]. The genetic basis for this complementation phenomenon remains to be established, but could result from complementation between non-allelic genes or from the heterosis effect of an allelic system.

In trying to map the number and the position of the NK-regulating gene(s) the offspring from various backcrosses were analyzed including high reactive F_1 hybrids crossed back to the low reactive parental strain. For this purpose, the low reactive A strain and a high reactive A F_1 hybrid were used. When spleens from mice of this cross were tested in the NK system, their activities were distributed over a wide range, covering the range of the high and the low reactive parental strains [203]. A linkage analysis was then performed in the same backcross population in order to localize the genes controlling NK activity. Spleen cells from mice of this cross were tested individually for NK activity. The same mice were also typed for 10 different markers segregating in this cross; they were typed for

H-2 haplotype and immunoglobulin heavy chain allotype (Ig-1 locus), and registered for two-coat color genes, the black-brown locus (B vs b) and albino-colored locus (c vs C). Five different tissue isozymes from kidney and liver, together with the C5 serum complement component, were also typed. Cytolytic activity of heterozygotes was compared to that of homozygotes for each of these markers. Highly significant linkage was established only with regard to the H-2 marker.

It is likely that linkage with the H-2 complex is quite weak, and that the H-2^{b}-linked factor(s) could be situated on chromosome 17 at some distance from the H-2 complex. Results supporting this alternative have come from studies of congenic resistant (CR) strains of mice. Upon analyzing the reactivity of various CR strains on A/Sn background (A.CA, A.SW, A.BY), carrying other H-2 donor haplotypes, their acitivities do not significantly exceed that of the low reactive A/Sn strain and never reach the level of reactivity of their H-2 donor strain [203]. This again argues for the view that at least some of the H-2^{b}-linked factors are situated outside the H-2 complex. Is there then any evidence that some of the 'NK genes' can map *within* the H-2 when studying H-2 haplotypes other than H-2^{b}? Recent results from *Harmon* et al. [104] and confirmed by *Klein* et al. [148] indicate an H-2^{d}-linked reactivity gene(s) mapping within the H-2 complex. Evidence for an H-2^{d}-linked factor stems from analysis of segregating backcross populations as well as from studies using CR and H-2 recombinant mice. In the backcross test, the high reactive DBA × C57Bl F_1 hybrid was crossed with the less reactive C57Bl strain. Again, H-2 heterozygotes proved to be more reactive than homozygotes. Here, however, the significant difference in reactivity between heterozygotes and homozygotes became apparent with considerably fewer mice than was necessary to use in the (C57Bl × A) × A backcross test mentioned earlier. Tentatively, one could ascribe this difference in variation to linkage strength between these two crosses: the H-2^{d}-linked factor(s) could be closer to, or even within, the H-2 complex in contrast to what might be the case for the H-2^{b}-linked gene(s). Results well in accordance with this have in fact been obtained using CR or H-2 recombinant strains. Here, *Harmon* et al. [104] initially showed, using CR or H-2 recombinant mice on B10·A background that heterozygosity within or near the D-end of the H-2 complex was sufficient to give high NK reactivity against the C57Bl lymphoma El-4. These results have been confirmed using a variety of other tumor targets of other genotypes by *Klein* et al. [148]. We must then conclude that the H-2^{d}-linked reactivity factor(s) maps within the H-2 complex, a fact which is probably in contrast to the previously discussed H-2^b-linked gene(s). Apparently, at least two different H-2-linked loci on chromosome 17, and probably several other non-H-2-linked genes as well, are involved in the control of NK activity.

In the earlier genetic analysis we used the highly NK-sensitive target YAC-1 only, but later a panel of different tumor targets of variable etiological and histological origin have been used [148], also including some human cell lines [89]. A crucial question has been whether a different strain distribution of reactivity is seen using different tumor targets. This question would seem to be of considerable importance, since the answer may help us to understand how the genetic factors control differences in reactivity. Could these 'NK genes' selectivity influence the reactivity to certain target cells but not to others, giving a different genotype distribution when different targets are tested? Alternatively, is the same genetic distribution seen regardless of target cell type, which would rather indicate a genetic influence of the 'overall' NK reactivity of a particular mouse strain? Most of the evidence supports the second alternative. Thus, carrying out NK analysis against a variety of different tumor targets, including MuLV-induced lymphomas of strain A origin, RadLV and Rauscher-induced C57Bl tumors, Balb/C plasmocytomas as well as certain human cell lines, the same distribution of reactivities between various strains was seen regardless of target cell type [89, 148]. This was even more striking when studying segregating backcross populations. Here high reactivity against all tested lines segregated together, and the H-2-linked component came through irrespective of target cell type [148]. In contrast, others [114, 264] have found, using other targets, that the A strain mice had levels of reactivity similar to those of C57Bl/6 mice, in contrast to the previously mentioned classification of the C57Bl/6 strain, as more reacitve than the A strain against YAC. The authors concluded that the genetic factors determining NK reactivity did not do this by regulating the overall levels of reactivity, but would rather regulate the specificity of the NK cells, a conclusion in contrast to ours. Whether this discrepancy depends on the RL♂1 tumor being unique from the tumors used in our studies remains to be established. The classification of the C57Bl strain as low reactor against some targets could rather be due to the general insensitivity of these tumors, and not caused by a selective NK unresponsiveness of this strain.

The relationship between the NK system and the ADCC lysis of antibody-coated tumor targets has also been studied from a genetic point of view. As discussed alsewhere, NK cells are endowed with dual functions: one being the capacity to lyse antibody-coated tumor targets via an Fc receptor-mediated mechanism and the other the ability to selectively destroy tumor cells via some hitherto evasive recognition system. Yet both functions seem to be regulated by the same genetic factors regarding effector cell lytic capacity. This was apparent from results showing that those strains which were the most efficient in NK killing also lysed antibody-coated 'NK-resistant' tumor targets the best [187]. Furthermore, both traits seem to segregate together under H-2-linked influence in populations using high responsive F_1 hybrids backcrossed to low reactive parental strains [190]. When NK cells function as cells in the ADCC system, specificity is not an intrinsic property of the effector cell itself, but is provided by addition of antibodies. Thus, it must be concluded that in this ADCC system the different levels of reactivity cannot be due to regulation of specific recognition structures as these are passively added. A major extrinsic factor in the regulation of NK activity is endogenous interferon. The induction of interferon by various agents such as Newcastle disease virus is known to be under genetic control in the mouse [45]. It has also been found [*Orn* et al., to be published] that the agent tilorone is under genetic control in the mouse as to its interferon-inducing ability. When making a comparative analysis of tilorone in relation to other interferon inducers, it was noted that the AKR strain was particularly unresponsive with regard to tilorone. This was noted both at the level of interferon induction and by a parallel failure to display significant increases in NK activity of this low NK strain. Interferon itself, however, could induce significant NK activity. The underlying genetics of this was found to be regulated at the level of the bone marrow stem cells as judged by the following set of experiments: H-2K mice of strains CBA/H (= high NK) and AKR (= low NK) were lethally irradiated followed by repopulation with bone marrow cells from either of the two strains. The mice were then left for reconstitution for 4–5 weeks, whereafter they were treated with tilorone. It was then found that all mice repopulated with CBA marrow, irrespective of recipient genotype, produced interferon and subsequent increases in NK activity, whereas the recipients of AKR marrow all failed to respond to tilorone treatment with either interferon synthesis or increases in NK activity. This would strongly suggest that there exists a genetic regulation at the level of bone marrow-derived cells with regard to ability to produce interferon to certain defined substances and probably also viruses. As the NK cells themselves are bone marrow-derived cells, and under genetic control as to levels of activity, it is at present not possible to exclude that normally occurring, endogenous interferon levels may vary in different strains and thus be responsible for the variations in NK levels seen. It is thus conceivable that the observed linkage between NK levels and H-2-associated genes in certain strains may in part have as an underlying basis an H-2-associated susceptibility to a particular virus which may then regulate NK activity in a secondary manner.

This mechanism could account for some but not all of the H-2 linkage since the frequency of cells recognizing and binding to the target (TBC) is also under dominant, H-2-linked genetic control [216] whereas interferon has no effect on the frequency of TBC [218]. It is most likely therefore that the H-2-linked NK genes in various strains control either the NK population size and/or the expression of a recognition receptor which binds to determinants on the target.

Although NK lysis of most or all tumor cells is under the same genetic control, recent evidence [146] suggests that different genes are involved in NK acitivtiy against some primary tissues such as peritoneal cells. Whereas tumor cells (YAC-1), normal cortical thymocytes from young mice, and embryonic fibroblasts exhibited the same patterns of lysis against a panel of LCMV-activated NK cells from low and high reactive genotypes, normal peritoneal target cells showed a different pattern. These results could most readily be explained if various subpopulations of NK cells were under separate genetic control; as suggested by the work of *Stutman* et al. [264].

In addition to the H-2-linked NK genes on chromosome 17 discussed above, at least one other gene (bg) on a different chromosome (chromosome 13) has been shown to be intimately involved in NK cytolysis in both mice [219, 221, 222] and man [86, 152, 228]. A mutant of the C57Bl/6 mouse, called beige, was found to have a selective defect in NK cytolysis and ADCC against tumor targets whereas effector functions mediated by T cells, promonocytes, macrophages, and B cells were normal. The mutant gene is recessive and showed no penetrance in the heterozygous condition. Bg/bg mice exhibited low NK activity against a variety of tumor cell and non-transformed targets compared to heterozygous, +/bg, or wild type, +/+, littermates which showed normal NK function [219]. The other known abnormalities in the mutants include defective lyosomal granules and a pigmentary defect. These various abnormalities could be related or unrelated pleiotropic effects of a single gene since the genetic lesion is thought to be a spontaneous point mutation [159] which has been mapped to the centromeric end of linkage group IV (chromosome 13) [160, 239]. Several independent investigations have now confirmed the NK defect in beige mice [21, 146] and an even more profound impairment has recently been found in the human [228]. As shown in figure 4; both NK cytolysis and ADCC against tumor targets was depressed > 100-fold compared to age- and sex-matched controls in terms of lytic units. This defect was apparent against a variety of different targets and was not due to a delayed kinetics of response or suppressor cells. Furthermore, the defect was selective since cytolysis by monocytes, neutrophils, and T cells was normal. The Chediak-Higashi (CH) syndrome was recognized originally as a familial trait [33, 117]. Several reports of consanguinity [156] and extensive family pedigrees [238] indicate that the CH syndrome is autosomal recessive [reviewed in 13]. The precise location of the CH gene is not yet known.

B. In vivo Tumor Resistance

As outlined above the primary basis for the genetic analysis of NK activity is the observed differences in in vitro NK reactivity among various mouse strains. It was natural to investigate whether the same genotype differences would appear when analyzing the levels of in vivo resistance to the growth of the same tumor as used as target in the in vitro NK asay. In other words, would the same mouse strain as displayed high in vitro NK activity also show high in vivo tumor resistance? To reduce the possibility that mice of some strains would mount significant, conventional trans-plantation rejections towards histocompatibility antigens expressed on the tumor graft, all strains to be used in this comparison were mated in an F_1 cross with the parental strain of tumor origin. Initially YAC cells were inoculated at small cell numbers (10^3 – 10^4) into semisyngeneic A/SnF_1 hybrids. Here a good positive correlation was seen between levels of NK activity and the in vivo resistance in all tested genotypes by *Kiessling* et al. [140]. Thus, those F_1 hybrids which in vitro displayed high NK activity were also relatively more resistant to tumor cell challenge in vivo, and vice versa. This strain distribution of resistance was seen when a relatively small tumor cell inoculum was given (10^3 – 10^4 cells), but resistance was invariably broken down at higher cell doses. The parallelism between NK acitivity and in vivo F_1 tests has also been noted in several other tumor systems. *Sendo* et al. [252] observed this, using a radiation induced Balb/C lymphoma tested in various Balb/C F_1 hybrids. Likewise, *Harmon* et al. [104] noted a similar correlation with the EL-4 tumor of C57Bl origin tested in various C57Bl F_1 hybrids. A good positive correlation between

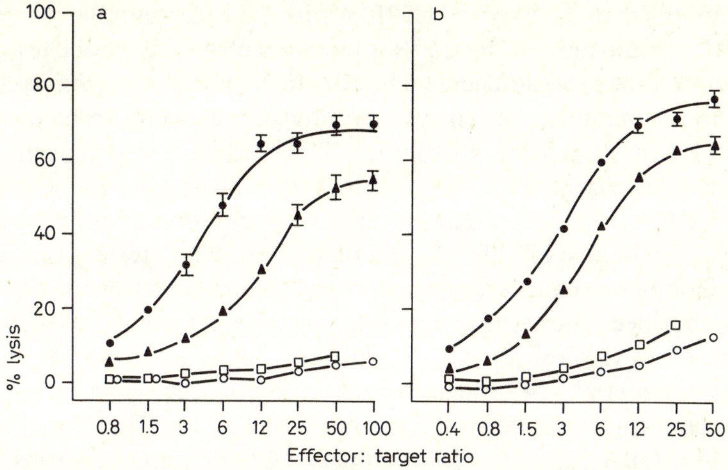

Fig. 4. Impairment of NK and ADCC activity in lymphocytes from Chediak-Higashi (CH) patients. Ficoll-Hypaque-separated, peripheral blood lymphocytes were depleted of monocytes on microexudates and incubated with ^{51}Cr-labelled K562 cells, a human myeloid leukemia cell line (a), or antibody-coated Chang cells, a human hepatocarcinoma cell line (b) at various effector:target ratios in a 4-hour cytolytic ^{51}Cr release assay. ●, ▲ = normal donors; □, ○ = Chediak-Higashi patients. Values represent the mean percent specific lysis ± SE in triplicate wells. Background lysis of non-antibody-coated Chang cells at a 50:1 effector: target ratio was 18, 7, 2, and 1.5% for ●, ▲, □, ○, respectively.

NK reactivity and in vivo resistance to a variety of other tumors has also been found by *Klein* et al. [148] and others. One of the limitations of the in vivo test compared to the in vitro assay, in that only one of the parentals can be tested in the former assay. To exemplify this problem we can consider results from two different tumor host systems. In the YAC tumor system it has been shown that the A × C57Bl F_1 hybrid is highly reactive as judged by in vitro NK activity and in vivo tumor resistance compared to the low reactive A/Sn strain. The C57Bl strain cannot be tested for in vivo tumor resistance with the (in this case) allogeneic A tumor cells, although the NK assay would indicate that this strain is more reactive than the A strain. This would then mean that high reactivity in the F_1 hybrids is due to dominance of high partner reactivity in the F_1 cross. The situation is somewhat different in the DBA/2 × C57Bl F_1 cross as already discussed in the previous chapter. In this F_1 hybrid, NK activity exceeds that of either of the parental strains, indicating some complementation phenomenon. In vivo, the C57Bl × DBA hybrid is more resistant to the challenge of various

C57Bl tumors than the parental C57Bl strain, but here again the resistance of the DBA strain against the C57Bl tumor can obviously not be tested. It is therefore impossible to decide whether the higher in vivo resistance of this hybrid is due to some type of 'hybrid vigor', as seems to be the case with the NK reactivity, or whether it is inherited in a dominant manner from the DBA parent.

As mentioned above, we have thus far failed to find any clear indication for the existence of different genetic patterns when testing tumors of widely different origin as targets in vitro. A further approach to this question has been the testing of a variety of different tumors of the same genotype for in vivo resistanc in various semisyngeneic F_1 hybrids. In one study by *Klein* et al. [149] the same genotype distribution was seen with all lymphoid tumors of A/Sn origin used in this study. The A × CBA, A × C3H, A × DBA/2, A × C57Bl and the A × C57Leaden F_1 hybrids were all significantly more resistant than the parental A strain, while the A × A.CA hybrid did not show this increased resistance. It can therefore be concluded that the classification of high and low resistant genotypes resembled the previous grouping of these genotypes in the YAC system by *Kiessling* et al. [140], and is in agreement with the in vitro NK reactivities of the same strains. Although all the lymphoid tumors of A origin showed the same genetic pattern, there were some clear indications for a different genetic control in resistance to sarcomas and carcinomas [149] although the participation of NK cells in in vivo resistance to nonlymphoid tumors is not yet established.

In view of the finding that F_1 hybrids are in general more resistant to tumor growth than the parental strain of tumor origin it will be instructive here to consider the phenomenon of hybrid resistance. According to the conventional laws of transplantation, grafts from an inbred strain to an F_1 hybrid, of which the inbred strain is one parent, are accepted [259]. The phenomenon of hybrid resistance is apparently an exception, and violates the usual laws of transplantation. *Snell* [258] was the first to note that C57Bl lymphomas grew less well in certain F_1 hybrid mice compared to the syngeneic parental strain. A host of similarly related phenomena has later been described [260]. One question of some controversy is whether this damage to the tumor cells is a passive or an active process with regard to the tumor cells. According to the interpretation of *Hellstrom and Hellstrom* [105], a form of self-destruction or suicide of the injected tumor cells due to H-2 incompatibility of some non-immunological type may be responsible for the phenomenon that they termed 'allogeneic inhibition'.

Mostly, however, this phenomenon is interpreted as an active process involving immunological effector mechanisms of the host animal.

In numerous genetic studies of hybrid resistance, or similarly related phenomena such as radiation-resistant rejection of hematopoietic grafts or allogeneic inhibition, a major role for the H-2 complex was established [260]. Originally, *Snell and Stevens* [259] pointed out that in F_1 hybrids the hybrid resistance increased in proportion to the histocompatibility disparity of the parents.

If one assumes that the tumor destruction in the F_1 versus parental tumor combination is an active process, there are many different ways of interpreting how this H-2-linked activity is generated. One question is whether it can be explained by the uniqueness of the F_1 effector cell or by the uniqueness of the parent strain antigen. Thus, one explanation would be that the high reactivity in the F_1 mice is due to introduction of a new set of H-2-linked Ir genes from the high reactive parent, enabling the F_1 hybrid via conventional immunity to recognize tumor-associated transplantation antigens (TATA) expressed on the tumor target, which are not recognized by the poorly reactive parental strain. Another way of interpreting the increased tumor resistance in H-2 heterozygote mice would be to assume similar recognition events as in resistance to bone marrow grafts. In this system extensively studied by *Cudkowicz* [41], heavily irradiated F_1 hybrids are injected with parental bone marrow cells. In certain genetic combinations, the parental bone marrow cells are rejected. As an explanation of hybrid resistance, *Cudkowicz and Hochman* [42, 43] suggested the existence of H-2-linked recessive histocompatibility genes. Such genes would generate an antigen or antigens in each parent which would be lacking in the F_1, provided the F_1 was heterozygous at the postulated locus. *Cudkowicz* [41] has called the proposed loci hybrid histocompatibility or Hh loci. In linkage studies conducted by *Cudkowicz and Hochman* [42, 43], evidence has been found for several such loci, all of which map at chomosome 17.

To return in our discussion to the NK in vitro system we should recall the H-2 linkage of NK activity as described in the previous chapter in this context. This is particularly noteworthy as the H-2 linkage of NK activity was seen in the same tumor host systems, where in vivo H-2-linked hybrid resistance phenomenon was observed. Taken as an example, the higher NK activity of the H-2 heterozygotes against the YAC-1 tumor target in the (A × C57Bl/6) × A backcross corresponds in vivo to a higher resistance of the H-2 heterozygotes to the same tumor [203]. the same was true also for tumors of other types. e.g. for both the Rauscher virus-induced tumor RBL-5 of C57Bl/6 origin or the chemically induced C57Bl/6 lymphoma El-4; a parallel H-2-linked influence of NK activity as well as hybrid resistance in vivo was observed in the (C57Bl × DBA) × C57Bl backcross [148]. Clearly, the fact that both phenomena are under the influence of H-2-linked factors in no way proves that NK cells are the only mechanism responsible for the H-2-linked hybrid resistance in the above-mentioned backcross mice. More evidence for this, however, stems from the backcross analysis, where the mice were compared individually for in vitro NK activity and in vivo tumor growth resistance. Here, spleen cells from spleenectomized mice were tested for NK activity, and the same were then challenged with living YAC tumor cells. These studies revealed that mice with high in vitro NK activity generally also had better tumor growth resistance [203]. However, this again may merely show that both phenomena are H-2-linked but they may still operate by different mechanisms. The elimination of conventional immune mechanisms, while maintaining the positive correlation between NK activity in vivo resistance as discussed in the next chapter would, however, support the importance of NK-dependent mechanisms in the above-described phenomena.

Knowing that the NK activity as well as the hybrid resistance phenomenon both are linked to the H-2 complex, it has been of utmost importance to compare the precise position of these two traits on chromosome 17, and ask whether they would map in the same region or even be identical. With regard to the hybrid resistance to bone marrow grafts, *Cudkowicz* [41] found evidence for multiple H-2-linked loci, but that D-end disparities seem to be most conducive to rejection. *Cudkowicz* [41] has designated the D-end-linked locus, Hh-1. Furthermore, he has shown that heterozygosity at this locus is sufficient for the expression of hybrid resistance to two radiation-induced C57Bl lymphomas. In confirmation of this, *Harmon* et al. [104] lately established that heterozygosity within or near the H-2- D region is sufficient to obtain a successful resistance to the EL-4 lymphoma of C57Bl origin. Interestingly, these authors also confirmed the finding of an H-2-linked influence on NK activity. By using CR or H-2-recombinant mice on B10 background, they showed that heterozygosity within or near the D-end of the H-2 complex was necessary for high NK activity against this tumor, as already mentioned in the previous chapter. The possibility was therefore considered that the homozygous C57Bl tumor cells expressed antigenic determinants analogous to the aforementioned Hh antigens, which would be recognized by the hybrids. Could NK cells then actually be directed against products of such Hh loci on the tumor target? At present, we do not consider this explanation the most plausible one. From what has already been discussed in the previous chapter, the genetic differences in F_1 crosses or the H-2-linked influence on NK activity do not seem to depend on the uniqueness of the target cell used in the assay. Thus, the H-2-associated factors determine high NK activity not only to tumor targets which could be expected to express products of the Hh-1 locus, such as various C57Bl lymphomas, but also to a variety of widely different tumor targets, including lymphomas of Balb/C and A/Sn origin, that is genotypes which should be 'silent' at the Hh-1 locus [148]. A more likely explanation would thus be a regulation of effector cell levels rather than specificities.

There is one aspect which adds some complexity to any comparison between NK activity and in vivo resistance to tumor cell growth. We have in a previous section described how NK activity can be augmented or boosted in vivo by various agents. therefore, one should bear in mind that, although the rejection tests of small tumor inocula are performed in non immunized animals, the possiblity that the tumor inoculum itself could induce NK activity must be considered. This is particularly likely if this induction has a very short latency period, as is indeed the case with the induction of NK cells. *Herberman* et al. [111] have in fact demonstrated how the inoculation of various types of tumor cells in vivo may lead to a rapid increase of in vitro NK reactivity. A clue to the mechanism behind this induction was originally provided by *Trinchieri* et al. [274], who demonstrated in a human NK system that certain tumor cells in vitro can induce interferon production in human PBL, and that this interferon can augment NK activity. The existence of the interferon induction system as a modifying system for NK activity clearly adds another dimension to the genetic control of tumor cell growth by NK cells. The knowledge of this may in fact suggest that when comparing in vitro NK activity and in vivo tumor resistance, the in vitro assay should also be done using mice which have been boosted with the relevant tumor in vivo. This should then be done very shortly before assay in order to reduce the possibility of measuring conventional immune mechansisms as well.

If one accepts that NK cells participate in the rejection of transplantable tumors, does this also mean that NK cells act in surveillance of newly arising neoplastic cells? It is quite likely that the inoculation of small numbers of tumor cells into unprimed animals in many ways does resemble the situation the organism encounters when faced with transformed aberrant cell clones developing into tumors. Nevertheless, a more appropriate model would be to actually study the genetic control of primary tumor induction in comparison to the NK system. To our knowledge, no such study has yet been performed, a fact which might appear somewhat surprising. Having in mind the multiplicity of factors known to influence oncogenesis, the difficulty in estimating the role of one particular component in tumor protection is obvious. Consequently a comparison between tumor incidence in various strains of mice with the efficiency of a certain component of the immune system, such as NK activity, would not seem very meaningful. Therefore, the possible importance of NK cells is not contradicted by the fact that no obvious correlation exists between tumor incidence and grouping of mouse strains into NK high or low reactive strains. Ideally, such a comparison should be performed using mice of the same genetic background, differing at the genes determining NK reactivity only.

The beige mutant provides such a strain since NK activity is deficient whereas all other immune and macrophage functions are normal. Initial studies in two separate laboratories show that beige mice are deficient in rejecting low-dose inoculas of one virally induced and one chemically induced synegeneic leukemia [134] as well as a malignant melanoma [269]. Homozygous bg/bg mice had an increased incidence of tumor takes, a shorter latency period and a larger tumor mass than heterozygous +/bg littermate controls. In addition the ability of B16 melanoma cells to metastasize to the lung was greatly increased in bg/bg mice [269]. This is in agreement with earlier studies showing that nude mice resisted metastasis of allogeneic tumor cells [56, 256] most likely by an NK mechanism [98]. Finally the most pertinent observations come from NK-deficient CH patients surviving early childhood infections. These patients almost invariably develop a spontaneous lymphoproliferative disorder which has been called malignant [reviewed in 13]. Hence, the evidence is becoming very strong that NK cells are indeed involved in surveillance against neoplasia. The universality of this surveillance mechanism awaits the screening of a broader range of tumor cells and inducing agents.

V. Ontogeny and Senescence

One of the characteristic features of the NK system in rats and mice is the rapid rise and fall in NK activity with age (fig. 5). NK-mediated cyto-

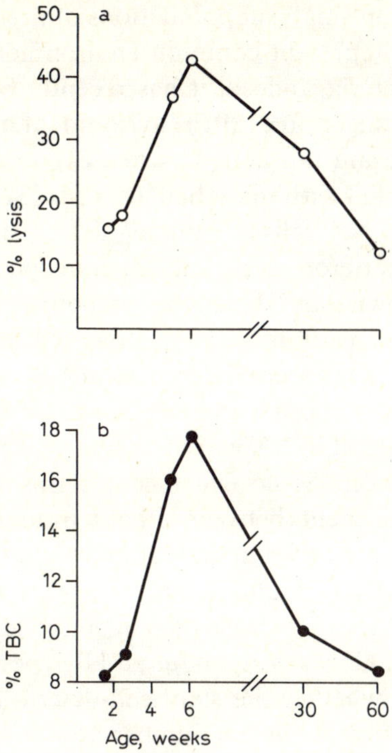

Fig. 5. Age-related changes in target-cell binding and lysis by CBA spleen cells. Nylon wool-passed spleen cells from 6 to 8 mice in each group were tested for cytolysis (a) and target-cell binding (b) at effector to YAC target ratios of 30:1 and 1:10, respectively. Adapted from *Roder and Kiessling* [216].

lysis is low or absent in fetal liver and the spleens of newborn mice, appears at 3–4 weeks of age (which is later than immunocompetent T cells appear), and shows a peak at 5–8 weeks followed by a slow decline to low levels in mice 6–12 months old [109, 138, 252]. This age distribution of in vitro activity is well correlated with resistance to growth of the same tumors in vivo [87]. The question which arises then is whether the characteristic age curve of NK activity reflects an intrinsic, pre-programmed developmental pathway in NK cell ontogeny or whether extrinsic factors are also involved. Due to the relatively sharp peak of NK activity at 6 weeks of age it is conceivable that the response reflects prior immunization against some ubiquitous antigens present early in life [109] since responses to alloimmunization or tumor antigens [161] follow similar kinetics. These antigens could be autoantigens found on small subpopulations of thymocytes [100, 101] or alternately might represent common environmental agents in the form of viruses and bacteria. Indeed, it has recently been found that mice raised in a specific pathogen-free (SPF) environment have little or no NK activity but rapidly regain this activity when exposed to conventional conditions or when injected with microbial antigens [37]. It cannot be concluded yet however that NK cells are antigen driven since microbial agents may also induce interferon, a potent but non-specific stimulator of NK activity [62]. It is known that NK function in both young (1 week) and old (1 year) mice is boosted by tilorone [194]. However, it has not yet been reported if the relative age-dependent differences in NK cytolytic activity are maintained in interferon-boosted animals. Such information would help to decide if the rise and fall in NK activity with age is preprogrammed at the precursor cell level. Some evidence for this was provided by bone marrow transfer experiments between lethally irradiated young or old syngeneic recipients [90]. As shown in table II, the recipients always expressed the level of NK activity found in the bone marrow donors. Therefore, the decline in NK activity with age is an intrinsic defect of precursor cells in the bone marrow. However, without an NK-specific marker it was not possible to decide whether the stem cell-determined decline in NK function was due to a decline in the number of NK cells generated or rather a defect in the function of those NK cells which did arise (table III). With the development of a sensitive technique for assessing the frequency of NK cells by their capacity to selectively adhere to target cells, we found that very young (2 week) and very old (1 year) mice possessed only one half the total number of NK cells present in 6-week-old mice (fig. 5). This observation, together with the fact that interferon inducers increase intrinsic NK cytolytic capacity but not total numbers of NK cells, suggests that both the maturation and senescence of NK activity can largely be explained on the basis of population size. Confirmation of this conclusion will depend on the development of cell surface markers which are independent of the NK recognition receptor.

The possibility of intrinsic defects at the receptor level in very young and old NK cells was investigated by comparing the relative NK receptor density and avidity at different ages. Purified target antigens were used in quantitative absorption studies or competitive inhibition assays in which increasing quantities of antigen inhibited the formation of target-effector conjugates. The results suggested that NK receptors 'mature' during ontogeny to higher avidity binding whereas the decline in NK function

Table II. NK activity in irradiated mice reconstituted with syngeneic bone marrow from young or old mice

Bone marrow donor	Bone marrow recipient	
	young	old
Young	high	high
Old	low	low

3×10^7 bone marrow cells from young (8 weeks) or old (1 year) CBA mice were transferred into lethally irradiated (800 R) recipients of the same strain. NK-mediated cytolysis of YAC cells was measured several weeks later at 50:1 effector:target ratio. Adapted from *Haller* et al. [90].

Table III. Possible mechanisms for the sensescence in NK function

Size of NK population	Intrinsic cytolytic function of each NK cell
Decline	normal
Decline	decline
Normal	decline
Normal	normal[1]

[1] In this case one would postulate an age-dependent increase in suppressor cells which could inhibit NK function. An increase in suppressor cells with age has previously been shown to result in the immunosenescence of B cell function [217].

during senescence can solely be attributed to a decrease in population size with no change in receptor avidity or density [230].

It should be pointed out that in autoimmune susceptible strains such as NZB, NK activity declines at a much later time than in conventional strains [*Roder,* unpublished observations]. In all other conventional strains tested, however, the age pattern is normal regardless of genotype. In contrast, the target cell chosen may be of primary importance since NK activity to cell lines derived from solid tumors does not show any of the age effects [264] characteristic of the response to lymphomas and other tumors as discussed above. Rats exhibited age-related changes in NK activity similar to that observed in mice [183, 253] although the results in the human are less clear with authors reporting both the presence [29] or absence of age effects [126, 192, 265].

Table IV. NK cells in *Xenopus laevis*

Organ	% lysis	
	Xenopus laevis	*Mus musculus* (CBA)
Spleen	14.6 ± 4	40.2 ± 3
Bone marrow	5.1 ± 1	17.6 ± 2
Thymus	4.8 ± 1	0.9 ± 1

2 individuals were pooled for each species. ^{51}Cr-labelled YAC cells were used at a 20:1 effector:target ratio in a 6-hour cytolytic assay at 20°C.

VI. Phylogeny

One important but virtually ignored aspect of NK cells regards their evolutionary appearance as revealed in the phylogenetic tree. Since the phylogeny of immunoglobulins and specific cell-mediated immunity have been well delineated [118], it would be instructive to compare the development of NK activity. If NK cells appeared very early in phylogeny before immunoglobulins and T cells in the primitive chordates or fishes then this would suggest that NK recognition structures are not coded for by V region genes but possibly histocompatibility-like genes [118] which would distinguish self from non-self and allow the destruction of non-self. Preliminary results [*Roder,* unpublished observation] suggest that the African toad, *Xenopus laevis* may have detectable levels of NK activity against murine tumors such as YAC (table IV). This species and a wide variety of other poikilotherms have also been shown to lyse chicken erythrocytes by an ADCC mechanism [130]. One preliminary report showed that hemocytes from a higher invertebrate, the lobster, *Homarus americanus,* spontaneously lysed xenogeneic tumor cells [P815×2, L5178^{-}1, and SL2) but not sheep erythrocytes or autologous cells [67]. It will be important to look at a variety of other lower species since this observation suggests that NK cells evolved prior to the appearance of highly specific and differentiated T and B cells.

VII. Environment

Early experiments with SPF mice [illegible] to show major differences in NK reactivity compared to mice or rats reared

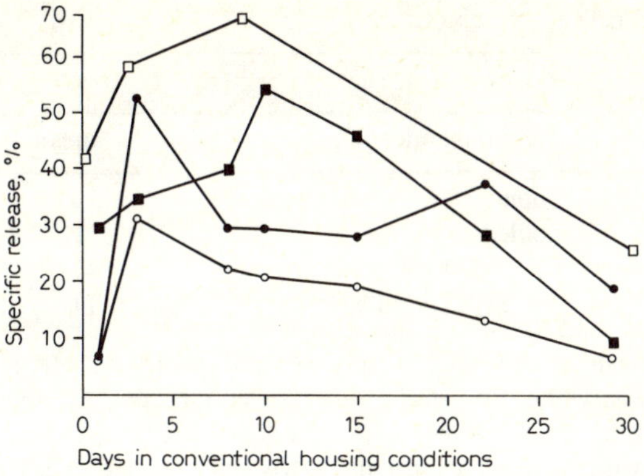

Fig. 6. Kinetics of NK cell activation by naturally acquired infection. Age-matched male 7- to 9-week-old CBA mice (●), 7-month-old CBA mice (○), or 9-week-old nude mice (■) were removed at intervals from SPF environment and placed in conventional animal quarters containing endemic Sendai virus. Spleen cell NCMC against ^{51}Cr-labelled YAC-1 target cells was measured. A 33:1 effector:target cell ratio is shown (spontaneous release: 13.6%). In another experiment 9-week-old female nude mice were tested (□) using a 66:1 effector:target cell ratio (spontaneous release 9.0%). Adapted from *Clark* et al. [37] with permission.

in conventional environments. However, recent studies by *Clark* et al. [37] clearly show that SPF mice have very low or non-existent NK activity which is rapidly augmented within 36–48 h after exposure to conventional environments (fig. 6). Our own studies with litters of outbred, gnotobiotic mice delivered by cesarean section into completely sterile barrier unit conditions revealed that these germ-free animals maintain some (> 10%) NK activity against YAC even though mice reared under identical conditions possess undetectable serum γ-globulins or lymphoid cells in the gastrointestinal tract [*Kiessling, Roder, Karre, Gustavsson,* unpublished observations]. The frequency of NK cells as measured in the TBC assay was also significant (table V) and cytolytic T cells generated by Con A or in mixed lymphocyte cultures were highly active (100 LU/10^6 cells). However, due to the possibility of genetic drift between the mice inside and outside the barriers, our studies do not yet allow us to make comparisons between the gnotobiotic and conventionally reared mice. The conclusion at present must be that NK activity in germ-free mice may be much lower than in mice reared in conventional environments, as shown by *Clark* et al. [37], although significant levels of NK activity persist in the absence of any microbial flora or exogenous viruses.

Table V. The frequency of NK cells in gnotobiotic mice

Strain	Environment	% TBC	
		individual mice	mean
CBA	conventional	pooled	20
A/Sn	conventional	pooled	9
SW	conventional	16, 16.5, 17, 17.2, 17.5	17
SW	germ-free	7.5, 10, 11.1, 13.2, 16	11

Nylon-passed spleen cells from mice of inbred, high (CBA) and low (A/Sn) NK-reactive strains or from an outbred Swiss white (SW) strain were assayed for TBC. All mice were raised in a conventional environment except the Swiss whites designated as germ-free. This latter group was born and raised in a sterile environment. SW mice were assayed individually whereas 6 CBA and 5 A/Sn mice were pooled. Note that the SW and SW-GF groups were single litters from two groups of mice which had been bred separately for several generations. Therefore, these two groups are not strictly comparable in the genetic sense.

In the *Clark* et al. [37] study the kinetics of NK cell activation by SPF mice placed in conventional environments (fig. 6) bears a striking resemblance to curves showing the natural, age-related rise and fall in NK activity. Further evidence suggested that the age curves may reflect the acquisition of a naturally occurring capsular polysaccharide of *Klebsiella pneumoniae* [37]. In addition, these animals were exposed to endemic Sendai virus as detected by serotyping. This particular virus [111] as well as others [169, 185, 186, 289] are known to activate NK cells. *Clark* et al. [37] have further postulated that the H-2D^d gene which regulates NK activity may do so by controlling the expression or presentation of virus.

Further observations provide indirect support for the hypothesis that NK activity is generated in an immunological manner to some ubiquitous environmental agent. Immunoglobulin from the serum of normal, nonimmunized rabbits exhibited the same specificity as murine NK cells when tested against a panel of targets [83]. Preincubation of targets with this serum blocked subsequent NK cytolysis [*Roder,* unpublished observation]. One interpretation of this finding is that the rabbits and mice used in this study were exposed to environmental agents which share cross-reacting determinants common to all NK-sensitive target cells.

The situation in the human is less clear. Some studies of laboratory workers and relatives or contacts of cancer patients show a higher incidence of NK-like activity [27, 191, 265] whereas other studies have failed to detect a difference [192, 206, 232].

VIII. Modulation

A. Interferon

Purified interferon derived from fibroblasts or leukocytes has recently been shown to augment NK cytolysis after in vivo and in vitro administration [62, 80] and anti-interferon antibodies block the NK enhancement observed in both mice [62] and humans [109, 116]. Inducers of interferon such as tilorone, statalon, Newcastle disease virus [62], polyinosinic-polycytidylic acid [185, 186], and tumor-derived or virus-infected cells [242, 276] also enhance NK-mediated cytolysis in humans, rats, and mice.

In the mouse the enhancement of endogenous NK activity appears to reflect an increase in the intrisic lytic potential of individual effector cells rather than recruitment or clonal expansion [218]. Spleen cells from tilorone-injected mice exhibited a marked increase in cytolysis of YAC targets with no detectable increase in the frequency of NK cells as judged in the TBC assay (fig. 7). Furthermore, since interferon effects are observed sometimes within minutes in the in vitro cytolytic assay, it is unlikely that proliferative effects would be significant. The actual mechanism of interferon-induced enhancement is still unknown but could conceivably involve an increase in the cycling of NK cells so that the total number of targets contacted and lysed with time is greater. Interferon is known to increase the rate of lysis, which would fit with this prediction. In comparison interferon increases the number of antigen-binding T cells in unstimulated murine lymphocyte population [166]. Therefore, interferon regulates endogenous T and NK cells by different mechanisms.

It is not known if interferon augments the triggering signal provided by the NK target or whether interferon itself is the signal. The latter possibility is supported by the observations that (1) anti-interferon antibody inhibits NK activity in normal, non-interferon-activated populations [116], and (2) certain patients with a profound depression in their ability to produce interferon have a concomitant depression in NK activity [281]. On the other hand, many targets have been found which do not induce any detectable interferon but are nevertheless lysed efficiently by NK cells which suggests that interferon is not an obligatory requirement for NK-mediated cytolysis [283]. It is possible that interferon simply blocks target-derived inhibitory products, such as prostaglandin E [51, 52], by competing for the same ganglioside receptor on the NK cell. If interferon actively induced cAMP, which it does in most cell types [60], then NK cytolysis would be greatly inhibited rather than enhanced since it has been

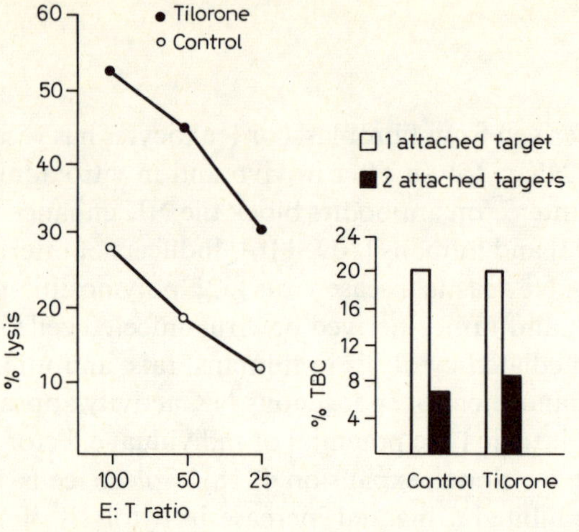

Fig. 7. The effect of an interferon inducer on target cell binding and lysis. Spleen cells were pooled from mice given 2 mg tilorone parentally, 24 h previously. Aliquots of these cells were tested in a cytolytic assay and for their ability to bind YAC targets at 0°C. The frequency of cells binding one or two attached targets was determined. Adapted from *Roder* et al. [218].

shown that increasing cAMP suppresses NK activity [225]. However, several hours prior to the increase in intracellular cAMP in a synchronized lymphocytic cell line, interferon induces a transient increase in intracellular cGMP [273]. Since cGMP also causes a slight enhancement of NK activity [225] then it is possible that interferon enhances NK activity by virtue of its cGMP-inducing effect. The species specificity of NK enhancement by interferon [*Gidlund,* personal communication] could support either the active or passive model depending on whether this restriction is determined by membrane receptors or other factors.

In comparison to the endogenous NK cells, the LCMV-activated NK cell behaved differently [*Kiessling and Welsh,* unpublished observation]. As shown in table VI, NK cells obtained 3 days after LCMV infection have a broader target selectivity, greater adherence to nylon wool, greater avidity FcR and much larger size than endogenous or 'resting' NK cells in non-infected mice. In addition the frequency of TBC was increase in LCMV-activated NK cells compared to non-activated NK cells. It is likely that LCMV acts through the agency of interferon and therefore these results taken together suggest that NK cells are heterogenous with respect to their response to interferon.

Table VI. Comparison of virally activated or unactivated NK (cells 218, 288, 289; *Kiessling and Welsh,* unpublished observations)

Property	Endogenous NK cells	LCMV-activated NK cell[1]
Target selectivity	narrow	broad
Nylon adherence	none	partially adherent
FcR	± weak	++
Cell size	small to medium 'resting' lymphocyte	large blast-like cell
TBC frequency (L cell target)	–	increase (10–30%)

[1] 3 days following infection of mice with LCMV.

In the human system the cells producing interferon in response to virus-infected or tumor-derived cells fall into the same lymphocyte subpopulation as NK cells since they possess Fc receptors but lack surface Ig or high affinity E receptors [276]. Therefore, it is possible but not yet proven that the cell which produces and responds to interferon is the NK cell.

Effector cells participating in ADCC against nucleated targets were boosted in a parallel fashion to NK activity in one study [109, 116] but not in another [275]. Since the NK cell and K cells participating in ADCC are probably identical then interferon would be expected to boost both activities in common.

The cytolytic activity of T cells [164] and the phagocytic activity of macrophages are also boosted by interferon [50, 123] and, therefore, its augmenting effects are non-selective. Many activities are inhibited rather than enhanced [54]. For example antibody responses [16, 125], proliferative responses to antigens or mitogens [198, 235, 257], delayed type hypersensitivity, allograft survival, and graft-versus-host reactions [44] are all suppressed by interferon.

In the human system interferon has been shown to protect certain target cells, such as fibroblasts, from lysis, in direct antagonism to its effect on NK cells [276]. Protein and RNA synthesis are required for this protective effect. Interferon-treated targets are also less efficient competitors in a cytolytic assay which suggests that some surface determinant, such as the NK-TS, is masked or absent. These results have been confirmed in the mouse in experiments showing that both fetal murine fibroblasts and adherent peritoneal cells are protected from NK lysis by interferon treatment [291]. In *Trinchieri's* [277] original studies interferon protected fibroblasts but not virus-infected or tumor-derived target cells, and it was suggested that during the process of immune surveillance interferon optimizes the conditions necessary for the lysis of virus-infected or tumor cells while at the same time protecting normal tissues from NK-mediated damage [243]. This result may be an oversimplification [291] and it remains to be determined whether the putative anti-tumor effects of interferon in vivo [34, 77–79, 262] are due to the boosting of NK cells.

B. 'Adjuvants'

A myriad of NK stimulants have been described, some or all of which may act through the agency of interferon. The term 'adjuvant' is used since no data has been presented which convincingly shows that the NK-boosting effect is immunologically specific. Various bacterial preparations such as live bacille Calmette-Guérin (BCG) [245] or *Corynebacterium parvum* [111, 185, 188] have been shown to enhance NK activity depending on the mode of administration [188]. In all of these studies, the cytolytic effector cells induced exhibited the characteristics of classical NK cells. As little as 1–100 ng of heat-killed *C. parvum* injected i.p. caused marked enhancement of NK activity in the peritoneal cavity of CBA/H mice whereas i.v. administration markedly suppressed NK cytolysis by an unknown mechanism [188]. The ability of *C. parvum* and BCG to boost NK activity could be due to their capacity to induce interferon [147; *Tracey,* personal communication] whereas the suppressive activity of *C. parvum* could be due to its ability to induce PGE [82]. *C. parvum* as well as BCG has been used as an inhibitor or tumor growth in vivo [95, 296].

Virus-infected or tumor-derived cells also induce interferon [274] and consequently augment NK-mediated cytolysis [111] as do natural or synthetic interferon inducers, such as tilorone, statalon [62], endotoxin, pyran, and poly I:C [114, 115, 185, 186]. Virus infections with Semliki Forest virus, Junjin virus [169], lymphocytic choriomeningitis virus [288, 289], Newcastle disease virus [62, 185, 186], Kilham rat virus [185, 186], Sendai virus, and murine sarcoma virus [111] also augment NK activity within the first few days of infection. Some viruses (HSV-1, PR 8, HK) were also effective at augmenting NK activity by in vitro co-cultivation [242].

C. Suppressor Cells

It has been suggested that suppressor cells may regulate NK activity in a variety of conditions which are known to inhibit NK function. Studies by *Savary and Lotzova* [245] show that NK activity against YAC is depressed in neonatal mice, mice tolerant to bone marrow grafts and in mice injected i.v. with *C. parvum*. Co-cultivation experiments suggest a suppressor mechanism in all three instances. Similar results have been obtained in mice treated with a number of agents, such as ^{89}Sr, carrageenan, hydrocortisone acetate or whole body irradiation (700 R) [42, 43]. The suppressor cell is not fully characterized as yet and has been described in both systems as 'macrophage-like' since: (1) it can be induced in nude mice and therefore might not be a T cell; (2) it is radioresistant (2000 R); (3) it shows no MHC restriction; (4) it adheres to Sephadex G-10, and (5) it serves as an accessory cell for the in vitro induction of killer cells. Some attempts to remove suppressor cells by fractionation (carbonyl iron, Sephadex G-10) showed an increase in NK activity but this could also be explained as a simple enrichment of NK cells. No attempt was made to recover suppressor cells and characterize them by positive criteria. The suppressor could simply be an NK cell which binds to the target but fails to induce lysis as shown in the beige mouse [219]. Such a cell would compete with cytolytic NK cells for targets. Alternatively, the suppressed animals might simply possess an increased number of natural targets such as those described in the thymus [99, 100]. These cells would compete with isotope-labelled tumor cell targets and reduce the observable cytolysis. Prozone and crowding effects have not been rigorously excluded since normal 'filler' cells also caused a significant inhibition of NK activity at the suppressor: effector ratios reported. If the suppressor cells were larger in size than the filler cells then more of this artifactual inhibition would be expected. Finally, the biological relevance of these forms of suppression is open to question, with the exception of the neonatal suppressors, Attempts to demonstrate NK suppressor cells in neonatal or aging mice in vitro in other laboratories have been unsuccessful. In summary, the existence of 'classical' suppressor cells in the NK system must be treated with caution and further experiments are needed before definitive conclusions can be reached.

Another kind of suppressor cell may in fact be the tumor cell itself via prostaglandin release. In the mouse [225] and human system [51, 52] PGE_1 and PGE_2 inhibit cytotoxicity by a direct effect on the NK cell which may be mediated by a PGE-induced increase in cAMP [225]. Studies by *Droller* et al. [51, 52] with indomethacin and other inhibitors of prostaglandin synthetase revealed that pretreatment of the target cells led to enhanced NK activity presumably by preventing the release of PGE. As suggested by these authors this could provide a mechanism whereby certain tumors escape immune surveillance. It is interesting in this regard that *Plescia* et al. [204] showed tumor cells producing high amounts of PG that were rejected more rapidly in vivo if pretreated with indomethacin. Macrophages also release PGE during phagocytosis and therefore it would be expected that inflammatory macrophages at the tumor cite might also inhibit cytolytic function in infiltrating NK cells. This could explain the decreased activity in NK cytolysis exhibited by cells recovered from the tumor site in some studies [240].

D. Tumor Growth

In mice carrying a variety of tumors (transplanted methyl-cholanthrene-induced sarcoma, primary MSV sarcoma, reticulum cell sarcoma or human lymphoblastoid cell lines in nude mice) NK cells were recovered from the tumor site or draining lymph nodes [5, 59]. It is not known if these NK cells were activated via interferon, recruited locally or homed from the lymphoid organs. The spleens from tumor-bearing mice were depleted of NK activity in the absence of a suppressor mechanism [5, 109], possibly as a result of migration to the tumor site or perhaps inhibition by PGE release from the tumor.

In humans the predominance of NK cells in inflammatory infiltrates varied greatly depending on the type of tumor examined [240, 272]. In a recent study by *Saksela* et al. [240], NK cells separated from several carcinomas and testicular seminomas were unable to lyse the usual leukemia targets in the human NK system (MOLT-4, K562) or the autochthonous tumor cells but were able to lyse anchorage-dependent, fetal fibroblasts. NK cells from the peripheral blood of these same patients efficiently lysed all of these targets which again points to a heterogeneity or polyclonality within the NK system. These findings are supported in part by earlier studies [126, 127, 282–284]. Without NK-specific markers it is not possible to comment on the meaning of the shifted NK reactivity at the tumor site. NK cells could be present but blocked from expressing their function by inhibitory factors such as PGE released from the tumor cells or infiltrating macrophages [22]. It is equally possible that subcellular target antigens [220] from parts of a necrotic tumor could inhibit cytolysis. A mere 2% contamination by tumor cells would yield at 2:1 competition/target ratio in this study and such ratios have previously been shown in many other systems [99, 100] to cause a marked reduction in NK cytolysis. In the study reported [240] one would expect that the normal

NK function towards autochthonous tumor cells in the peripheral blood of these cancer patients would at least prevent blood-borne metastasis.

IX. Specificity

A number of criteria must be fulfilled in order to designate a cytolytic phenomenon as specific. Most investigators would agree that NK cells are at least selective since some target cell types are highly sensitive to cytolysis whereas others are not [109, 138]. NK-resistant targets were sensitive to alloimmune T cells thereby demonstrating that they were not simply more resistant to cytolysis in general. The concept of specificity in the context of an immune surveillance system requires that (1) a lock and key, or enzyme substrate model is appropriate to describe the target-effector interactions, and (2) the system is sufficiently polymorphic to deal with the wide range of target cells sensitive to NK-mediated lysis. The list of NK-sensitive targets now includes cells of xenogeneic, allogeneic or syngeneic origin which may be transformed or non-transformed and derive from normal tissues or tumors from diverse histogenic origins (melanocytes, lymphocytes, mast cells, fibroblasts, macrophages, nerve cells, and others). A priori it would seem unlikely, although it cannot be excluded, that this plethora of diverse targets would share a common determinant and hence the requirement for a polymorphic recognition system if NK cells are to be of general significance in an immune surveillance system.

Previous work has shown that various unlabelled competitor cells could inhibit isotope release from labelled target cells [197], and similar observations in the NK system were used to infer that the target and competitor shared common determinants which were recognized by surface receptors on the NK cell [109, 138]. Those cells most sensitive to NK lysis proved to be the best competitors and a good correlation was found between susceptibility to direct lysis and competition among a wide variety of targets [89, 99, 109, 138]. However, simple competition tests alone are not enough to designate an NK interaction as specific. What are the non-specific effects that could occur during a lytic reaction in a competitive inhibition assay? When examining real data one sometimes observes that potent effectors against a given target also give higher background levels against target cells that are usually insensitive. Likewise, highly sensitive targets are usually sensitive to effectors from a broader range of genotypes. This general susceptibility of targets to lysis could represent a membrane property (possibly) and would therefore be a threshold event, unrelated to the presence or absence of particular antigens. Similarly, high levels of effector activity could reflect (1) their degree of activation over a more specific resting state, if activation of NK cells causes non-specific lysis as occurs in the macrophage system, or (2) some general membrane property (charge, hydrophobicity, fluidity, cell size, motility, glycolayer, microvilli) that allows mor intimate contact with certain targets regardless of the target antigens they carry [8]. It is not possible at present to distinguish between these various alternatives.

Recently a powerful statistical method, called interaction analysis, has become available to help distinguish the underlying specificities involved in NK-mediated lysis [153, 266–268]. Interaction analysis starts from the premise that both specific and non-specific reactions occur during NK-mediated lysis. Testing one effector against a target panel is not very informative and says nothing about specificity except how broad the reactions are. Likewise testing many effectors against one target says nothing about specificity. However, if effectors from different donors show different patterns against a panel of targets then different recognition receptors and target antigens are involved in the system. Therefore, in the Takasugi system a panel of effectors and targets are measured for the degree of lysis and then a two-way analysis of variance is set up to test the hypothesis that no specific reactions have occurred.

The analysis is rigorous because it tends to underestimate specific effects. In addition, if 'non-selective' lysis really represents (1) a sum of many specificities shared between targets, or (2) one antigen shared on many targets which is equally recognized by all the effectors, then interaction analysis in reality will only pick out interactions involving very unique and infrequent antigens. The method is probably an underestimate of the real number of specificities involved. If the analysis indicates significant selective reactions by one effector on different targets this could be interpreted as a sharing of antigens or as one heterogeneous effector suspension reacting to two different antigens. The only method for distinguishing these possibilities is a cross-competition assay in which unlabelled targets are added to the reaction mixture of effectors and ^{51}Cr-labelled targets. The ability of different cell lines to compete infers the existence of cross-reacting target antigens and a complementary receptor on the effector cell. The data obtained can also be analyzed by interaction analysis to distinguish specific from non-specific inhibitory effects.

The discovery that certain target cells can induce interferon [274] which can subsequently enhance the cytolytic activity of NK cells [62, 274, 277], and at the same time protect certain targets from cytolysis [277, 280], will require a reevaluation of all the competition experiments reported to date. *Trinchieri* et al. [275] have shown that when target cells and competitor cells are both interferon inducers, target cells are less efficiently lysed than in systems in which target and competitor cells differed in their ability to induce interferon. Interaction analysis of this data revealed two classes of target cells most likely on the basis of interferon-inducing activity rather than the presence or absence of antigens recognized by NK cells. The problem becomes more apparent in comparing intraspecies NK restrictions [99] since both the NK-augmenting and target-protecting actions of interferon are species-specific, whereas interferon induction itself is not. Some of these complications could be resolved (1) using targets that are relatively refractory to the inhibitory effects of interferon, (2) using short competition assays (stopped with EDTA and cytochalasin B) in the presence of anti-interferon antibody.

The most direct approach to elucidating NK specificity may lie in the isolation of target 'antigens' and recognition structures. The recognition receptor on the NK cell, which is necessary for target-effector binding, was found to be a de novo synthesized, cell surface protein based on trypsin sensitivity and the inability to regenerate in the presence of cycloheximide or at 0°C (fig. 8). It is not yet known if the receptors are coded by the V region genes as in the case of T cell receptors [11] or if they represent remnants of a more primitive surveillance mechanism. At the target cell level an attempt was made to isolate the target antigens recognized by NK cells.

Preincubation of NK cells with detergent-solubilized, cell surface proteins of YAC lymphoma cells prevented subsequent binding to intact YAC targets [220]. The inhibitory material bound to Con A-Sepharose columns and could be eluted with the specific sugar, suggesting that the target structures may be glycosylated. The target structures consisted of 3 molecular species tentatively assigned molecular weights of 130, 160, and $240 \times 10^3 \pm 10 \times 10^3$ based on electrophoretic mobility in SDS-polyacrylamide gels as shown in figure 9 and 10. These molecules could not be detected in gels of relatively NK-insensitive target cells such as P815, A9HT, ywt, YWA or EL-4 and the quantity obtained from the gels varied directly with the NK sensitivity of YAC [220] which, like other tumor cells, is more sensitive when grown in vitro than in vivo [7, 10]. The NK target structures from YAC were selective since they inhibited the binding

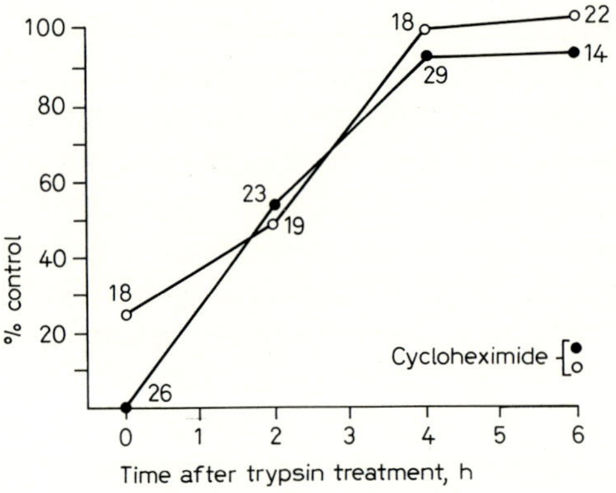

Fig. 8. The recovery of target binding and cytolytic potential following trypsinization. Spleen cells passed through nylon wool columns, anti-immunoglobulin columns, or treated with anti-θ serum and complement were then treated 30 min with 1% trypsin or trypsin buffer as control and incubated for varying periods of time at 37°C in culture medium. For controls, cycloheximide (20 μg/ml) was added to some cells at time 0. The number of Ig+ cells remaining in these column passed cells was 0 in 1,000 counted cells as determined by direct staining with fluoresceinated anti-mouse Ig (polyvalent). The values represent the percent of control TBC (○) or percent control lysis (●) for trypsin-treated cells compared to cells treated with trypsin buffer only. The actual control values for percent lysis or percent TBC are given beside the plotted values. Adapted from *Roder* et al. [218].

of NK cells but not alloimmune T cells to their appropriate targets. Conversely, H-2 antigens specifically inhibited alloimmune T cells, but not NK cells, from binding to targets expressing the appropriate H-2 haplotype (figure 9) [220, 227]. Therefore, the receptors on NK cells have a specificity distinct from the receptors on T cells. The conclusion is also supported by the observation that interferon protects fibroblast targets from lysis by NK cells but not CTL [277, 291], presumably by shutting off synthesis of the NK-TS. Immune T cells therefore must recognize a different determinant. Additional cell lines (F9, K562) which completely lack MHC-related structures are also sensitive to NK cells which further argues that MHC products are not the target of NK cells. In addition, NK-sensitive targets have been found which lack Fc, C3 and SRBC receptors as well as surface Ig [89], thereby suggesting that none of these surface entities serve as the NK target structure.

Additional NK-sensitive tumor cells also expressed some or all of the target molecules exhibited by YAC and some of these structures shared

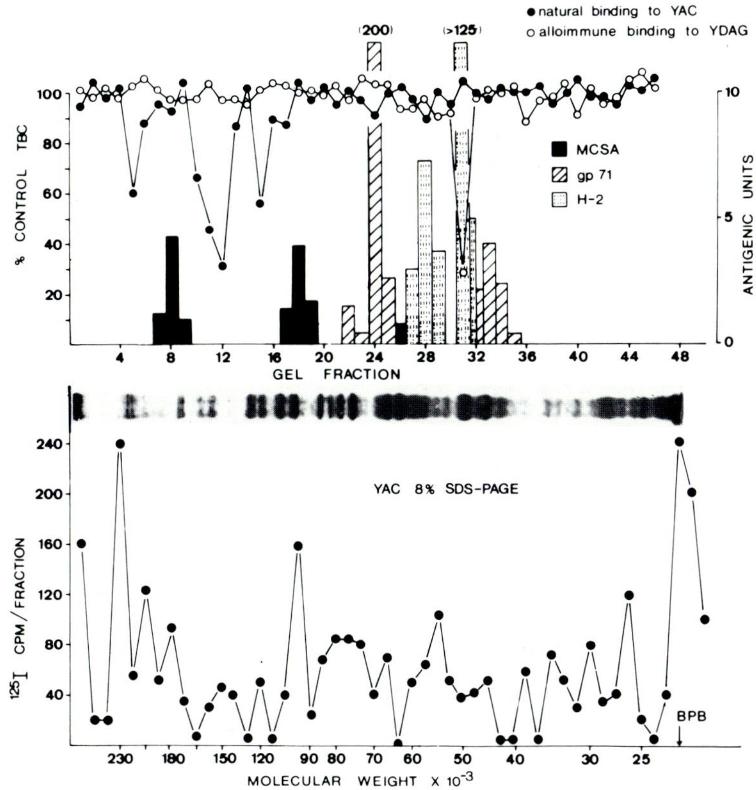

Fig. 9. YAC target structure binds selectively to NK cells. Triton X-100 solubilized and reduced proteins from ^{125}I-labelled YAC cells were electrophoresed in 8% SDS-polyacrylamide gels. The proteins of adjacent channels were stained with Coomassie blue (middle) and gel slices were assayed in a gamma counter (bottom). Proteins were recovered from individual gel slices and assayed for antigenic profiles by inhibition of the appropriate antiserum-mediated cytotoxicity (bars, top panel). In addition, proteins from each fraction were tested for their capacity to inhibit binding of NK cells (nylon-passed CBA spleen) to intact YAC (●) or binding of alloimmune T cells to YDAG targets expressing the appropriate H-2 specificity (○). The T cells were derived from carbonyl iron-treated, peritoneal exudate cells of CBA mice immunized with DBA/2 spleen, thymus, and liver 13 days earlier. The frequency of T cell binders was reduced 81% by pretreatment with rabbit anti-Thy-1 plus complement. the frequency of control TBC in NK or T cell populations was 26 and 16.2%, respectively. Adapted from *Roder* et al. [220].

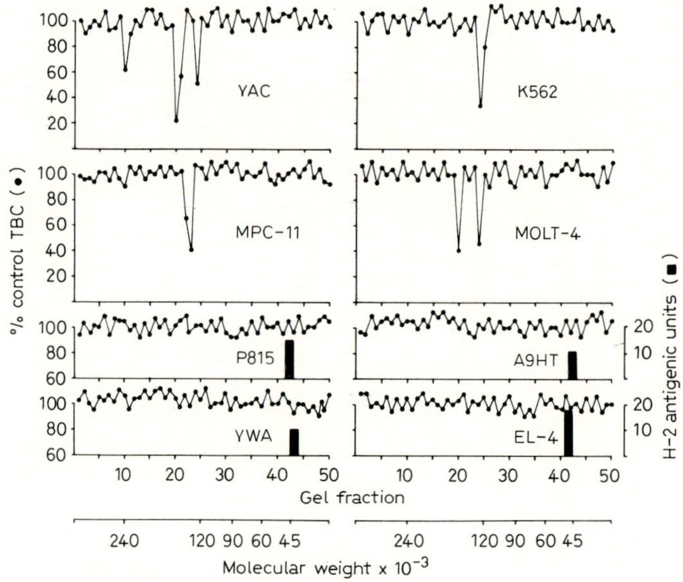

Fig. 10. The expression of NK cell target structures on NK-sensitive but not insensitive cell lines. Triton X-100-solubilized and reduced proteins from various target cells were electrophoresed in 10% SDS-polyacrylamide gels. Proteins were removed from gel fractions in adjacent channels and tested for their capacity to inhibit the binding of NK cells (nylon-passed CBA spleen cells) to intact YAC targets in the case of gels from the NK-insensitive cell lines (P815, EL-4, YWA, A9HT). Gels from NK-sensitive targets (YAC, MPC-11, K562, MOLT-4) were tested for inhibition of NK cell binding to the intact, homologous target. The position of the H-2 antigens (closed bars) was determined by inhibition of antiserum-mediated cytotoxicity. Adapted from *Roder* et al. [220].

specificities in the case of MPC-11, 136-6, and X-63 or were unique in the case of MOLT-4 and K562 as shown by cross inhibition studies [220]. It should be pointed out that this method may underestimate NK-TS specificities and we cannot be certain that all NK-TS molecular species are isolated from each different tumor cell line. However, the results suggest that NK cells recognize more than one specificity. In addition murine NK cells recognized a different spectrum of NK-TS molecules than human NK cells [226]. Although the nature of the NK-TS is not fully understood, it does not possess determinants common to gp71, Moloney cell surface antigen, p30, p15E or H-2. This provides the first direct evidence that in some instances, the NK-TS may not be related to murine leukemia virus products. Several lines of indirect evidence also support the conclusion.

Hence, there was no correlation in 11 different cell lines between expression of virion gp71, p30, p15 or p12, and sensitivity to NK cell-mediated lysis by CBA spleen cells. In addition, immunoresistant sublines of YAC selected for low expressions of MCSA were equally sensitive to NK cells [5, 55]. In further studies, NK-sensitive human cell lines were superinfected with xenotropic murine leukemia virus and were shown to express virion gp71, p30, and p12. The sensitivity of these cells to NK-mediated lysis, however, was not influenced [143]. These data, taken together, strongly argue against the involvement of endogenous virus products in the target cell specificity of NK cells from some strains of mice such as CBA. In different strains, however, gp71 was shown to inhibit NK cell-mediated cytolysis of an AKR tumor by C57Bl × C3H, F_1 mice but not NIH Swiss mice [162], and an NK clone was inhibited by gp71 [*Cantor,* personal communication]. The response to exogenous virus may be a more universal phenomenon.

It has recently been suggested that NK cells in nude mice possess some degree of virus specificity against a variety of exogenous RNA viruses [177, 209]. NK-insensitive, xenogeneic tumor cells (HeLA) became susceptible to lysis by NK cells when infected with measles, VSV or mumps virus. In cold target competition experiments, NK-mediated lysis of isotope-labelled HeLa-measles was inhibited only by unlabelled HeLa-measles target and not HeLa-mumps or HeLa-VSV which were equally susceptible to cytolysis. Reciprocal combinations of targets and competitors were not reported. In view of the possible interferon induction in these 18-hour ^{51}Cr release assays and the limitations of competition assays (above), it will be important to determine if target-effector conjugates are specifically inhibited by virus or virus-containing membranes from infected targets. Other investigators found no differences in susceptibility to NK lysis between virus-infected or uninfected targets in a short-term, 4-hour ^{51}Cr release assay [291]. In the same experiments, the frequency of TBC did not increase following virus infection of the target cell by a number of different viruses. These results suggest that non-specific, interferon mediated boosting of NK cells may occur in 18-hour assays rather than increased NK recognition of specific viral structures.

The widespread nature of common NK-TS might suggest that the NK-TS are (1) de-repressed fetal or embryonic antigens of a highly conserved nature, (2) differentiation antigens, or (3) environmentally determined antigens. Although it is not possible to distinguish between these alternatives at present it is interesting to note that normal unimmunized rabbits have naturally occurring IgG and other unimmunoglobulins in their serum with similar specificity as mouse NK cells, which suggests that an infectious agent common to rabbit and mouse may generate antibody in the rabbit and NK cells in the mouse [83]. The NK-like specificity of the rabbit antibody was shown by: (1) a positive correlation between those targets in a panel which were lysed both by NK cells and natural rabbit antibody plus complement; (2) a correlation between NK sensitivity and the ability of a given target to absorb the rabbit antibody, and (3) target cells pretreated with natural rabbit antibody were less sensitive to NK-mediated lysis in a short-term ^{51}Cr release assay, whereas anti-H-2 antibody had no effect [*Roder,* unpublished observation].

It has also been shown that immature cells in the normal thymus [100, 101] and bone marrow are sensitive to lysis by NK cells. Therefore, the NK-TS may be an autoantigen which is not directly dependent on neoplastic transformation for expression, although it may be expressed more abundantly on transformed cells, which are often more sensitive to cytolysis. Finally, it cannot be ruled out that NK-TS represents a lectin-like substance or glycosyltransferase since models of cell-cell contact involving at least a few specificities are possible under these circumstances [61, 170, 254].

Factors which may be important in the expression of NK-TS have also been examined. A somatic cell hybrid between NK-sensitive and NK-insensitive murine parental cell lines did not express the NK-TS and consequently was not bound or lysed by NK cells [226]. Several additional hybrids in NK high/low combinations were also low in NK sensitivity [1]. Since alloantigens are usually codominantly expressed in somatic hybrids, whereas differentiation states or markers are usually suppressed, the low NK sensitivity in this and other similar hybrids had led to the suggestion that NK sensitivity is related to differentiation [1]. The finding that NK-TS is not expressed on the YAC-IR/A9HT hybrid further suggests that the NK-TS itself may be expressed according to differentiation regulatory mechanisms, thus leading to low NK sensitivity in the YAC-IR/A9HT hybrid. This phenomenon is not universal however since NK sensitivity was intermediate in human high (K562) × low (P3H-R1) hybrids [2, 151].

A different form of NK-TS modulation is also shown to occur. YAC cells, when removed from the selective pressures of an NK containing milieu in vivo and grown in vitro gradually exhibit increasing sensitivity to NK-mediated lysis over a period of weeks [7]. We have previously shown a concomitant increase in the quantitative expression of NK-TS in these in vitro explanted cells [220]. Cell cycle effects may not be important in the expression of NK–TS since NK cells bind to targets equally well which are at various stages in the cell cycle [*Roder,* unpublished observation].

The available evidence supports the concept of limited heterogeneity in the putative NK 'recognition' receptor. Hence, NK cells from high (CBA) and low (A/Sn) NK-reactive strains exhibited identical avidities for NK-TS from YAC cells [230]. That some heterogeneity in the receptor exists is suggested by the observation that receptor avidity increases in early postnatal development in CBA mice [230]. In addition, NK cells from mice recognized different patterns of NK-TS than did human NK cells [226]. Since several unique NK-TS specificities have also been found [220, 226] then these observations taken together suggest that the NK cell pool is polyspecific and has some heterogeneity in the recognition structure, albeit much less than would be expected of an antibody-combining site.

There are no convincing studies yet which show that NK specificities are polyclonally distributed and indeed one recent study suggests they are not. A number of long-term cell lines established from cloned NK cells (Lyt 1^-2^-, $Ly5^+$, thy 1^+) all exhibited the same rank order of cytolysis against a panel of diverse target cells [46]. This suggests either that individual NK cells are polyspecific or that all NK targets share at least one NK target antigen in common. The work of *Ortaldo* et al. [to be published], examining NK depletion on target cell monolayers, suggests that most tumor cell lines share one specificity and maintain others which are unique in support of our work on the isolated NK-TS [220, 226]. It remains to be determined if different subpopulations of NK cells recognize different target specificities.

X. Mechanisms

A. The Lytic Cycle

NK-mediated cytolysis can be divided into 4 discrete stages: (1) recognition and binding to the target; (2) triggering and activation of the lytic mechanism; (3) release of NK lymphotoxin and lethal hit, and (4) target cell death as shown schematically in figure 11 and table VII. We have recently described a TBC assay to visualize target-effector contact [216, 218], based on earlier studies in the CTL system [87, 237, 299].

Investigations into the nature of the nylon non-adherent lymphocyte

Table VII. Model of NK cytolysis

Steps	Relevant molecules structures or processes	Genes	Supporting system	Stimulators	Inhibitors
1 Contact	surface protein recognition moiety (cation requirement)	chrom. 17 H-2 linked (mouse)	protein synth. (E. metabolism not required)		trypsin/pronase/papain puromycin/cycloheximide EDTA/DMSO
Binding	microvilli engulfment	?	cyto-skeleton		cytochalasin B colchicine
2 Trigger	cAMP/cGMP	?	adenylate cyclase	cGMP carbamyl-choline	cAMP/PGE$_1$/theophylline histamine/epinephrine
	ATP?	?	glycolysis oxidative phosphorylation		DNP NaN$_3$ NaF/iodoacetate
Regulator (interferon sensitive)	IF production IF receptor IF target	chrom. 5 (2) and chrom. 21 (human)	protein synthesis ganglioside synthesis	interferon poly I:C NDV tilorone	anti-IF antibody
3 Lethal hit	lysosomal enzymes	chrom. 13 ch/bg gene	Golgi and secretory vesicles	cGMP carbamyl choline	chloroquine PMSF/DIFP glutaraldehyde
4 Target death	colloid osmotic lysis				

This table combines data obtained in both human and mouse experimental systems. References are given in text.
PMSF = Phenylmethyl sufonylfluoride; DIFP = diisopropyl fluorophosphate are inhibitors of serine proteases.

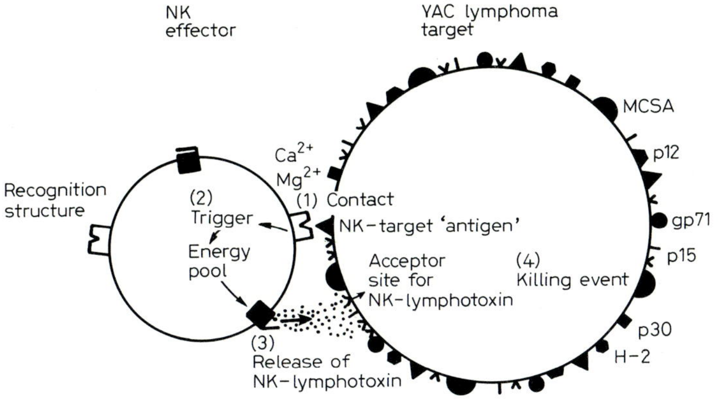

Fig. 11. A Schematic model of NK cytolysis (working hypothesis). The target structure has been isolated and described [220] whereas the NK receptor is inferred. Step 1 involves a divalent cation-dependent and specific contact between the target and effector which is inhibited by EDTA. Step 2 involves an interaction between the NK receptor and target structure which triggers (cyclic nucleotide-dependent?) an energy-dependent intrinsic arming of the NK cell which is inhibited by DNP or NaN_3. In step 3 the lytic moiety or 'black box' is activated and brought into close opposition with the target cell membrane by a dynamic process (fig. 2) which is inhibited by colchicine. The actual killing event may involve a lysosome-derived, serine esterase activity (NK-lymphotoxin) which is inhibited by phenylmethyl sulfonylfluoride and diisopropyl fluorophosphate. These lytic molecules bind to a putative acceptor site on the target cell membrane. Target cell death follows and (step 4) is thought to result from colloid osmotic lysis (fig. 3).

binding to target cells (TBC) revealed that the frequency of the TBC and the level of lysis in a population were closely correlated in kinetics experiments and under varying biological conditions of age, organ site, genotype, and target selectivity. The genes controlling the frequency of TBC were inherited in a dominant fashion and were linked to the H-2 complex as shown in figure 12 [216]. In addition, cell populations enriched or depleted in TBC by velocity sedimentation showed a corresponding increase or decrease in the ability to lyse YAC targets and the majority of single TBC isolated in droplets or monodispersed in agar killed the target to which they were attached [218]. This suggests that (1) the majority of TBC detected are indeed NK cells, and (2) the lytic event is independent of helper cells although the involvement of soluble factors and helper or suppressor cells in the formation of TBC cannot be excluded. It is not yet known if NK cells can recycle to lyse several targets as in the CTL system

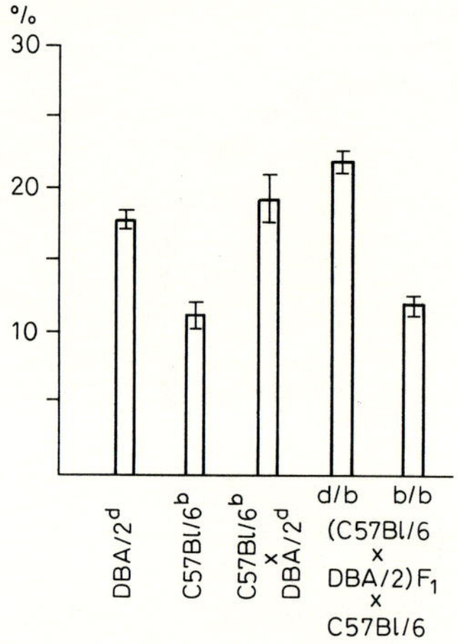

Fig. 12. H-2 linkage of target cell binding. Spleen cells from parents or offspring of a C57Bl/6 x (C57Bl/6xDBA/2) F_1 backcross were recovered from nylon wool-passed fractions. Values represent the mean ± SE for 20 individually tested mice, which were grouped as homozygous or heterozygous at the H-2 locus by prior histocompatibility testing. The difference in the frequency of TBC in H-$2^{b/b}$ backcross mice is significant ($p < 0.01$). Adapted from *Roder and Kiessling* [216].

[299]. Binding to the target was found to be a necessary prerequisite for lysis to occur. If cell contact was inhibited by EDTA, lysis was not observed and in kinetics experiments target cell binding was detectable 5–10 min prior to measurable cell damage. In addition, when both target cell binding and lysis were abolished by trypsinization of the effector (or by papain and pronase) both functions regenerated in a parallel time course (figure 8) in the absence of cycloheximide which is compatible with the suggestion that binding occurs prior to lysis and is mediated by 'protein like' cell surface 'recognition' structures. Prolonged (3 h) treatment with puromycin and cycloheximide alone also prevented TBC formation possibly by blocking the synthesis of the recognition receptor [*Argov and Roder,* unpublished observation]. H-2-linked genes controlled the frequency of detectable TBC and therefore these genes could conceivably code for the NK recognition receptor although other explanations are possible.

Target-lymphocyte contact alone, irrespective of the effector cell properties, was not a sufficient condition for lysis since nylon retained spleen cells did not lyse the targets to which they attached. These results are reminiscent of findings in the CTL system [32, 58, 81, 171, 172, 237, 299] but TBC in the NK system are unique. Whereas CTL are totally prevented from binding to target cells at low temperatures or in the presence of metabolic inhibitors [9, 172], NK cells do form stable contacts with the target in similar drug concentrations and binding is also evident at 0°C. This suggests that energy is required for binding by CTL but not NK cells.

Membrane and cytoskeleton integrity may also be important in the NK target-binding step. The aprotic dipolar solvent, dimethylsulfoxide (DMSO), inhibited NK-mediated binding [229]. Although the mechanism of drug action is not entirely clear, this agent would be expected to destabilize the plasma membrane and lead to a disorganization of integral membrane proteins, including recognition receptors [285]. Secondary inhibition of the lytic phase could also occur if free radicals are involved since DMSO is also a free radical scavenger. In the CTL system, DMSO inhibits the CTL-target contact phase [285] but its effects on the lytic step are controversial [68, 285]. Cytochalasin B also inhibited TBC and lysis in the NK system but it was not possible to assess if the lytic stage was inhibited independently or by virtue of inhibition of the earlier TBC stage. In the CTL system, target binding but not the lytic step is inhibited [reviewed in 32] and the available evidence favors the view that the relevant drug action involves disruption of microtubules [131] rather than secondary effects, such as inhibition of glucose transport [24]. Since tightly opposed interdigiting microvilli are observed in the area of NK-target contact (figure 2) [218], it is conceivable that such tertiary strengthening of the adhesion is dependent on cell motility and therefore requires intact microtubules. These results raise the possibility that membrane and cytoskeleton organization may be important for NK-target binding. However, the free movement of integral membrane proteins may not be necessary since the protein cross-linking agent, glutaraldehyde, completely abolished lysis with only partial inhibition of the target-binding phase [229].

The target cell requirements for step 1 in this target-effector interaction in the NK system are less well studied. However, biochemical isolation procedures have established that the NK-target 'antigen' on the surface of target cells consists of up to three glycoprotein molecules of MW 130,000–240,000 [220], as discussed above in the section IX. The expression of these target antigens may be blocked by interferon [242] and may be controlled by differentiation genes as suggested by studies of somatic cell hybrids between NK-sensitive and insensitive targets [226]. These structures are also cleaved by trypsin and pronase and turn over at a low rate ($>$3 h) [*Argov and Roder,* unpublished observation]. Energy metabolism does not appear to be important and in other systems it has been shown that neither protein synthesis, RNA or DNA synthesis, is required and, indeed, enucleated targets are lysed equally well by NK cells [255].

The triggering of NK cells (step 2) subsequent to making contact with the target cell appears to involve two pathways which may or may not be linked, namely energy metabolism and the cyclic nucleotide system. The lytic phase of NK cytolysis was blocked by DNP, NaN_3 or low temperature [218] as observed for CTL-mediated cytolysis [9, 229]. CTL cytolysis was more temperature-dependent than NK cytolysis [229] although as previously shown, CTL responding to a secondary challenge are less temperature-dependent than primary CTL [32]. Inhibitors of glycolysis such as iodoacetate and NaF also block the NK lytic phase which suggests that the energy necessary for NK lysis may be derived from both glycolysis and oxidative phosphorylation as in the CTL system [9, 68, 168]. Dibutyryl cAMP (dB-cAMP) and the cAMP-elevating agents, prostaglandin E_1, theophylline, and histamine markedly suppressed NK cytolytic function with no effect on TBC formation (figure 13), [225]. Conversely, dB-cGMP and carbamylcholine accelerated the rate of lysis to a small but significant degree and could compete with inhibitory doses of dB-cAMP to reduce the level of suppression, thereby suggesting that the cAMP/cGMP ratio might be important in NK-mediated lysis [225].

The mechanism by which cyclic nucleotides modulate NK cytolysis is unclear but perhaps the most attractive hypothesis is a 'stimulus-secretion' model. Under this model, target-effector contact, via specific recognition structures, leads to an alteration in intracellular cyclic nucleotide levels which in turn triggers a lytic event in the effector cell. The observation that cAMP-or cGMP-active agents alter NK cytolysis with no effect on the target cell contact phase favors this sequence of events. Alternatively cAMP is known to inhibit cell locomotion [84] and would therefore simply block recycling of NK cells.

The nature of the lytic events (step 3) is entirely unclear but within the 'stimulus-secretion' model of NK cytolysis we postulate that lysosomal enzymes ('NK-lymphotoxin') may be the actuel lytic moiety since: (1) chloroquine, a selective lysosomal inhibitor, inhibited NK cytolysis as well as lysosomal enzymes [229]; the mutation in chromosome 13 of the mouse, beige, results in both defective lysosomal enzyme function and a partial impairment in the NK lytic mechanism, whereas the level of TBC

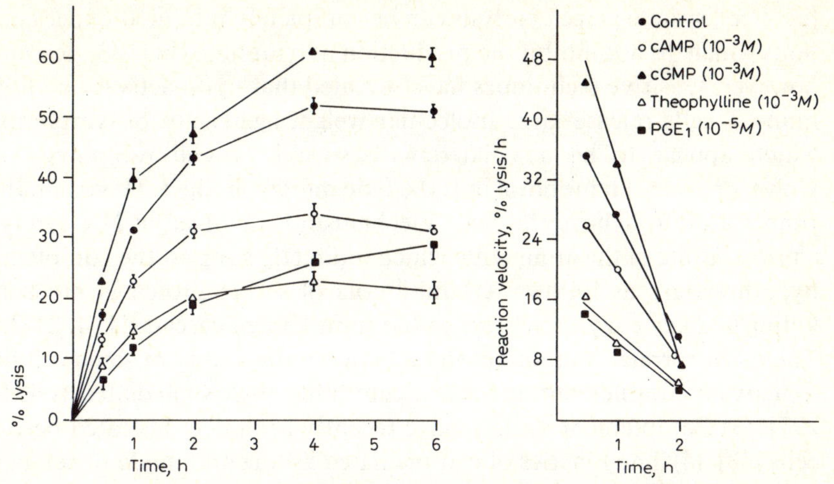

Fig. 13. Inhibition of NK-mediated cytolysis by cAMP-active agents. Nylon wool column-passed CBA spleen cells were pretreated 5 min with the various agents shown and ^{51}Cr-labelled YAC target cells were then added. At various times after centrifugation supernatants were assayed for ^{51}Cr release. The right panel is a re-plot of the data in the left panel. Adapted from *Roder and Klein* [225].

are normal [219]; (3) the same concentrations and variety of agents which inhibit the lytic phase of NK cytolysis (cAMP, PGE_1, theophylline, histamine, high-dose epinephrine) also inhibit lysosome secretion in cells [287] and conversely agents which increase lysosomal discharge (cGMP, carbamylcholine) significantly augment NK lysis [225]; (4) inhibitors of serine esterases, phenylmethyl sulfonylfluoride and diisopropyl fluorophosphate, inhibited NK cytolysis but not TBC formation [218], and (5) interferon enhances NK cytolysis [62] and is also known to augment IgE-mediated histamine release from basophils [124] and therefore might also cause the release of lysosomal enzyme-containing granules. If lysosomes are involved in target destruction, then a special localized mode of delivery is implicated since 'innocent bystanders' were not killed in NK cytolytic experiments [229]. These observations suggest that the lysosome-dependent, 'stimulus-secretion' model, shown in table VII, is worthy of further investigation in the NK system.

Although we favor a serine protease as the actual lytic moiety, others have argued that 'classical' lymphotoxin [15, 74, 120, 278] or complement [174] may be involved. Lymphotoxin is an unlikely candidate at least in the T cell system since cytolysis can be completely inhibited under conditions which do not inhibit the production of lymphotoxin [108]. Recently, however, sensitive techniques have revealed that upon activation alloimmune T cells release large molecular weight aggregates of lymphotoxin which appear to be associated with specific T cell receptors [120]. However, even if lymphotoxin is the lytic moiety, in the CTL system there is no reason to believe the same mechanism is involved in NK cytolysis, a priori. Some of the arguments which indirectly support the complement hypothesis are as follows: (1) Inhibitors of serine proteases block NK killing and key enzymatic steps in the complement cascade [174, 218]. (2) There is a correlation (not absolute) between the ability of some cell lines to activate complement and its susceptibility to NK-mediated lysis [99, 297]. (3) Complement bridges have recently been demonstrated between cells [48]. (4) The kinetics of cell-mediated killing and mode of cell death strongly resemble complement-mediated lysis [174]. (5) Some unpublished observations suggest that anti-complement antibody may block killing by cytotoxic T lymphocytes. (6) Some forms of cytolysis such as ADCC are enhanced by complement [234]. It is clear that more direct experiments are required to elucidate the lethal hit stage.

Recent results [231] lead to the inference of an acceptor site on the target cell which binds the putative NK-lymphotoxin released during the lethal hit (figure 11). Clones of NK-resistant (NK^R) mutants (YAC-6, YAC-6.28) derived from YAC expressed normal NK-TS and were bound by NK cells but not lysed (table VIII). The NK^R clones however were equally susceptible to lysis mediated by cytotoxic T lymphocytes or antibody and complement thereby suggesting that they were selectively resistant to NK cytolysis only. It is reasonable to assume therefore that this mutation has occurred in an acceptor site for the NK-lymphotoxin-like substance.

The effect of hormones on NK activity has also been investigated. Insulin had no effect in vitro on NK cells whereas CTL-mediated cytolysis was boosted [225]. Insulin is thought to act through the cGMP system [119, 263] and we have shown that cGMP causes a small but significant enhancement of NK cell activity (figure 13). Therefore, NK cells, which are small 'resting' lymphocytes, may lack the insulin receptors found on activated CTL [119]. Adrenalin had both augmenting and suppressive effects depending on the concentration [*Klein,* unpublished observation] whereas β-estradiol administered in vivo was suppressive [248], due to a non-physiological effect on bone marrow.

Table VIII. Selective NK resistance in a clone of mutagenized YAC cells

	YAC parent	YAC-19	YAC-6	YAC-6.28
Lysis (LU/10^7)[a] by				
(a) NK	85	100	10	5
(b) CTL	3000	2800	2500	2700
(c) Lectin-CTL	1100	800	1000	900
Competition vs YAC				
parent[b]	0.6	0.5	0.8	0.6
% TBC[c]	20	22	23	24

Effectors were derived from nylon passed fresh spleen (NK), mixed lymphocyte cultures (CTL) or spleen cells cultured in the presence of Con A (lectin-CTL).
[a] Data is expressed as the number of lytic units/10^7. 1 LU was defined as the number of effector cells required to yield 25% cytolysis.
[b] Data is expressed as the unlabelled competitor/labelled target cell ratio required for 50% inhibition of ^{51}Cr release from YAC parent cells in an NK assay.
[c] The frequency of nylon-passed CBA spleen cells binding to the various targets.
These data are summarized from *Roder* et al. [231].

A number of genes are almost certainly involved in the lytic cycle although their precise location in the scheme of events is unknown. H-2 linked genes on chromosome 17 determine the size of the NK pool and possibly code for the recognition receptor on NK cells although this remains speculative [216]. Genes on chromosomes 2, 5 and 21 in the human code for interferon production as well as elements involved in interferon action [270]. In addition a gene, bg, on chromosome 13 in the mouse is involved in some post-recognition event in the lytic cycle (see below), possibly related to lysosome function. Since interferon does not completely restore the bg defect [229] then the interferon sensitive event probably occurs earlier in the lytic cycle than the bg-sensitive step.

B. The Beige Mutation

Germ line mutations affecting defined cell populations are often valuable tools in elucidating the function of these cells in complex biological systems such as cytolysis. We have recently found that the mutant gene in the mouse called beige (bg^J), leads to a profound and selective impairment of naturally occurring killer lymphocytes, whereas other forms of cytolysis-mediated by T cells, macrophages, and promonocytes are apparently normal [219]. This defect in NK cytolysis was predetermined at the level of progenitor cells in the bone marrow as revealed in radiation chimeras [221]. This impairment in NK function could not be accounted for by an altered organ distribution, target selectivity or ontogenesis. Interferon could improve but not fully restore the response which suggests that the defect is not totally due to a lack of endogenous interferon stimulation in beige mice [229]. Furthermore, bg mice synthesize normal amounts of interferon in response to viral challenge [290]. The frequency of target-binding cells was normal in all lymphoid organs [221] which suggests that the defect is intrinsic to the NK cell and does not involve an altered population size or an inability to recognize and interact with the target. Rather, the defect may lie within the lytic pathway subsequent to target cell contact. The bg gene provides the first defined genetic marker for sequencing the events involved in the cytolytic pathway.

The bg mutant also provided an opportunity to reexamine the relationship between NK and ADCC effector cells. ADCC against chicken erythrocytes was found normal in bg/bg mice (figure 14). This form of cytolysis is mediated by a variety of different effectors including FcR^+, adherent cells in the mouse [142, 187] and monocytes or granulocytes in the human system [181]. Therefore, it is likely that the cytolytic function of these cell populations in the beige mouse is normal. Some of the ADCC against tumor cells is mediated by different lymphocytic effectors, called K cells [201], and this population is defective in beige mice (figure 15). However, ADCC against tumor cells mediated by promonocytes from beige mice was normal [222] which indicates that the NK/ADCC defect in beige lymphocytes is not simply due to a generalized failure in the ADCC mechanism against tumor cells. Since ADCC against P815 tumor cells was deficient in beige mice these observations strongly suggest that NK and K cells are identical. This view is supported by recent work showing NK and K cells have similar properties [244]. In addition antibody-coated, NK-insensitive P815 cells compete with NK cytolysis of ^{51}Cr-labelled YAC cells in a cold target inhibition assay [187]. It would seem, therefore, that NK cells in the mouse are capable of lysing tumor cells by two independent recognition mechanisms; one antibody-dependent and the other antibody-independent.

C. The Chediak-Higashi Gene

The beige mouse has been used for over a decade as a model for the CH syndrome in man [293]. Both species exhibit abnormal leukocyte granulation and skin pigmentation, whereas their immune responsiveness is intact (table IX). If the defective NK gene in the beige mouse was closely

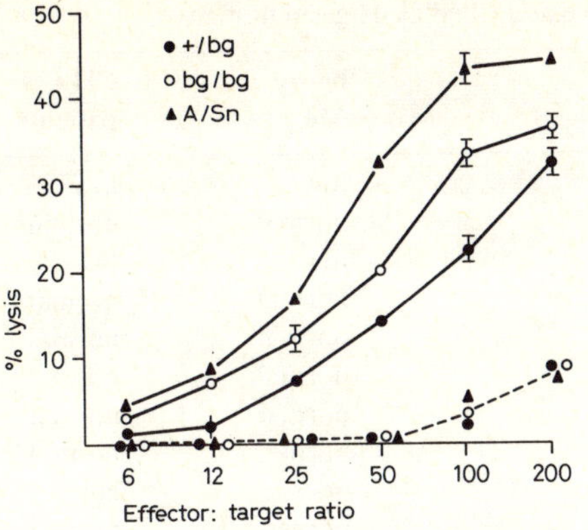

Fig. 14. ADCC against chicken erythrocytes (CRBC) in beige mice. Spleen cells were pooled from 4 mice in each group and tested in a 4-hour ^{51}Cr release assay against CRBC which had been precoated with 1/200 mouse anti-CRBC serum (solid lines). In some wells 200 μg/ml aggregated γ-globulin was included (dotted lines). Lysis of non-antibody-coated CRBC was <1% at all effector:target ratios. Adapted from *Roder et al.* [222].

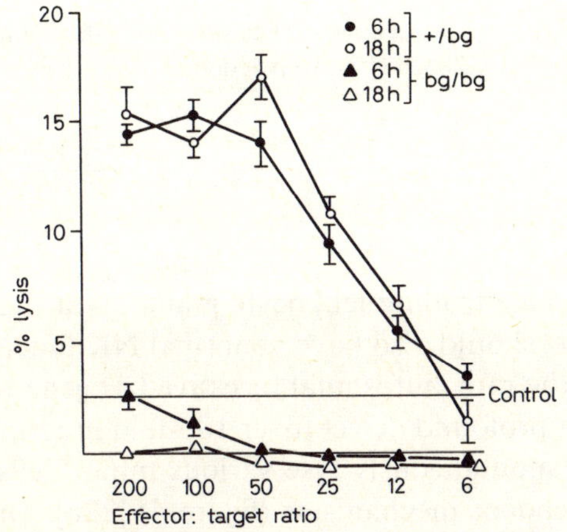

Fig. 15. ADCC against tumor cells in beige mice. ^{51}Cr-labelled P815 target cells were coated with a predetermined optimum concentration (1/60) of A.BY × C57Bl anti-A/Sn serum and mixed with spleen cells in a 6- or 18-hour cytolytic assay. Control lysis of uncoated P815 cells was <2% at the highest effector:target ratio. Adapted from *Roder et al.* [222].

Table IX. Immune function in beige mice and Chediak-Higashi humans[1]

Function	bg/bg Mice	CH patients
NK cytolysis	low	low
Frequency of NK target-binding cells	normal	normal
ADCC vs tumor cells	low	low
ADCC vs erythrocytes	normal	normal
CTL-MLC	normal	normal
In vivo alloimmune	normal	ND
Lectin induced	normal	normal[2]
Skin graft rejection	normal	normal
Delayed type hypersensitivity	normal	normal
T cell-mitogens	normal	normal
B cell-mitogens	normal	normal
Ig production	normal	normal
Activated macrophage cytolysis	normal	normal
Promonocyte cytolysis		
Spontaneous	normal	normal
ADCC	normal	ND
NK interferon boost	yes[3]	yes[3]
cGMP restoration of NK function	no	yes[2]
Lysosomal enzyme function	low	low

[1] Homozygous bg/bg mice were compared to heterozygous +/bg littermate controls. Human CH patients were compared to age- and sex-matched normal donors or unaffected siblings.
Cytolytic assays were performed with tumor cells as targets. Data supporting this summary are referenced elsewhere [86, 152, 219, 221, 228, 293]. ND = Not done.
[2] Not in all patients.
[3] Does not approach level of untreated or IF-boosted normal cells.

linked or identical to the gene controlling leukocyte granulation then it was predicted that CH patients should also have abnormal NK function.

Several patients carrying the rare, autosomal recessive CH gene were examined and found to have a profound defect (over 100-fold in terms of lytic units) in their ability to spontaneously lyse various tumor cells by antibody-dependent or independent mechanisms (figure 4) [228]. Other forms of killing mediated by monocytes, neutrophils, and T cells were relatively normal [152]. As found in the beige mouse, the ability of CH lymphocytes to bind to target cells was not impaired and therefore the defect in the human may also effect a post-recognition event [86].

These two genes, bg and CH, may not only be important as defined genetic markers for sequencing the NK cytolytic pathway but they also provide an opportunity to unequivocally assess the importance of the NK cell in surveillance against neoplasia. Recent studies show that the beige mouse is more susceptible to the growth and dissemination of transplantable tumors [134, 269] and CH patients surviving childhood infections almost invariably develop a lymphoproliferative disorder which may be malignant [47]. It is becoming increasingly likely therefore that NK cells are involved in surveillance against at least some tumors.

D. Conclusions

A picture of the NK cytolytic mechanism is emerging (table VII) in which various genes control and regulate discrete steps in the lytic cycle. H-2-linked genes on chromosome 17 control the initial stage of target-effector recognition either through their regulation of population size and/or the putative recognition receptor itself. Surface glycoproteins, which are important for recognition, have been isolated from the target cell membrane. Target recognition is postulated to trigger the cyclic nucleotide system which turns on those energy pathways important in the activation and release of the actual lytic moiety (lysosomal enzymes?). These intervening steps are susceptible to the action of genes regulating the production of, or response to, interferon. Finally the terminal stages of the cycle, close to the lethal hit, are regulated by the bg gene in the mouse and the CH gene in the human. Finally, there is most likely a second receptor on the target cell which binds the lytic molecules released by the NK cell. This model is amenable to further investigation, and shows that the mechanism of NK cytolysis is distinct from that in T cells, promonocytes, macrophages/monocytes or neutrophils.

XI. In vivo Relevance

A. Tumor Resistance

Following the discovery that NK cells comprise a new set of lymphocytes which can mature in the absence of a thymus, a major effort in this field has been to understand the in vivo relevance of this T cell-independent effector mechanism. A number of different results, listed in table X, now seem to indicate that NK cells may be the in vitro correlate of a biologically important surveillance system.

Table X. Evidence for in vivo relevance of mouse NK cells in rejection of tumor cells

1	Strong positive correlation between in vitro NK activity and in vivo tumor resistance in rejection studies of tumor transplants in various genotypes [140, 252]
2	T cell-deficient F_1 hybrid mice have an increased resistance to the growth of an NK-sensitive semi-syngeneic lymphoma [141]
3	NK-sensitive tumors grow less well in nude mice than normal littermates [141, 286]
4	B cell-deprived mice have elevated NK activity and are better able to resist NK-sensitive, syngeneic tumors [17]
5	Lymphocytes enriched for NK cells efficiently delay tumor growth upon transfer into irradiated syngeneic recipients [30, 141]
6	Bone marrow chimeras between NK high- or low-reactive F_1 genotypes show the same in vivo resistance to semi-syngeneic tumor transplant as the bone marrow donor strain [87, 88]
7	Newly hatched chickens can be passively protected against Marek's disease and leukemia induction by transfer of an 'NK like' cell type from adult animals [158]
8	Good correlation between in vitro NK activity and the ability to reject prelabelled tumor cells in a 4-hour in vivo assay [115, 214]
9	Aging mice (>6 months) exhibit a marked decline in NK-mediated tumor rejection in vivo or in vitro [87, 94] and an increasing incidence of spontaneous tumors. Young mice repopulated with NK cells from old bone marrow also have diminished NK function [90]
10	The mouse mutant beige, selectively impaired in NK function [219], rejects tumor transplants and metastasis less efficiently [134, 269]
11	Humans, carrying the analogous beige gene (Chediak-Higashi gene), are selectively impaired in NK function and exhibit a profound (85%) incidence of a spontaneous, 'lymphoma-like' lymphoproliferative disorder [228]

Previously, in vivo resistance against tumor growth has been attributed mainly to immune T cells [39, 65, 71]. However, there is a large body of growing data to support the hypothesis that T cells are not primarily involved in immune surveillance against all tumors [247]. Some of this data, in addition to that above, will be discussed here.

Several reports indicate that T cell-deficient mice are quite resistant to growth of certain allogeneic or syngeneic transplantable tumors [56, 64, 75, 233, 256]. It has also been observed that some human tumors do not grow well in nude mice [53, 246], a relevant point to bear in mind when considering the lytic effect of mouse NK cells on certain human T cell lines. In addition, *Bonmassar* et al. [14] observed increased resistance in allogeneic nude mice to the growth of a mouse lymphoma. It must be considered, however, that several T cell-independent mechanisms mediated for example by macrophages or K cells could be active in T cell-deprived mice and all the effects cannot be attributed to NK cells. The relative role of NK cells in relation to other non-T immune mechanisms must be established clearly for each tumor studied.

The first set of evidence arguing for an in vivo relevance of NK cells in tumor rejection came from a series of genetic experiments, showing a strong positive correlation between NK activity and in vivo resistance to tumor transplants in various genotypes [140]. The genetics of these experiments will be discussed at length elsewhere in this review (section IV). It should be stressed that although striking similarities in the genetic control of NK reactivity and that of in vivo tumor resistance to a number of different tumors was found previously, T cell and macrophage activity were not examined and therefore it cannot be ruled out that these cell types may also play a role in the observed rejections.

Most of the early studies of the in vivo relevance of NK cells were performed with the MLV-induced lymphoma YAC but the conclusions drawn have been confirmed in other host tumor systems. Three different lines of evidence supported the notion that NK cells were responsible for the rejection of the YAC tumor. One was the 'neutralization test' previously used to demonstrate the importance of immune effector cells [294]. In this 'semi-in vivo' system spleen cells were admixed with living YAC cells and inoculated subcutaneously into irradiated syngeneic recipients. The advantage of this method is that the same spleen cell fraction can simultaneously be tested both in vitro for NK activity and in vivo for protection against tumor growth. The results were clear: spleen cells depleted of T cells, B cells and monocyte-macrophages were the most efficient fraction both in delaying tumor growth and in the in vitro NK assay [141]. Lately, the same experimental model was used by others [30] to demonstrate the in vivo protective effect of spleen cells depleted of Thy-1$^+$ cells and then positively selected for Ly 5$^+$ Ig$^-$ cells. This highly selected population conferred complete protection against the RL♂1 tumor inoculated into sublethally irradiated Balb/c mice, and against YAC cells inoculated into A/J mice.

In another experimental series semi-syngeneic A, F_1 hybrids were thymectomized as adults, then irradiated and reconstituted with fetal liver cells. T cell depletion of these mice was demonstrated by their inability to produce antibodies to a T cell-dependent antigen [141]. These mice showed an increased in vivo resistance to challenge with small numbers of syngeneic YAC cells when compared to the untreated control mice. In contrast, larger doses of allogeneic tumor cells histoincompatible at the H-2 locus showed the expected preferential growth in the thymectomized and reconstituted groups of mice. Also nude mice on a Balb/c background used in the same study showed strong resistance to YAC tumor growth. Since these animals were H-2-incompatible with regard to the YAC tumor used, a reaction against H-2-determined antigens cannot be excluded. Others, however, have demonstrated high resistance to syngeneic tumors transplanted into nude mice, as will be discussed below.

The last but probably strongest of the indirect arguments for NK cells but not T cells being involved in the in vivo YAC rejections comes from experiments with bone marrow chimeras. Thus, it has been shown that NK reactivity in bone marrow chimeras between NK high and low reactive strains follows the reactivity of the bone marrow donor [90]. This chimera model provided a unique chance to further investigate the role of NK cells in the in vivo resistance of T cell-depleted mice. Consequently, H-2 compatible mice of either high or low NK strains were mated with low reactive A mice syngeneic to the A strain lymphoma YAC. The resulting F_1 hybrid offspring were again either high or low NK-reactive according to the parental haplotype. Furthermore, all offspring of F_1 mice were semi-syngeneic to YAC and were 'identical' at the H-2 region. Thus, transfers of high or low bone marrow precursor cells into lethally irradiated recipients of low or high NK type were feasible. In order to exclude possible influence of the T cell system on in vivo tumor resistance the bone marrow recipients were thymectomized prior to reconstitution. For the first time T cell-deficient hosts which expressed either high or low in vitro NK activity and yet were genetically identical, with the exception of their hemopoietic tissue, were available for in vivo tumor challenge. The results of subcutaneous challenge with small tumor cell doses showed a striking positive correlation between in vivo tumor resistance and NK activity in vitro [87], thereby demonstrating a real tumor-protective role of NK cells in vivo. These experiments have recently been repeated in (AKR × CBA) F_1 hybrid mice repopulated with high NK (CBA) or low NK (AKR) bone marrow and challenged with an NK-sensitive AKR lymphoma [215]. High NK chimeras but not low NK chimeras were efficient in tumor rejection. No differences in tumor growth were observed in chimeras challenged with an NK-insensitive AKR lymphoma.

This series of experiments has demonstrated that an organism without a functioning T cell system can efficiently protect itself against growing tumor cells, and that NK cells most probably play an important role in this thymus-independent defence mechanism. A further question which could be asked is whether only those tumors which are sensitive to NK effects in vitro would be efficiently rejected in T cell-deficient mice. This was analyzed by *Warner* et al. [286], who compared the growth of a number of Balb/c tumors in Balb/c-nu/nu and Balb/c nu/+ heterozygous littermates. They found that tumors which were lysed in vitro by spleen cells from normal syngeneic nude mice showed a reduced growth rate in syngeneic nude compared to syngeneic normal mice, whereas tumors which were resistant to in vitro NK lysis failed to display this selectively reduced growth rate in nude mice. This approach for analyzing the possible involvement of NK cells in T cell-independent resistance would seem a useful one, and may be less laborious than the in vitro manipulation methods described above in the YAC system.

It is also important to consider the role of the B cell system in tumor resistance. Recent studies by *Gordon* and collaborators have shown that elevated in vitro NK activity against YAC or syngeneic, methylcholanthrene (MCA)-induced fibrosarcomas, in mice deprived of B cells by chronic anti-μ treatment, correlated well with an increased resistance to (1) induction of primary tumors by MCA [19], (2) transplantation of syngeneic MCA-induced fibrosarcomas [17], and (3) spontaneous metastasis from the primary tumor site [18]. An important control in these studies was the observation that the incidence and growth rate of Moloney sarcoma virus-induced progressing tumors did not differ in μ-suppressed or normal control mice that were untreated or treated with normal γ-globulin [*Gordon,* unpublished observation]. Since host defence in the MSV system is clearly T cell-dependent [70] this indicates that B-deprived mice have normal T function and augmented NK activity does not influence a T cell-dependent rejection system.

A major difference between NK cells and other effector mechanisms lies in the fact that NK cells preexist in the non-primed host whereas T cells and activated macrophages cannot act until after a certain latency period of several days. A recent experimental model [31] adapted by *Riccardi* and collaborators [115, 214] takes advantage of these 'kinetic' differences in effectors and should provide a useful system for studying NK cells in vivo. Mice were inoculated i.v. with tumor cells prelabelled with ^{125}I-iododeoxyuridine and the amount of isotopes in various organs was measured after 4 h. The authors found good correlation between the ability to eliminate the isotope-labelled tumor cells and the levels of host NK reactivity, since young mice of high NK-reactive strains were found more efficient in eliminating the prelabelled tumor cells than low NK-reactive strains. An increased elimination was also seen in mice which had been treated with poly I:C or *C. parvum,* treatments which augment NK activity. Thus, this assay could provide a very convenient method for rapid estimation of host NK activity and its relevance for in vivo tumor protection.

In the pre-NK era of tumor immunology the possibility to assess the in vivo significance of T cells in immune surveillance was greatly enhanced by the development of the athymic mouse mutant nude, which is selectively impaired in its T cell functions [236]. In this, the NK era, a similar opportunity presents itself with the recent discovery that the mouse mutant beige is selectively deficient in NK cell function [219]. This finding will be extensively discussed elsewhere in this review, and the potential usefulness of this model is obvious. The fact that this mutant has normal levels of cytolytic T cell function, activated macrophage killers, and ADCC-dependent promonocytes (table IX) provides the first opportunity to asses the role of NK cells alone in immune surveillance. In addition, the role of T cells or macrophages in tumor graft rejection can be assessed in the absence of NK 'interference'. In a recent study homozygous bg/bg mice had significantly decreased natural resistance to the growth of low dose inoculas of one virally and one chemically induced syngeneic leukemia compared to phenotypically normal, heterozygous bg/+ littermates [134]. This decreased resistance was manifested in a higher tumor incidence, shorter latent period, and greater tumor diameter (table XI). In another study [269] B16 melanoma cells [57] demonstrated an increased growth rate, faster induction time and an increased metastatic potential in bg/bg mice compared to +/bg controls [269]. Cell lines established from metastatic foci have a decreased sensitivity to NK cells compared to cells from the primary tumor as would be predicted if NK cells exert negative selection on the tumor population in vivo [72]. Further studies revealed that humans homozygous for the analogous CH gene have a profound increase in spontaneous lymphoproliferative malignancy [47]. It is clear therefore that NK cells play an important role in tumor growth and metastasis and possibly surveillance against neoplasia.

All the in vivo experiments discussed above have been performed using a very small tumor inocula. With inoculation of high tumor cell numbers, the NK-mediated in vivo resistance is easily broken down [140]. It is therefore possible that NK cells represent only a first line of defence against newly arising neoplastic cells, and that other effector mechanisms such as ADCC, T cells, and macrophages are active later in tumor devel-

Table XI. Tumor growth and metastasis in an NK-deficient mutant

Parameter measured	Injection site	Tumor			bg/bg	+/bg
1 Latency	s.c.	EL-4,	P52,	$B16^{s}$	short	long
2 Incidence of takes	s.c.	EL-4,	P52,	$B16^{s}$	high	low
3 Tumor size	s.c.	EL-4,	P52,	$B16^{s}$	large	small
4 Time to death	s.c.	EL-4,	P52	–	short	long
5 Spontaneous pulmonary metastasis	s.c.	–	–	$B16^{s}$	many	none
6 Lung colonies	i.v.	–	–	$B16^{s}$	many	none
7 Clearance of ^{125}IUdR-labelled cells	i.v.	–	P52	–	slow	rapid
8 In vitro NK activity (3 days)	s.c.	EL-4,	P52	–	low	high
9 In vitro NK activity	none	EL-4,	P52,	$B16^{s}$	low	high

Data for the EL-4 and P52 leukemias are summarized from *Karre* et al. [134]. The $B16^{s}$ melanoma data is from *Talmadge* et al. [269].
There were no differences between bg/bg and +/bg mice in parameters 1–5 when NK-resistant $B16^{R}$ tumor cells were inoculated [269]. $B16^{s}$ is an NK-sensitive line derived by prolonged (>32 days) in vitro culture [269]. The EL-4 and P52 tumors were maintained as in vivo, ascites-passed lines [134].

opment. The fact that NK activity has been found depressed in tumor-bearing mice [5, 110], a situation in which NK cells might have been overwhelmed in early surveillance, does not therefore argue against their in vivo relevance.

When discussing the in vivo relevance of NK cells it should be noted that all studies have been performed with rejection of tumor transplants. It is quite likely that the inoculation of small numbers of tumor cells into unprimed animals resembles the situation the organism encounters when faced with transformed aberrant cell clones developing into tumors. Nevertheless, a more appropriate model would be to actually study models of primary tumor induction by carcinogenic agents, oncogenic viruses or irradiation. It is imperative also to investigate the role of NK cells in protection against spontaneous tumor development. No such study has yet been performed, and this must clearly be a major aim in further investigations. Our working model of NK involvement in immune surveillance, in relation to other effector mechanisms, is presented in table XII and

Table XII. The placement of NK cells in a hypothetical sequence of anti-tumor surveillance

1	A neoplastic transformation occurs via somatic mutation, viral or chemical induction [150]
2	Proliferating tumor cells are held in check by NK cells, which preexist at high numbers (1–2% of total lymphocyte pool)
3	After a sensitizing period of several days, some of the remaining tumor cells are lysed by immune T cells assuming that the tumor is antigenic (most spontaneous tumors are not)
4	Immune T cells are triggered by tumor antigens to release macrophage-activating factors which induce macrophages to lyse remaining tumor cells
5	Finally, small amounts of anti-tumor antibody may appear in the circulation and bind to any remaining tumor cells thus rendering them susceptible to ADCC by K cells or NK cells (which may be identical), promonocytes, macrophages, and polymorphonuclear leukocytes
6	Tumor cells escaping from the primary tumor during metastasis are held in check by NK cells during transport to, or at, the metastatic foci (assuming NK-resistant variants have not arisen from the selection process in step 2)

All of these mechanisms (fig. 16) have been shown to be important in the rejection of some experimental tumors. However, it remains to be determined whether any of them play a role in the prevention of spontaneous tumors and some effectors (T cells, B cells, antibody) have been shown not to be important in this regard. The important point is that NK cells could potentially provide the first level of defence. There is also some evidence to suggest that such a model would be appropriate in surveillance of virally infected cells.

figure 16 as a temporal sequence of hypothetical events following a neoplastic transformation.

One can hypothesize that since NK cells preexist at relatively high frequency (0.6–2.4% of total lymphocytes [216]) and do not require lengthy preactivation, they would be expected to provide a first line of defence against newly arising malignancies or their metastasis as outlined in table XII. Other potential effectors, such as T cells, activated macrophages or K cells participating in ADCC, if they are important at all, would only be effective after a specific priming or activation period requiring days or weeks.

It is not unreasonable to assume that under the selective pressure of such a lethal condition as malignancy, several independent defence mechanisms may have evolved. In principle the theory is an adaptation of the one originally proposed by *Burnett* [26]. However, the role of the T cell in immune surveillance has not withstood critical tests, as reviewed above, and in the scheme shown in table XII and figure 16 we have placed the NK cell as the foremost barrier, but not necessarily the only one, to tumor

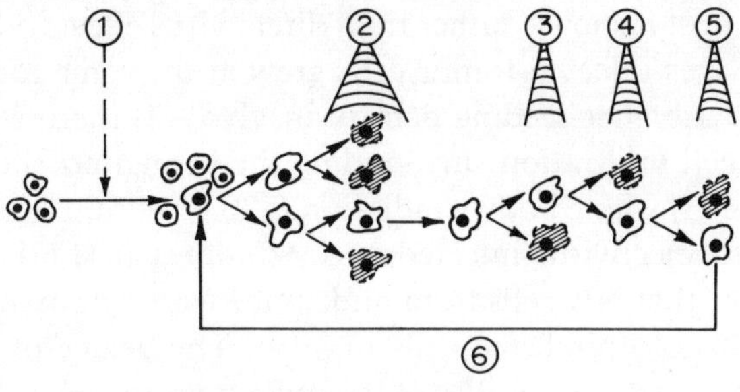

Fig. 16. The NK hypothesis of anti-tumor surveillance. See table XII for a description of steps 1–6.

development and metastasis. The mechanism whereby the NK cell recognizes and destroys the developing tumor has been discussed in section IX. It is however instructive to consider some recent work by *Collins* et al. [40] which shows a correlation between tumorigenic potential of cloned chemically transformed cell lines and their susceptibility to NK-mediated lysis in vitro. N cells, a cloned fetal fibroblast line which does not grow in agarose, was not tumorigenic in vivo (in normal mice or lethally irradiated adult thymectomized, fetal liver reconstituted, AT×FL) and was not lysed by NK cells in vitro. When these cells were transformed by MCA they grew in agarose and after cloning had the potential to grow as tumors in AT×FL mice [200] only, which were NK-deficient [40] in this study. This same mutation, N→I, also caused the cells to become susceptible to NK-mediated cytolysis in vitro. Cell lines, derived from these I tumors, were capable of growth in normal mice. This I→C mutation, or selection, was accompanied by a loss of NK sensitivity, thereby allowing the tumor to escape NK surveillance and grow in normal mice. Although this is a useful model it would be misleading to suggest NK cells exclusively lyse tumor cells, since virus-infected cells, hemopoietic grafts and even some normal tissues are NK-sensitive as discussed below.

B. Virus Infection

Recent work by *Reid* et al. [209, 210] has shown that nude mice efficiently reject xenogeneic tumor cells (HeLa, BHK 21) which are infected by persistent, non-cytopathic RNA viruses (VSV, measles, mumps, influenza, rabies) whereas non-infected tumor cells grow in 100% of injected animals, often with as few as 10 cells in the inoculum. The rejection seemed to be an active host response rather than direct virus effects on the tumor cells since: (1) virus-infected tumor cells grew at the same rate in vitro and did not have a higher seeding density in vivo; (2) there was a marked mononuclear cell infiltration surrounding the benign nodules of virus-infected cells, and (3) whole body irradiation (500 R) eliminated the capacity of nude mice to reject virus-infected cells. *Minato* et al. [177] used the same system to show that NK cells from nude mice were cytotoxic for these virus-infected but not uninfected cells in vitro. The degree of NK activity in vitro correlated with the ability of the nude mice to reject these same virus-infected tumors in vivo. As expected, NK cells taken from mice as early as 2 days after injection of virus-infected cells were non-specifically enhanced in cytolytic activity possibly due to the effects of interferon. Unlike the boosting of NK in normal strains which diminishes to normal levels within a week [111], NK boosting of nude mice persisted for a month or more. The authors suggest that increased viral shedding in nude mice may be responsible for this phenomenon. It is interesting to note in this regard that *Clark* et al. [37] found that nude mice retained much higher NK activity in SPF conditions than normal strains. These observations taken together raise the possibility that T cells may regulate virus release in normal strains.

Other investigators have also reported that virus-infected cells are more susceptible to NK-mediated cytolysis [3, 12, 103, 241] whereas some [291], using different virus-target combinations, find that virus infection reduces the target cells susceptibility. The potential importance of NK cells however in some systems was revealed in a study of EBV-infected human targets [12]. Since NK cells lysed virus-producing cells at an early stage of the cycle, before the virus particles are assembled, it was suggested that they may restrict the in vivo spread of this virus from latently infected cells [12].

C. Hemopoietic Grafts

Although this review primarily has dealt with the anti-tumor activity of NK cells, recent results have also indicated that NK cells can interact with normal non-malignant tissue. Striking similarities were found between the NK system and the rejection of bone marrow grafts [167], as studied in a system developed by *Cudkowicz* [41]. In this experimental system, heavily irradiated F_1 hybrids are injected with parental bone marrow cells, and in certain genetic combinations the parental bone marrow cells are rejected. As an explanation of hybrid resistance, the author suggested the existence of H-2-linked recessive histocompatibility genes. Such genes would generate an antigen, or antigens, in each parent which would be lacking in the F_1, provided the F_1 was heterozygous at the postulated locus. *Cudkowicz* [41] has called the proposed loci hybrid histocompatibility or Hh loci. In extensive linkage studies evidence was found for several such loci, all of which map on chromosome 17.

The reason why NK cells should be considered when discussing the rejection of bone marrow grafts stems from two different sets of information based on (a) similarities of the genetics in the two systems and (b) similarities in the nature of the effector mechanism regulating bone marrow rejection and NK activity. With regard to the genetic similarities, both phenomenon are examples of H-2-linked traits, therefore it is of importance to compare the precise position of these two traits on chromosome 17.

Cudkowicz [41] found evidence for multiple H-2-linked loci, regulating resistance to bone marrow grafts, but that D-end disparities seem to be most conducive to rejection. He designated the D-end-linked locus Hh-1. Furthermore, he has shown that heterozygocity at this locus is sufficient for the expression of hybrid resistance to two radiation-induced C57Bl lymphomas [41]. In the NK system, *Harmon* et al. [104] established that heterozygosity within or near the H-2D region is sufficient to obtain a successful resistance to the EL-4 lymphoma of C57Bl origin. Interestingly, these authors also confirmed the finding of an H-2 linked influence on NK activity. By using CR or H-2 recombinant mice on B10 background, they showed that heterozygosity within or near the D-end of the H-2 complex was necessary for high NK activity. The possibility was therefore considered that the homozygous C57Bl tumor cells expressed antigenic determinants analogous to the aforementioned Hh antigens, which would be recognized by the hybrids [36]. Could NK cells then actually be directed against products of such Hh loci on the tumor target? At present, we do not consider this explanation the most plausible one. From what has already been discussed in the previous chapter, the genetic differences in F_1 crosses or the H-2-linked influence on NK activity do not seem to depend on the uniqueness of the target cell used in the assay. Thus, the H-2-associated factors determine high NK activity not only to tumor targets which could be expected to express products of the Hh-1 locus, such as various C57Bl lymphomas, but also to a variety of widely different tumor targets, including lymphomas of Balb/c and A/Sn origin, that is, genotypes which should be 'silent' at the Hh-1 locus. A more likely explanation would be that the genes regulate effector cell levels rather than specificities.

Apart from similarities in the genetic control of bone marrow resistance and the NK system, a very close relationship appears to exist between the active mechanisms in the two systems. The notion that similar mechanisms might be active in resistance to hemopoietic grafts as in control of leukemogenesis has previously been suggested [14]. Results from collaborative studies between our laboratory and others have suggested that NK cells might constitute this mechanism [144], and recently similar results were obtained also by others [31, 167]. Briefly, the similarities between the two systems include: (a) independence of the presence of mature T cells; (b) a similar distribution of effector cells in different lymphoid organs, the spleen being the major site; (c) suppression of reactivity by a single injection of rabbit anti-mouse bone marrow serum; (d) suppression by a single injection of the anti-macrophage agents silica and carrageenan; (e) maturation of reactivity relatively late during the fourth week of life; (f) suppression of reactivity by multiple injections of parental spleen cells into F_1 mice in certain genetic combinations, and (g) dependency of the integrity of bone marrow, as demonstrated by reduction of reactivity by prolonged exposure of the animals' bone marrow to local irradiation from ^{89}Sr incorporated into bone tissue.

Taken together these common features of the two systems point to a very intimate association between NK cell activity and resistance to hemopoietic grafts. Recent results from *Hochman and Cudkowicz* [122], however, have demonstrated an exception to the parallelism. They have shown that whereas NK activity is sharply decreased after in vivo administration of hydrocortisone acetate, hybrid resistance to parental marrow grafts is not impaired by this treatment. The authors stress, however, that the association between NK activity and resistance to hemopoietic grafts was not contradicted by their finding, and that both activities were probably generated from a single differentiation pathway, but that they differ with respect to sensitivity to cortisone. It must therefore be concluded that although striking similarities seem to exist between the mechanisms in these two systems, further studies are necessary to elucidate the exact nature of this relationship.

D. Autoreactivity

A more direct indication that NK cells can interact with normal tissue is the observation that low levels of sensitivity can be seen with normal non-transformed cells as targets in cytotoxicity assays. It is of particular interest in relation to a possible surveillance function of NK cells that lymphoid cells from thymus and bone marrow have been reported to display sensitivity to NK lysis [100, 101, 184, 193], although peritoneal cells have also been used by some authors [184, 289]. The question arises to what extent the NK cells can interact in vivo with such NK–sensitive 'natural target cells', and whether one important function of NK cells would be to regulate or interact with other cell types in the organism. With thymocytes as target cells in a short-term NK assay, our groups have recently reported an inverse relationship between the occurrence of NK-sensitive thymocytes and the NK activity of the host. This inverse relationship was found with regard to the four different parameters studied: age, genotype, organ site, and boosting by interferon-inducing agents [101]. Thus, thymocytes from young mice of NK low reactive strains displayed higher lysis sensitivity than thymocytes from older mice or from NK high reactive strains. Also, by giving interferon inducers, thereby strongly increasing NK activity of the host, the thymocyte sensitivity of these mice was largely abrogated.

One possible interpretation of these results could therefore be that the observed inverse relationship between thymocyte sensitivity and host NK activity is due to an elimination of the NK-sensitive target from the organism by a direct killing affected by the NK cells themselves. We have recently characterized the NK-sensitive thymus target as a large, immature cortisone-sensitive cell type (table XIII) [100]. One hypothesis would be that NK cells could have a normal physiological role to control these immature cells in the bone marrow, during their migration to the thymus or when 'accidently' released in the peripheral lymphoid system. As T cell lymphomas in general are known to be NK-sensitive, and to originate as immature pre-leukemic cells in the bone marrow and differentiate into overt leukemias in the thymus, this theory could have considerable implications in explaining the surveillance role of NK cells in control of leukemogenesis. It should be stressed here, however, that other alternative theories than a direct elimination of 'natural NK targets' by NK cells could explain the above-mentioned inverse relationship between these cells and host NK activity. Considering that NK cells have been suggested to be on the T cell lineage [115], the possibility that NK cells and NK thymus targets have a common progenitor cell could also explain this inverse relationship.

According to this theory both the NK cell itself and the NK-sensitive

Table XIII. Characteristics of the NK-sensitive subpopulation in the thymus of young strain A/Sn mice. From *Hansson* et al. [100].

Characteristics	NK-sensitive population[1]	Majority of cortical cells	Medullary thymocytes[2]
Cortisone sensitivity	+	++	–
Killed by treatment with fresh rabbit or guinea pig serum	+	+	–
Expression of receptors for peanut agglutinin	+	+	–
Mobility in 1 × g velocity sedimentation, mm/h	5	3.5	n.d.
Morphology	large, rapidly dividing	small	n.d.
Rate of mitotic activity	high	low	n.d.
Electrophoretic mobility	intermediate	low	high

+ or – signs refer to the presence or absence of the characteristics.

[1] As measured in a 3-hour NK assay where fractionated or unfractionated thymocytes were used as target cells with nylon wool-passed CBA effector spleen cells.

[2] As measured using anti-H-2-educated cytotoxic T cells as indicators for the medullary cells, known to express high amounts of H-2 antigens.

thymocyte target arise from a common progenitor cell and various influences (age, genetics, organ environment, interferon) can drive the progenitor more or less on one limb (i.e. NK) at the expense of the alternative limb (i.e. thymocyte target).

The following evidence indirectly supports the common progenitor theory: (1) If one drives the maturation of the thymus limb, by using thymopoietin or other thymic hormones then NK activity should decrease by a competition effect. Thymic hormone has been reportet to decrease NK [114]. (2) Nude mice would have an impaired thymic limb since they lack a thymic environment and this explains why NK activity is higher in nu/nu than +/nu littermates. If syngeneic thymus is grafted to nu/nu, NK activity decreases [114]. (3) Similarly, thymectomizing young mice should also increase NK which it does [88]. (4) Beige mice have normal frequency NK but deficient lytic machinery [219], whereas thymocytes from beige mice are similarly NK-sensitive compared to normal littermates [102]. This argues against direct modulation (i.e. killing) of thymocytes by NK and rather supports a differentiation hypothesis as outlined.

A third alternative explanation to consider is that both NK cells and the NK-sensitive thymus target are under the influence of a common 'modulator', which concomitantly would increase the activity or frequency of one of these cell types while decreasing that of the other. Interferon is clearly a candidate to consider in this context. In support of this theory, *Trinchieri and Santoli* [277] have recently demonstrated that interferon can influence both NK activity and target sensitivity in a diametrically opposed manner, increasing the lytic activity of NK cells while protecting the target cells for lysis, particularly normal tissue. We have recently found that interferon also confers such a direct protective effect on NK-sensitive thymocytes [102].

NK lysis of primary normal tissue is not restricted to cells from lymphoid organs, since NK sensitivity has been described using peritoneal macrophages as targets [184, 288]. In the system described by *Welsh* [289] and *Welsh and Zinkernagel* 288] primary adherent peritoneal cells (PC) were used as target cells with effector cells from LCMV-induced mice. An interesting feature of this system is that here cells from different strains of mice showed unique ranges of target specificities. Thus, NK cells from most mouse strains lysed allogeneic PC at much higher efficiency than syngeneic ones. This contrasted to what the authors as well as our group has observed using continuous cell lines as targets, where the same strain distribution of high and low reactive mouse strains will be seen regardless of the genotype of the target cells. What is even more interesting is that this also clearly distinguishes the PC target system from the results obtained with primary thymocytes as targets where, similar to the results with continuous cell lines, no evidence for heterogeneity at the effector cell level has been seen, using thymocyte targets from different genotypes [146]. It should be stressed, however, that some caution must be taken in comparing the PC assay system with a standard NK assay since some differences exist between the two assays. Thus, although the effector cell in the PC assay has the characteristics of an 'NK-like' cell it has been obtained from LCMV-boosted mice, and very little if any effects are seen with non-induced spleen cells [*Welsh*, personal communucation]. The assay system is also an overnight incubation, and no effects are seen in short-term assays, in contrast to when thymocytes or continuous cell lines are used as targets. The possibility that the PC lysis is mediated by another mechanism or cell type than the one active in a standard NK assay must therefore be further investigated.

XII. Concluding Remarks

The evidence suggests that NK cells play an important role in resistance to some tumors and possibly virus-infected cells in vivo. Similar or identical cells may also be important in the rejection of hemopoietic grafts and the regulation of T cell subpopulations within the thymus. These lymphocytes arise de novo in the bone marrow, develop in an age-dependent manner independent of thymic influence, and lack the characteristics of mature macrophages, B cells or T cells. NK cells are under H-2-linked genetic control and can be activated in vivo by interferon, bacterial adjuvants, and tumor cells or viruses.

Hence, NK cells have been added to a growing list consisting of T lymphocytes, K cells participating in an antibody-dependent mechanism, and activated macrophages or promonocytes all capable of exerting cytolytic or cytostatic, anti-tumor effects. This multiplicity of effector mechanisms is not unexpected since several levels of defence against a lethal condition such as malignancy would clearly have a selective advantage during evolutionary history.

NK cells are unique since there is no H-2 restriction of killing, no classical immunological memory and these cells kill a wide range of unrelated cells with some degree of specificity, as revealed in inhibition studies with isolated target cell antigens. NK-mediated cytolysis can proceed by both an antibody-dependent or antibody-independent mechanisms and is distinct from the cytolytic mechanism utilized by T cells, macrophages or promonocytes. In addition, NK cells appear to be a heterogeneous population and preexist at high levels in the host in contrast to other effector mechanisms (T cells, macrophages, ADCC) which require days or weeks to be primed. NK cells therefore might be expected to provide a first line of defence against newly arising malignancies. This hypothesis is supported by deficient tumor rejection in the NK-defective mutant, beige, and by the high incidence (85%) of spontaneous 'lymphoma-like' disorders in patients with selective NK deficiencies (CH).

Several important questions remain only partially resolved at best, including the specificity, clonality, control, and mechanism of target-effector interaction. The NK cell still lacks a specific serological marker and little is known regarding the recognition structure. We believe that a concerted biochemical and phylogenetic approach will help determine whether the NK receptors are coded by variable region genes or whether they represent remnants of a more primitive surveillance mechanism.

Recent developments in the cloning of NK cells, the isolation of target structures and the ability to selectively deplete NK cells in vivo (beige mice) will provide valuable tools with which to approach these fundamental questions.

References

1 Ahrlund-Richter, L.; Klein, ; Masucci, G.: Somatic cell hybrids between high NK-sensitive (YAC-1R) and several low sensitive sarcoma or L-cell-derived mouse lines exhibit low sensitivity. Som. Cell Genet. *6:* 89 (1980).

2 Ahrlund-Richter, L.; Masucci, G.; Klein, E.: A human somatic cell hybrid of the high natural killer sensitive line, K562, and the low sensitive line, P3HR-3, has intermediated sensitivity (submitted).

3 Anderson, M.J.: Innate cytotoxicity of CBA mouse spleen cells to Sendai virus-infected L cells. Infect. Immunity*20:* 608 (1978).

4 Archer, R.: Inheritance of the macrophage alloantigenic marker (mph-1) in inbred mice. Genet. Res. *26:* 213 (1975).

5 Becker, S.; Klein, E.: Decreased 'natural killer' effect in tumor-bearing mice and its relation to the immunity against oncorna virus-determined cell surface antigens. Eur. J. Immunol. *6:* 892 (1976).

6 Becker, S.; Fenyo, E.M.; Klein, E.: The 'natural killer' cell in the mouse does not require H-2 homology and is not directed against type or group-specific antigens of murine C-viral proteins. Eur. J. Immunol. *6:* 882 (1976).

7 Becker, S.; Kiessling, R.; Lee, N.; Klein, G.: Modulation of sensitivity to natural killer (NK) cell lysis after in vitro explanation of a mouse lymphoma. J. natn. Cancer Inst. *61:* 1495 (1978).

8 Becker, S.; Stendahl, O.; Magnusson, K.: Physico-chemical characteristics of tumor cells susceptible to lysis by natural killer (NK) cells. Immunol. Commun. *8:* 73 (1979).

9 Berke, G.; Gabison, D.: Energy requirements of the binding and lytic steps of T lymphocyte-mediated cytolysis of leukemic cells in vitro. Eur. J. Immunol. *5:* 671 (1975).

10 Biddison, W.E.; Palmer, J.C.: Development of tumor cell resistance to syngeneic cell-mediated cytotoxicity during growth of ascitic mastocytoma P815Y. Proc. natn. Acad. Sci. USA *75:* 329 (1977).

11 Binz, H.; Wigzell, H.: Antigen-binding, idiotypic T-lymphocyte receptors; in Stutman, Contemp. Top. Immunobiol., pp. 113 (Plenum Press, New York 1977).

12 Blazer, B.; Patarroyo, M.; Klein, E.; Klein, G.: Increased sensitivity of human lymphoid lines to natural killer cells after induction of the Epstein-Barr viral cycle by superinfection or sodium butyrate. J. exp. Med. *151:* 614 (1980).

13 Blume, R.S.; Wolff, S.M.: The Chediak-Higashi syndrome: studies in four patients and a review of the literature. Medicine, Baltimore *51:* 247 (1972).

14 Bonmassar, E.; Campanile, F.; Houchens, D.; Crino, L.; Goldin, A.: Impaired growth of a radiation induced lymphoma in intact or lethally irradiated allogeneic athymic (nude) mice. Transplantation *20: 343 (1975).*

15 Bonnard, G.D.: Models for the mechanism of natural cell-mediated cytotoxicity: in

Riethmuller, Wernet, Cudkowicz, Perspectives in immunology (Academic Press, New York 1978).

16 Brodeur, B.R.; Merigan, T.C.: Mechanism of the suppressive effect of interferon on antibody synthesis in vivo. J. Immun. *114:* 1323 (1975).

17 Brodt, P.; Gordon, J.: Anti-tumor immunity in B lymphocyte-deprived mice II. In vitro studies. J. Immun. (in press).

18 Brodt, P.; Gordon, J.: Anti-tumor immunity in B lymphocyte-deprived mice. I. Immunity to a chemically induced tumor. J. Immun. *121:* 359 (1978).

19 Brodt, P.; Gordon, J.: Heightened resistance of B lymphocyte deprived mice to carcinogenesis by 3-methylcholanthrene (submitted for publication).

20 Brooks, C.G.; Rees, R.I.; Leach, R.H.: High non-specific reactivity of normal lymphocytes against mycoplasma-infected target cells in cytotoxicity assays. Eur. J. Immun. *9:* 159 (1979).

21 Brunda, M.J.; Herberman, R.B.; Holden, H.T.: Natural killer cell activity in beige mice: spontaneous reactivity and augmentation of cytotoxicity by interferon and interferon inducers (submitted).

22 Brune, K.; Glatt, M.; Kalin, H.; Peskar, B.: Pharmacological control of prostaglandin and thromboxane release from macrophages. Nature, Lond. *274:* 261 (1978).

23 Brunner, T.K.; Engers, H.D.; Cerottini, J.C.: The ^{51}Cr release assay as used for the quantitative measurement of cell-mediated cytolysis in vitro; in Bloom, David, in vitro methods in cell-mediated and tumor immunity, pp. 423 (Academic Press, New York 1976).

24 Bubbers, J.E.; Henney, C.S.: Studies on the mechanism of lymphocyte mediated cytolysis. V. The use of cytochalasin A and B to dissociate glucose transport from the lytic event. J. Immun. *115:* 145 (1975).

25 Bunting, W.L.; Kiely, J.M.; Owen, C.A.: Radiochromium labelled lymphocytes in the rat. Proc. Soc. exp. Biol. Med. *113:* 370 (1963).

26 Burnett, F.M.: Immunological surveillance (Pergamon Press, Oxford 1970).

27 Byers, V.S.; Levin, A.S.; Hackett, A.J.; Fudenberg, H.H.: Tumorspecific cell-mediated immunity in household contacts of cancer patients. J. clin. Invest. *55:* 500 (1975).

28 Callewaert, D.M.; Lightbody, J.J.; Kaplan, J.; Jaroszewski, J.; Peterson, W.D.; Rosenburg, J.C.: Cytotoxicity of human peripheral lymphocytes in cell-mediated lymphocytes, antibody-dependent, cell mediated lympholysis and natural cytotoxicity assays after mixed lymphocyte culture. J. Immun. *121:* 81 (1978).

29 Campbell, A.C.; Waller, C.; Wood, J.; Aynsley-Green, A.; Yu, V.: Lymphocyte subpopulations in the blood of newborn infants. Clin. exp. Immun. *18:* 469 (1974).

30 Cantor, H.; Kasai, M.; Shen, F.W.; Leclerc, J.C.; Glimcher, L.: Immunogenetic analysis of 'natural killer' activity in the mouse. Immunol. Rev. *44:* 3 (1979).

31 Carlson, G.A.; Wegmann, T.G.: Rapid in vivo destruction of semi-syngeneic and allogeneic cells by non-immunized mice as a consequence of non-identity at H-2. J. Immun. *118:* 2130 (1977).

32 Cerottini, J.C.; Brunner, K.T.: Mechanism of T and K cell-mediated cytolysis; in Loor, Roelants, B and T cells in immune recognition, (pp. 319 (Wiley & Sons, Chichester 1977).

33 Chediak, M.: Nouvelle anomalie leucocytaire de caractère constitutionnel et familial. Revue Hémat. *7:* 362 (1952).

34 Chernyakhovskaya, I.Y.; Slavina, E.G.; Svet-Moldavsky, G.J.: Antitumor effect of lymphoid cells activated by interferon. Nature , Lond. *288:* 71 (1970).

35 Claesson, M.H.; Olsson, L.: Autoreactive natural killer-like cells from agar-cloned murine bone marrow cells. Nature, Lond. *283:* 578 (1980).

36 Clark, E.A.; Harmon, R.C.; Wicker, L.S.: Resistance of H-2 heterozygous mice to parental tumors. II. Characterization of Hh-lcontrolled hybrid resistance to syngeneic fibrosarcomas and the EL-4 lymphoma. J. Immun. *119:* 648 (1977).

37 Clark, E.A.; Russell, P.H.; Egghart, M.; Horton, M.A.: Characteristics and genetic control of NK cell-mediated cytotoxicity activated by naturally acquired infection in the mouse. Int. J. Cancer *24:* 688 (1979).

38 Cohen, A.M.; Burdick, J.F.; Ketcham, A.S.: Cell mediated cytotoxicity: an assay using ^{125}I-iododeoxyuridine-labelled target cells. J. Immun. *107:* 895 (1971).

39 Collavo, D.; Colambatti, A.; Chieco-Bianchi, L.; Davies, A.J.S.: T. lymphocyte requirement for MSV tumor prevention or regression. Nature, Lond. *249:* 169 (1974).

40 Collins, J.L.; Patek, P.Q.; Cohn, M.: The relationship of tumorigenicity to NK recognition and lysis. J. Immun. (in press, 1980).

41 Cudkowicz, G.: Hybrid resistance to parental grafts of hematopoietic and human lymphoma cells. The proliferation and spread of neoplastic cells, p. 661 (Williams & Wilkins, Baltimore 1968).

42 Cudkowicz, G.; Hochman, P.: Regulation of natural killer activity by macrophage-like and other types of suppressor cells; in Siskind, Developmental immunobiology (Grune & Stratton, New York 1978).

43 Cudkowicz, G; Hochman, P.: Do natural killer cells engage in regulated reactions against self to ensure hemeostasis? Imunol. Rev *44:* 13 (1979).

44 DeMayer, E.; DeMayer-Guignand, J.: Effect of interferon on cell-mediated immunity; in Baron and F. Dianzani, The interferon system: a current review to 1978. Tex. Rep. Biol. Med. *35:* 370 (1977).

45 DeMayer, R.E.; DeMayer-Guignand, J.; Bayly, D.W.: Effect of mouse genotype on interferon production. Immunogenetics *1:* 438 (1975).

46 Dennert, G.: Cloned lines of natural killer cells. Nature, Lond. *287:* 47 (1980).

47 Dent, P.B.; Fish, L.A.; White, J.F.; Good, R.A.: Chediak-Higashi syndrome. Observations on the nature of the associated malignancy. Lab. Invest. *15:* 1634 (1966).

48 Dietrich, M.; Landen, B.: Complement bridges between cells: analysis of a possible cell-cell interaction mechanism. J. exp. Med. *146:* 1484 (1977).

49 Domzig, W.; Lohmann-Matthes, M.L.: Antibody-dependent cellular cytotoxicity (ADCC) against tumor cells. II. The promonocyte identified as effector cell. Eur. J. Immunol. *9:* 267 (1979).

50 Donahoe, R.M.; Huang, K.Y.: Interferon preparations enhance phagocytosis in vivo. Infect. Immunity *13:* 1250 (1976).

51 Droller, M.J.; Perlmann, P.; Schneider, M.U.: Enhancement of natural and antibody-dependent lymphocyte cytotoxicity by drugs which inhibit prostaglandin production by tumor target cells. Cell. Immun. *39:* 154 (1978).

52 Droller, M.J.; Schneider, M.U.; Perlmann, P.: A possible role of prostaglandins in the inhibition of natural and antibody-dependent cell-mediated cytotoxicity against tumor cells. Cell. Immun. *39:* 165 (1978).

53 Epstein, A.L.; Herman, M.M.; Kim, H.; Dorfman, R.F.; Kaplan, H.S.: Biology of the

human malignant lymphomas. III. Interacranial heterotransplantation in the nude, athymic mouse. Cancer *37:* 2158 (1976).

54 Epstein, L.B.: The effects of interferons on the immune response in vitro and in vivo: in Stewart, Interferons and their actions (CRC Press, Cleveland 1977).

55 Fenyo, E.M.; Klein, G.: Independence of Moloney virus-induced cell surface antigen and membrane associated virion antigens in immunoselected lymphoma sublines. Nature, Lond. *260:* 355 (1976).

56 Fidler, I.J.; Caines, S.; Dolan, Z.: Survival of hematogenously disseminated allogeneic tumor cells in athymic nude mice. Transplantation *22:* 208 (1976).

57 Fidler, I.J.; Nicolson, G.L.: Tumor cell and host properties affecting the implantation and survival of blood-borne metastatic variants of B16 melanoma. Israel. J. med. Scis *14:* 38 (1978).

58 Fishelson, Z.; Berke, G.: T lymphocyte-mediated cytolysis: dissociation of the binding and lytic mechanisms of the effector cell. J. Immun. *120:* 1121 (1978).

59 Fitzgerald, K.L.; Ponzio, N.M.: Natural killer cell activity in reticulum cell sarcomas. Cell Immunol. *43:* 185 (1979).

60 Friedman, R.M.; Grollman, E.F.; Chang, E.H.; Kohn, L.D.; Lee, G.; Jay, F.T.: Interferon and glycoprotein hormones; in Baron, Dianzani, The interferon system: a current review to 1978. Tex. Rep. Biol. Med. *35:* 326 (1977).

61 Galili, U.; Galili, N.; Vanky, F.; Klein, E.: Natural species restricted attachment of human and murine T lymphocytes to various cells. Proc. natn. Acad. Sci. USA *75:* 2396 (1978).

62 Gidlund, M.; Orn, A.; Wigzell, H.; Senik, A.; Gresser, I.: Enhanced NK cell activity in mice injected with interferon and interferon inducers. Nature, Lond. *273:* 759 (1978).

63 Gidlund, M.; Ojo, E.E.; Orn, A.; Wigzell, H.; Murgita, R.A.: Severe suppression of the B-cell system has no impact on the maturation of natural killer cells in mice. Scand. J. Immunol. *9:* 167 (1979).

64 Giovanella, B.C.; Stehlin, J.S.; Williams, L.J., Jr.: Heterotransplantation of human malignant tumors in 'nude' thymusless mice. II. Malignant tumors induced by injection of cell cultures derived from human solid tumors. J. natn. Cancer Inst. *52:* 921 (1974).

65 Glaser, M.; Lavrin, D.H.; Herberman, R.B.: In vivo protection against syngeneic gross virus-induced lymphoma in rats: comparison with in vitro studies of cell-mediated immunity. J. Immun. *116:* 1507 (1976).

66 Glimcher, L.; Shen, F.W.; Cantor, H.: Identification of a cell surface antigen expressed on the natural killer cell. J. exp. Med. *145:* 1 (1977).

67 Goldenberg, P.Z.; Greenberg, A.H.: Cell-mediated cytolysis of xenogeneic cells by hemocytes from a higher invertebrate (abstr.). Proc. Can. Fed. Biol. Soc. *21:* 109 (1978).

68 Goldstein, P.; Smith, E.T.: Mechanism of T-cell-mediated cytolysis: the lethal hit stage; in Stutman, Contemp. Top. Immunobiol, p. 273 (Plenum Press, New York 1977).

69 Gomard, E.; Leclerc, J.C.; Levy, J.: Spontaneous antilymphoma reaction of preleukemic AKR mice is a non-T-cell killing. Nature, Lond. *250:* 671 (1974).

70 Gorczynski, R.M.: Evidence for in vivo protection against murine-sarcoma virus-induced tumors by T lymphocytes from immune animals. J. Immun. *112:* 533 (1974).

71 Gorczynski, R.M.; Norbury, C.: Immunity to murine sarcoma virus induced tumors. III. Analysis of the cell populations involved in protection from lethal tumor progression of sublethally irradiated, MSV-inoculated mice. Br. J. Cancer *30:* 118 (1974).

72 Gorelik, E.; Fogel, M.; Feldman, M.; Segal, S.: Differences in resistance of metastatic

tumor cells and cells from the local tumor growth to natural killer cells. J. natn. Cancer Inst. *63:* 1397 (1979).

73 Goud, T.J.; Schotte, C.; vanFurth, R.: Identification and characterization of the monoblast in mononuclear phagocyte colonies grown in vitro. J. exp. Med. *143:* 1180 (1975).

74 Granger, G.A.; Daynes, R.A.; Runge, P.; Prieur, A.M.; Jeffes, E.: Lymphocyte effector molecules and cell-mediated immune reactions. Contemp. Top. mol. Immunol. *4:* 205 (1975).

75 Greenberg, A.H.; Greene, M.: Non-adaptive rejection of small tumor inocula as a model of immune surveillance. Nature, Lond. *264:* 356 (1975).

76 Greenberg, A.H.; Playfair, J.H.L.: Spontaneously arising cytotoxicity to the P-815-Y mastocytoma in NZB mice. Clin. exp. Immunol. *16:* 99 (1974).

77 Gresser, I.; Coppey, J.; Flacoff, E.; Fontaine, D.: Interferon and murine leukemia. I. Inhibitory effect of interferon preparations on development of Friend leukemia in mice. Proc. Soc. exp. Biol. Med. *124:* 84 (1967).

78 Gresser, I.; Coppey, J.; Bourali, C.: Interferon and murine leukemia. VI. Effect of interferon preparations on the lymphoid leukemia of AKR mice. J. natn. Cancer Inst. *43:* 1083 (1969).

79 Gresser, I.; Bourali, C.; Levy, J.P.; Brouty Boye, D.; Thomas, M.T.: Increased survival in mice inoculated with tumor cells and treated with interferon preparations. Proc. natn. Acad. Sci. USA *63:* 51 (1969).

80 Gresser, I.; De Maeyer-Guignard, J.; Tovey, M.G.; De Maeyer, E.: Electrophoretically pure mouse interferon exerts multiple biologic effects. Proc. natn. Acad. Sci. USA *76:* 5308 (1979).

81 Grimm, E.A.; Bonavida, B.: Studies on the induction and expression of T cell mediated immunity. IV. Heterogeneity of lytic efficiency exhibited by isolated cytotoxic T lymphocytes prepared from highly enriched populations of effctor-target conjugates. J. Immun. *119:* 1041 (1977).

82 Grimm, W.; Seitz, M.; Kirchner, H.; Gemsa, D.: Prostaglandin synthesis in spleen cell cultures of mice injected with *Corynebacterium parvum.* Cell. Immunol. *40:* 419 (1978).

83 Gronberg, A.; Hannson, M.; Kiessling, R.; Anderson, B.; Karre, K.; Roder, J.C.: Natural antibodies in normal rabbit serum with similar specificity pattern as mouse natural killer cells. J. natn. Cancer Inst. *64:* 113 (1980).

84 Gullin, J.I.; Sandler, J.A.; Clyman, R.I.; Manganiello, V.C.; Vaughn, M.: Agents that icrease cytclic AMP inhibit monocyte locomotion. J. Immun. *120:* 492 (1978).

85 Habu, S.; Hayakawa, K.; Okumura, K.; Tada, T.: Surface markers on natural killer cells of the mouse. Eur. J. Immunol. *9:* 938 (1979).

86 Haliotis, T.; Roder, J.; Klein, M.; Ortaldo, J.; Fauci, A.S.; Herberman, R.B.: The Chediak-Higashi gene in humans. I. Impairment of NK function. J. exp. Med. *151:* 1039 (1980).

87 Haller, O.; Hansson, M.; Kiessling, R.; Wigzell, H.; Nonconventional natural killer cells may play a decisive role in providing resistance against syngeneic tumor cells in vivo. Nature, Lond. *270:* 609 (1977).

88 Haller, O.; Kiessling, R.; Gidlund, M.; Wigzell, H.: Natural killer cells to leukemia: in vitro studies. Proc. Int. Symp. on Tumor-Associated Antigens and their Specific Immune Responses (Academic Press, London 1977).

89 Haller, O.; Kiessling, R.; Orn, A.; Karre, K.; Nilsson, K.; Wigzell, H.: Natural cytotoxicity to human leukemia mediated by mouse non-T cells. Int. J. Cancer *20:* 93 (1977).

90 Haller, O.; Kiessling, R.; Orn, A.; Wigzell, H.: Generation of natural killer cells: an autonomous function of the bone marrow. J. exp. Med. *145:* 1411 (1977).

91 Haller, O.; Wigzell, H.: Suppression of natural killer cell activity with radioactive strontium: effector cells are marrow-dependent. J. Immun. *118:* 1503 (1977).

92 Haller, O.; Gidlund, M.; Kurneck, J.T.; Wigzell, H.: In vivo generation of mouse natural killer cells: role of the spleen and thymus. Scand. J. Immunol. *8:* 207 (1978).

93 Haller, O.; Gidlund, M.; Hellstrom, U.; Hammarstrom, S.; Wigzell, H.: A new surface marker on mouse natural killer cells. Receptors for *Helix pomatia* A haemagglutinin. Eur. J. Immunol. *8:* 765 (1978).

94 Haller, O.; Orn, A.; Gidlund, M.; Wigzell, H.: In vivo activity of murine NK cells; in Herberman Natural cell-mediated immunity against tumors (Academic Press, New York, in press).

95 Halpern, B.N.; Biozzi, G.; Stiffel, G.; Monton, D.: Inhibition of tumor growth by administration of killed *Corynebacterium parvum.* Nature, Lond. *212:* 853 (1966).

96 Hammarstrom, S.: Structural specificity, binding properties, and some biological activities of a blood group A-reactive hemagglutinin from the snail *Helix pomatia.* Ann. NY. Acad. Sci. *234:* 183 (1974).

97 Hammarstrom, S.; Hellstrom, U.; Dillner, M.L.; Perlmann, P.; Perlmann, H.; Axelsson, P.; Robertsson, E.S.: in Hoffman-Ostenhof et al., Affinity chromatography. Proc. Int. Symp., Vienna (Pergamon Press, Oxford 1977).

98 Hanna, N.; Fidler, I.: Role of natural killer cells in the destruction of circulating tumor emboli. J. natn. Cancer Inst. *65:* 801 (1980).

99 Hansson, M.; Karre, K.; Bakacs, T.; Kiessling, R.; Klein, G.: Intra- and interspecies reactivity of human and mouse natural killer (NK) cells. J. Immun. *121:* 6 (1978).

100 Hansson, M.; Karre, K.; Kiessling, R.; Roder, J.C.; Andersson, B.; Hayry, P.: Natural NK cell targets in the mouse thymus; characteristics of the sensitive cell population. J. Immun. *123:* 765 (1979).

101 Hansson, M.; Kiessling, R.; Anderson, B.; Karre, K.; Roder, J.C.: NK-cell sensitive T-cell subpopulation in thymus: inverse correlation to host NK activity. Nature, Lond. *278:* 174 (1979).

102 Hansson, M.; Kiessling, R.; Andersson, B.; Welsh, R.: Effect of interferon and interferon inducers on the NK sensitivity of normal mouse thymocytes J. Immunol. *125:* 2225 (1980).

103 Harfast, B.; Andersson, T.; Perlmann, P.: Immunoglobulin independent natural cytotoxicity of Fc receptor-bearing human blood lymphocytes to mumps virusinfected target cells. J. Immun. *121:* 255 (1978).

104 Harmon, R.C.; Clark, E.; O'Toole, C.; Wicker, L.: Resistance of H-2 heterozygous mice to parental tumors. 1. Hybrid resistance and natural cytotoxicity to EL-4 are controlled by the H-2D-Hh-1 region. Immunogenetics *4:* 601 (1977).

105 Hellstrom, K.E.; Hellstrom, I.: Allogeneic inhibition of transplantated tumor cells. Prog. exp. Tumor Res., vol. 9, p. 40 (Karger, Basel 1967).

106 Hellstrom, U.; Hammarstrom, S.; Dillner, M.L.; Perlman, H.; Perlman, P.: Fractionation of human blood lymphocytes on *Helix pomatia* A haemagglutinin coupled to Sepharose beads. Scand. J. Immunol. *5:* suppl. 5, p. 45 (1976).

107 Henney, C.S.: Studies on the mechanism of lymphocyte-mediated cytolysis. II. The use of various target cell markers to study cytolytic events. J. Immun. *110:* 73 (1973).

108 Henney, C.S.; Gaffney, J.; Bloom, B.R.: On the relation of products of activated lymphocytes to cell-mediated cytolysis. J. exp. Med. *140:* 837 (1974).
109 Herberman, R.B.; Nunn, M.E.; Lavrin, D.H.: Natural cytotoxic reactivity of mouse lymphoid cells against syngeneic and allogeneic tumors. I. Distribution of reactivity and specificity. Int. J. Cancer *16:* 216 (1975).
110 Herberman, R.B.; Nunn, M.E.; Holden, H.T.; Lavrin, D.H.: Natural cytotoxic reactivity of mouse lymphoid cells against syngeneic and allogeneic tumors. II. characterization of effector cells. Int. J. Cancer *16:* 230 (1975).
111 Herberman, R.B.; Nunn, M.E.; Holden, H.T.; Staal, S.; Djeu, J.Y.: Augmentation of natural cytotoxic reactivity of mouse lymphoid cells against syngeneic and allogeneic target cells. Int. J. Cancer *19:* 555 (1977).
112 Herberman, R.B.; Bartram, S.; Haskill, J.S.; Nunn, M.E.; Holden, H.T.; West, W.H.: Fc receptors on mouse effector cells mediating natural cytotoxicity against tumor cells. J. Immun. *119:* 322 (1977).
113 Herberman, R.B.; Nunn, M.E.; Holden, H.T.: Low density of Thy-1 antigen on mouse effector cells mediating natural cytotoxicity against tumor cells. J. Immun. *121:* 304 (1978).
114 Herberman, R.B.; Holden, H.T.: Natural cell-mediated immunity; in Klein, Weinhouse, Advances in cancer research, vol. 27, p. 305 (Academic Press, New York 1978).
115 Herberman, R.B.; Dejeu, J.Y.; Kay, H.D.; Ortaldo, J.R.; Riccardi, C.; Bonnard, G.D.; Holden, H.T.; Fagnani, R.; Santoni, A.; Puccetti, P.: Natural killer cells: characteristics and regulation of activity. Immunol. Rev. *44:* 43 (1979).
116 Herberman, R.B.; Ortaldo, J.R.; Bonnard, G.D.: Augmentation by interferon of human natural and antibody dependent cell-mediated cytotoxicity. Nature, Lond. *277:* 221 (1979).
117 Higashi, O.: Congenital gigantism of peroxidase granules. J. exp. Med. *59:* 315 (1954).
118 Hildeman, W.H.: Specific immunorecognition by histocompatibility markers: the original polymorphic system of immunoreactivity characteristic of all multicellular animals. Immunogenetics *5:* 193 (1977).
119 Hilderman, J.H.; Strom, T.B.: Specific insulin binding site on T and B lymphocytes as a marker of cell activation. Nature, Lond. *274:* 62 (1978).
120 Hiserodt, J.C.; Tiango, G.; Granger, G.: The LT system in experimental animals. IV. Rapid specific lysis of ^{51}Cr-labelled allogeneic target cells by highly unstable high M.W. lymphotoxin receptor complex(es) released in vitro by activated alloimmune murine T lymphocytes. J. Immun. *123:* 332 (1979).
121 Hochman, P.S.; Cudkowicz, G.: Different sensitivities to hydrocortisone of natural killer cell activity and hybrid resistance to parental marrow grafts. J. Immun. *119:* 2013 (1977).
122 Hochman, P.; Cudkowicz, G.: Suppression of natural cytotoxicity by spleen cells of hydrocortisone-treated mice. J. Immun. *123:* 968 (1979).
123 Huang, K.Y.; Donahoe, R.M.; Gordon, F.B.; Dressler, H.R.: Enhancement of phagocytosis by interferon-containing preparations. Infect. Immunity *4:* 581 (1971).
124 Ida, S.; Hooks, J.J.; Siraganian, R.P.; Notkins, A.L.: Enhancement of IgE mediated histamine release from human basophils by viruses: role of interferon. J. exp. Med. *145:* 892 (1977).
125 Johnson, H.; Baron, S.: The nature of the suppressive effect of interferon and interferon inducers on the in vitro immune response. Cell. Immunol. *25:* 106 (1976).
126 Jondal, M.; Pross, H.: Surface markers on human B and T lymphocytes. VI. Cytotoxicity

against cell lines as a functional marker for lymphocyte subpopulations. Int. J. Cancer *15:* 596 (1975).
127 Jondal, M; Svedmyr, E.; Klein, E.; Singh, S.: Killer T cells in a Burkitt's lymphoma biopsy. Nature, Lond. *255:* 405 (1975).
128 Jondal, M.; Targan, S.: In vitro induction of cytotoxic effector cells with spontaneous killer specificity. J. exp. Med. *147:* 1621 (1978).
129 Jones, B.; Jones, T.C.: Allograft cytotoxicity cooperation between alloimmune T cells and macrophages. Immunology *34:* 247 (1978).
130 Jurd, R.D.; Doritis, A.: Antibody-dependent cellular cytotoxicity in poikilotherms. Dev. comp. Immunol. *1:* 341 (1977).
131 Kalina, M.; Hollander, N.: The effect of cytochalasin B on effector-target cell interaction. Quantitative and ultrastructural study. Immunology *29:* 709 (1975).
132 Karre, K.; Haller, O.; Becker, S.; Orn, A.; Andersson, L.; Ranki, A.M.; Kiessling, R.; Hayry, P.: Physical characteristics of mouse natural killer cells. Cell. Immunol. (in press).
133 Karre, K.; Seeley, J.K.: Cytotoxic thy 1.2-positive blasts with NK-like target selectivity in murine mixed lymphocyte cultures. J. Immun. *123:* 1511 (1979).
134 Karre, K.; Klein, G.O.; Kiessling, R.; Klein, G.; Roder, J.C.: Low natural in vivo resistance to syngeneic leukemias in natural killer deficient mice. Nature, Lond. *284:* 624 (1980).
135 Kasai, M.; Iwamori, M.; Nagai, Y.; Okumura, K.; Tada, T.: A glycolipid on the surface of mouse natural killer cells. Eur. J. Immunol. *10:* 175 (1980).
136 Keller, R.: Macrophage-mediated natural cytotoxicity against target cells in vitro. I. Macrophages from diverse anatomical sites and different strains of rats and mice. Br. J. Cancer *37:* 732 (1978).
137 Kerbel, R.S.; Roder, J.C.; Pross, H.F.: Induction of a unique promonocyte-like cell line with spontaneous killer cell activity Int. J. Cancer, in press (1981).
138 Kiessling, R.; Klein, E.; Wigzell, H.:Natural killer cells in the mouse. I. Cytotoxic cells with specificity for mouse Moloney leukemia cells. Specificity and distribution according to genotype. Eur. J. Immunol. *5:* 112 (1975).
139 Kiessling, R.; Klein, E.; Pross, H.; Wigzell, H.: Natural killer cells in the mouse. II. Cytotoxic cells with specificity for mouse Moloney leukemia cells. Characteristics of the killer cell. Eur. J. Immunol. *5:* 117 (1975).
140 Kiessling, R.; Petranyi, G.; Klein, G.; Wigzell, H.: Genetic variation of in vitro cytolytic activity and in vivo rejection potential of non-immunized semi-syngeneic mice against a mouse lymphoma line. Int. J. Cancer *15:* 933 (1975).
141 Kiessling, R.; Petranyi, G.; Klein, G.; Wigzell, H.: Non-T-cell resistance against a mouse Moloney lymphoma. Int. J. Cancer *17:* 1 (1976).
142 Kiessling, R.; Petranyi, G.; Karre, K.; Jondal, M.; Tracy, D.; Wigzell, H.: Killer cells: a functional comparison between natural, immune T-cell and antibody-dependent in vitro systems. J. exp. Med. *143:* 112 (1976).
143 Kiessling, R.; Haller, O.; Fenyo, E.M.; Steinitz, M.; Klein, G.: Mouse natural killer (NK) cell activity against human cell lines is not influenced by superinfection of the target cells with xenotropic murine C-type virus. Int. J. Cancer *21:* 460 (1978).
144 Kiessling, R.; Hochman, P.S.; Haller, O.; Shearer, G.M.; Wigzell, H.; Cudkowicz, G.: Evidence for a similar or common mechanism for natural killer cell activity and resistance to hemopoietic grafts. Eur. J. Immunol. *7:* 655 (1977).

145 Kiessling, R.; Haller, O.: Natural killer cells in the mouse; an alternative immune surveillance mechanism? Contemp. Top. Immunobiol. *8:* 171 (1978).
146 Kiessling, R.; Welsh, R.: Killing of normal cells by activated mouse natural killer cells: Evidence for two patterns of genetic regulation of lysis. Scand. J. Immunol. *11:* 363 (1980).
147 Kirchner, H.; Hirt, H.M.; Becker, H.; Munk, K.: Production of an antiviral factor by murine spleen cells after treatment with *Corynebacterium parvum.* Cell. Immunol. *31:* 172 (1977).
148 Klein, G.; Klein, G.; Kiessling, R.; Karre, K.: H-2 associated control of natural cytotoxicity and hybrid resistance against RBL-5. Immunogenetics *6:* 651 (1978).
149 Klein, G.; Klein, G.O.; Karre, K.; Kiessling, R.: 'Hybrid resistance' against parental tumors: one or several genetic patterns? Immunogenetics *7:* 391 (1978).
150 Klein, G.: Lymphoma dcvclopment in mice and humans: diversity of initiation is followed by convergent cytogenetic evolution. Proc. natn. Acad. Sci. USA *76:* 2442 (1979).
151 Klein, G.; Zeuthen, J.; Eriksson, I.; Terasaki, P.; Bernoco, M.; Rosen, A.; Masucci, G.; Povey, S.; Ber, R.: Hybridization of a myeloid leukemia-derived human cell line (K562) with human Burkitt's lymphoma line (P3HR-1). J. natn. Cancer Inst. *64:* 725 (1980).
152 Klein, M.; Roder, J.C.; Haliotis, T.; Korec, S.; Jett, J.; Herberman, R.B.; Katz, P.; Fauci, A.S.: The Chediak-Higashi gene in humans. II. the selectivity of the defect in NK and ADCC function. J. exp. Med. *151:* 1049 (1980).
153 Koide, Y.; Takasugi, M.: Determination of specificity in natural cell-mediated cytotoxicity by natural antibodies. J. natn. Cancer Inst. *59:* 1009 (1977).
154 Komuro, K.; Itakura, K.; Boyse, E.A.; Joh, M.: Ly-5: A new lymphocyte antigen system. Immunogenetics *1:* 452 (1975).
155 Koren, H.; Amos, B.; Buckley, B.: Natural killing in immunodeficient patients. J. Immun. *120:* 796 (1978).
156 Kritzler, R.A.; Terner, J.Y.; Lindebaum, J.; Magidson, J.; Williams, R.; Preisign, R.; Phillips, G.B.: Chediak-Higashi syndrome. Cytologic and serum lipid observations in a case and family. Am. J. Med. *35:* 583 (1964).
157 Kumar, V.; Ben-Ezra, J.; Bennett, M.; Sonnenfeld, G.: Natural killer cells in mice treated with ^{89}strontium: normal target-binding cell numbers but inability to kill even after interferon administration. J. Immun. *123:* 1832 (1979).
158 Lam, K.M.; Linna, T.J.: Protection of newly hatched chickens from Marek's disease (JMV) by normal spleen cells from older animals; in Bentvelzen et al., Advances in comparative leukemia research (Elsevier/North-Holland/Biomedical Press, Amsterdam 1977).
159 Lane, P.W.: Mouse News Lett. *25:* 35 (1962).
160 Lane, P.W.; Murphy, E.D.: Susceptibility to spontaneous pneumonitis in an inbred strain of beige and satin mice. Genetics N.Y., *72:* 451 (1972).
161 Lavrin, D.H.; Herberman, R.B.; Nunn, M.; Soares, N.: Studies of murine sarcoma virus-induced immunity in mice. J. natn. Cancer Inst. *51:* 1497 (1973).
162 Lee, J.C.; Ihle, J.N.: Characterization of the blastogenic and cytotoxic responses of normal mice to ecotropic C-type viral gp 71. J. Immun. *118:* 928 (1977).
163 Leibold, W.; Bridge, S.: $^{75}S_e$-release: a short-and long-term assay system for cellular cytotoxicity. Z. Immun. Forsch. exp. Ther. *155:* 287 (1979).

164 Lindahl, P.; Leary, P.; Gresser, I.: Enhancement by interferon of the specific cytotoxicity of sensitized lymphocytes. Proc. natn. Acad. Sci. USA *69:* 721 (1972).

165 Lohmann-Matthes, M.L.; Domzig, W.; Roder, J.C.: Promonocytes have the functional characteristics of natural killer cells. J. Immun. *123:* 1883 (1979).

166 Lonai, P.; Steinman, L.: Physiological regulation of antigen binding to T cells: role of a soluble macrophage factor and of interferon. Proc. natn. Acad. Sci. USA *74:* 5662 (1977).

167 Lotzova, E.; Savary, C.: Possible involvement of natural killer cells in bone marrow graft rejection. Biomédecine *27:* 341 (1977).

168 MacDonald, H.R.; Koch, C.J.: Energy metabolism and T-cell-mediated cytolysis. I. Synergism between inhibitors of respiration and glycolysis. J. exp. Med. *146:* 698 (1977).

169 MacFarlan, R.I.; Burns, W.H.; White, D.O.: Two cytotoxic cells in peritoneal cavity of virus-infected mice: antibody-dependent macrophages and nonspecific killer cells. J. Immun. *119:* 1569 (1977).

170 Marchase, R.B.; Voxbeck, K.; Roth, S.: Intercellular adhesive specificity. Biochim. biophys. Acta *457:* 385 (1976).

171 Martz, E.: Early steps in specific tumor cell lysis by sensitized mouse T lymphocytes. I. Resolution and characterization. J. Immun. *115:* 261 (1975).

172 Martz, E.: Mechanism of specific tumor cell lysis by alloimmune T lymphocytes: Resolution and characterization of discrete steps in the cellular interaction; in Stutman, Contemp. Top. Immunobiol., pp. 301 (Plenum Press, New York 1977).

173 Matthes, M.J.; Sharrow, S.O.; Herberman, R.B.; Holden, H.T.: Identification and separation of thy-1 positive mouse spleen cells active in natural cytotoxicity and antibody-dependent cell-mediated dytotoxicity. J. Immun. *123:* 2851 (1979).

174 Mayer, M.M.: Mechanism of cytolysis by lymphocytes. A comparison with complement. J. Immun. *119:* 1195 (1977).

175 Meerpohl, H.G.; Lohmann-Matthes, M.L.; Fisher, H.: Studies on the activation of mouse bone marrow-derived macrophages by the macrophage cytotoxicity factor (MCF). Eur. J. Immunol. *6:* 213 (1976).

176 Miller, R.G.; Phillips, R.A.: Sedimentation analysis of the cells required to initiate an in vivo immune response to sheep erythrocytes. Proc. Soc. exp. Biol. Med. *135:* 63 (1970).

177 Minato, N.; Bloom, B.R.; Jones, C.; Holland, J.; Reid, L.M.: Mechanism of ejection of virus-presistently infected tumor cells by athymic nude mice. J. exp. Med. *149:* 1117 (1979).

178 Moller, G. (ed.): Natural killer cells. Immunol. Rev. *44:* 44 (1979).

179 Mueller, J.; Brundell, G.; Buerki, H.; Keller, H.U.; Hess, O.W.; Goltier, H.: Nonspecific acid esterase activity; a criterion for differentiation of T and B lymphocytes in mouse lymph nodes. Eur. J. Immunol. *5:* 270 (1974).

180 Nelson, D.S. (ed.): Immunobiology of the macrophage, pp. 509 (Academic Press, New York 1976).

181 Nelson, D.L.; Bundy, B.M.; Pitchon, H.E.; Blaese, R.M.: The effector cells in human peripheral blood mediating mitogen induced cellular cytotoxicity and antibody-dependent cellular cytotoxicity. J. Immun. *117:* 1472 (1976).

182 Nelson, K.; Malmstrom, P.; Jonsson, A.; Sjogren, H.O.: Separation of rate leukocytes by countercurrent distribution in aqueous two-phase systems. II. Subpopulations which

mediate selective and nonselective lysis of normal and colon carcinoma target cells in vitro. Cell. Immunol. *37:* 422 (1978).

183 Nunn, M.E.; Djeu, J.Y.; Glaser, M.; Lavrin, D.H.; Herberman, R.B.: Natural cytotoxic reactivity of rat lymphocytes against syngeneic Gross virus-induced lymphoma. J. natn. Cancer Inst. *56:* 393 (1976).

184 Nunn, M.E.; Herberman, R.B.; Holden, H.T.: Natural cell-mediated cytotoxicity in mice against nonlymphoid tumor cells and some normal cells. Int. J. Cancer *20:* 381 (1977).

185 Oehler, J.R.;Herberman, R.B.: Natural cell-mediated cytotoxicity in rats. III. Effects of immunopharmacologic treatments on natural reactivity and on reactivity augmented by polyinosinic polycytidylic acid. Int. J. Cancer *21:* 221 (1978).

186 Oehler, J.R.; Lindsay, L.R.; Nunn, M.E.; Holden, H.T.; Herberman, R.B.: Natural cell-mediated cytotoxicity in rats. II. In vivo augmentation of NK-cell activity. Int. J. Cancer *21:* 210 (1978).

187 Ojo, E.; Wigzell, H.: Natural killer cells may be the only cells in normal mouse lymphoid cell populations endowed with cytophilic ability for antibody-coated tumor target cells. Scand. J. Immunol. *7:* 297 (1978).

188 Ojo, E.; Haller, O.; Kimura, A.; Wigzell, H.: An analysis of conditions allowing *Corynebacterium parvum* to cause either augmentation or inhibition of natural killer cell activity against tumor cells in mice. Int. J. Cancer *21:* 444 (1978).

189 Ojo, E., Haller, O.; Wigzell, H.: *Corynebacterium parvum*-induced peritoneal exudate cells with rapid cytolytic activity against tumor cells are non-phagocytic cells with characteristics of natural killer cells. Scand. J. Immunol. *8:* 215 (1978).

190 Ojo, E.; Karre, K.; Wigzell, H.: The genetic control of natural cell-mediated reactivity against antibody-coated murine tumors (submitted for publication).

191 Oldham, R.K.; Siwarski, D.; McCoy, J.L.; Plata, E.J.; Herberman, R.B.: Evaluation of a cell-mediated cytotoxicity assay utilizing ^{125}iododeoxyuridine-labelled tissue-culture target cells. Natn. Cancer Inst. Monogr. *37:* 49 (1973).

192 Oldham, R.K.; Djeu, J.Y.; Cannon, G.B.; Siwarski, D.; Herberman, R.B.: Cellular microcytotoxicity in human tumor systems: analysis of results. J. natn. Cancer Inst. *55:* 1305 (1975).

193 Ono, A.; Amos, D.B.; Koren, H.S.: Selective cellular natural killing against human leukaemic T cells and thymus. Nature, Lond. *266:* 546 (1977).

194 Orn, A.; Gidlund, M.; Wigzell, H.: Genetics of interferon induced NK-activity (abstr.) (submitted for publication).

195 Ortaldo, J.R.; Bonnard, G.D.; Kind, P.O.; Herberman, R.B.: Cytotoxicity by cultured human lymphocytes: characteristics of effector cells and specificity of cytotoxicity. J. Immun. *122:* 1489 (1979).

196 Ortaldo, J.R.; Bonnard, G.D.; McDermott, R.P.; Kind, P.H.; Herberman, R.B.: Cytotoxicity from cultured cells: analysis of precursors involved in generation of human cells mediating natural and antibody-dependent cell-mediated cytotoxicity (submitted).

197 Ortiz de Landazuri, M.; Herberman, R.B.: Specificity of cellular immune reactivity to virus-induced tumors. Nature new Biol. *238:* 18 (1972).

198 Pacheco, D.; Falcoff, R.; Catmot, L.; Flock, F.; Werner, G.; Falcoff, E.: Inhibitory effect of interferon on DNA and RNA synthesis in murine spleen cells stimulated by lectins. Ann. Immunol., Paris *127C:* 163 (1976).

199 Paige, C.J.; Figarella, E.F.; Cuttito, M.J.; Cahan, A.; Stutman, O.: Natural cytotoxic

cells against solid tumors in mice. II. Some characteristics of the effector cells. J. Immun. *121:* 1827 (1978).

200 Patek, P.Q.; Collins, J.L.; Cohn, M.: Transformed cell lines susceptible or resistant to in vivo surveillance against tumorigenesis. Nature, Lond. *276:* 510 (1978).

201 Perlman, P.; Cerottini, J.C.: Cytotoxic lymphocytes. The antigens, vol. 5 (Academic Press, New York, in press).

202 Petranyi, G.; Kiessling, R.; Klein, G.: Genetic control of 'natural' killer lymphocytes in the mouse. Immunogenetics *2:* 53 (1975).

203 Petranyi, G.; Kiessling, R.; Povey, S.; Klein, G.; Herzenberg, L.; Wigzell, H.: The genetic control of natural killer cell activity and its association with in vivo resistance against a Moloney lymphoma isograft. Immunogenetics *3:* 15 (1976).

204 Plescia, O.J.; Smith, A.H.; Grinwich, K.: Subversion of immune system by tumor cells and role of prostaglandins. Proc. natn. Acad. Sci. USA *72:* 1848 (1975).

205 Pollack, S.B.; Tam, M.R.; Nowinski, R.C.; Emmons, S.L.: Presence of T cell associated antigens on murine NK cells. J. Immun. *123:* 1818 (1979).

206 Pross, H.F.; Jondal, M.: Cytotoxic lymphocytes from normal donors. A functional marker of human non-T lymphocytes. Clin. exp. Immunol. *21:* 226 (1975).

207 Pross, H.; Baines, M.: Spontaneous human lymphocyte-mediated cytotoxicity against tumor target cells. Cancer Immunol. Immunother. *3:* 75 (1977).

208 Pross, H.; Gupta, S.; Good, R.A.; Baines, M.G.: Spontaneous human lymphocyte-mediated cytotoxicity against tumor targets. VII. The effect of immunodeficiency disease. Cell. Immunol. *43:* 160 (1979).

209 Reid, L.M.; Jones, C.; Minato, N.; Bloom, B.R.; Holland, J.: Rejection of virus persistently infected tumor cells. I. Rejection in athymic mice. J. exp. Med. *149:* 1117 (1979).

210 Reid, L.M.; Jones, C.L.; Holland, J.: Virus carrier state suppresses tumoriginicity of tumor cells in athymic (nude) mice. J. gen. Virol. *42:* 609 (1979).

211 Reisner, Y.; Linker-Israeli, I.; Sharon, N.: Separation of mouse thymocytes into two subpopulations by the use of peanut agglutinin. Cell. Immunol. *25:* 129 (1976).

212 Reisner, Y.; Ravid, A.; Sharon, N.: Use of soybean agglutinin for the separation of mouse B and T lymphocytes. Biochem. biophys. Res. Commun. *72:* 1585 (1976).

213 Reisner, Y.; Ikehara, S.; Hodes, M.; Good, R.A.: Allogenic hemopoietic stem cell transplantation using mouse spleen cells fractionated by lectins: In vitro study of cell fractions. Proc. natn. Acad. Sci. USA *77:* 1164 (1980).

214 Riccardi, C.; Puccetti, P.; Santoni, A.; Herberman, R.B.: Rapid in vivo assay of mouse NK cell activity. J. natn. Cancer Inst. *63:* 1041 (1979).

215 Riesenfeld, I.; Orn, A.; Gidlund, M.; Axberg, I.; Alin, G.; Wigzell, H.: Positive correlation between in vitro and NK acitivity and in vivo resistance towards AKR lymphoma cells. Int. J. Cancer *25:* 399 (1980).

216 Roder, J.C.; Kiessling, R.: Target-effector interaction in the natural killer cell system. I. Co-variance and genetic control of cytolytic and target-cell-binding subpopulations in the mouse. Scand. J. Immunol. *8:* 135 (1978).

217 Roder, J.C.; Duwe, A.K.; Bell, D.A.; Singhal, S.K.: Immunological senescence. I. The role of suppressor cells. Immunology *35:* 837 (1978).

218 Roder, J.C.; Kiessling, R.; Biberfeld, P.; Anderson, B.: Target effctor interaction in the natural killer cell system. II. The isolation of NK cells and studies on the mechanism of killing. J. Immun. *121:* 3509 (1978).

219 Roder, J.C.; Duwe, A.K.: The beige mutation in the mouse selectively impairs NK cell function. Nature, Lond. *278:* 451 (1979).
220 Roder, J.C.; Rosen, A.; Fenyo, E.M.; Troy, F.A.: Target-effector interaction in the natural killer cell system: the isolation of target structures. Proc. natn. Acad. Sci. USA *76:* 1405 (1979).
221 Roder, J.C.: The beige mutation in the mouse. I. A stem cell predetermined impairment in natural killer cell function. J. Immun. *123:* 2168 (1979).
222 Roder, J.C.; Lohmann-Matthes, M.L.; Domzig, W.; Wigzell, H.: The beige mutation in the mouse. II. Selectivity of the natural killer (NK) cell defect. J. Immun. *123:* 2174 (1979).
223 Roder, J.C.; Kiessling, R.: A model of target cell recognition and lysis by natural killer cells. Adv. exp. Biol. Med. *114:* 745 (1979).
224 Roder, J.C.; Lohmann-Matthes, M.L.; Domzig, W.; Kiessling, R.; Haller, O.: A functional comparison of tumor cell killing by activated macrophages and natural killer cells. Eur. J. Immunol. *9:* 276 (1979).
225 Roder, J.C.; Klein, M.: Target-effector interaction in the natural killer cell system. IV. Modulation by cyclic nucleotides. J. Immun. *123:* 2785 (1979).
226 Roder, J.C.; Ahrlund-Richter, L.; Jondal, M.: Target-effector interaction in the human and murine natural killer cell system. Specificity and xenogeneic reactivity of the solubilized target structure complex and its loss in a somatic cell hybrid. J. exp. Med. *150:* 471 (1979).
227 Roder, J.C.; Karre, K.: The partial isolation of subcellular MHC products which are recognized by alloimmune T lymphocytes. Immunogenetics *10:* 343 (1980).
228 Roder, J.C.; Haliotis, T.; Klein, M.; Korec, S.; Jett, J.; Ortlado, J.; Herberman, R.B.; Katz, P.; Fauci, A.S.: A new immunodeficiency disorder in humans involving NK cells. Nature, Lond. *284:* 553 (1980).
229 Roder, J.C.; Argov, S.; Klein, M.; Petersson, C.; Kiessling, R.; Andersson, K.; Hansson, M.: Target-effector interaction in the natural killer cell system. V. Energy requirements, protein synthesis and the possible involvement of lysosomal enzymes. Immunology *40:* 107 (1980).
230 Roder, J.C.: Target-effector interactions in the natural killer (NK) cell system. VI. The influence of age and genotype on target binding characteristics. Immunology *41:* 483 (1980).
231 Roder, J.C.; Beaumont, T.; Kerbel, R.S.; Haliotis, T.; Kozbor, D.: A mutation (NK^R) which renders YAC cells selectively resistant to cytolysis by NK cells (submitted for publication).
232 Rosenberg, E.B.; Herberman, R.B.; Levine, P.H.; Halterman, R.H.; McCoy, J.L.; Wunderlich, J.R.: Lymphocyte cytotoxicity reactions to leukemia-associated antigens in identical twins. Int. J. Cancer *9:* 648 (1972).
233 Rotter, V.; Trainin, N.: Inhibition of tumor growth in syngeneic chimeric mice mediated by a depletion of suppressor T cells. Transplantation *20:* 68 (1975).
234 Rouse, B.T.; Grenwal, A.S.; Babiuk, L.A.: Complement enhances antiviral antibody-dependent cytotoxicity. Nature, Lond. *266:* 456 (1977).
235 Rozee, K.R.; Lee, S.H.S.; Ngan, J.: Effect of priming on interferon inhibition of Con A-induced spleen cell blastogenesis. Nature new Biol. *245:* 16 (1973).
236 Rygaard, J.; Povlsen, C.O.: Is immunological surveillance not a cell-mediated immune function? Transplantation *17:* 135 (1976).

237 Ryser, J.E.; Sordat, B.; Cerottini, J.C.; Brunner, K.T.: Mechanism of target cell lysis by cytolytic T lymphocytes. I. Characterization of specific lymphcyte and target cell conjugates separated by velocity sedimentation. Eur. J. Immunol. *7:* 110 (1977).

238 Sadan, N.; Yaffee, D.; Rozenszajn, L.; Adar, H.; Soroker, B.; Efrat, P.: Cytochemical and genetic studies in four cases of Chediak-Higashi-Steinbrinck syndrome. Acta haemat. *34:* 26 (1965).

239 St. Amand, W.; Cupp, M.: Mouse News Lett. *19:* 38 (1958).

240 Saksela, E.; Timonen, T.; Ranki, A.; Hayry, P.: Morphological and functional characterization of isolated effector cells responsible for human natural killer activity to fetal fibroblasts and to cultured cell line targets. Immunol. Rev. *44:* 71 (1979).

241 Santoli, D.; Trinchieri, G.; Leif, F.S.: Cell-mediated cytotoxicity against virus-infected target cells in humans. I. characterization of the effector lymphocytes. J. Immun. *121:* 526 (1978).

242 Santoli, D.; Trinchieri, G.; Koprowski, H.: Cell-mediated cytotoxity in humans against virus-infected target cells. II. Interferon induction and activation of natural killer cells. J. Immun. *121:* 532 (1978b).

243 Santoli, D.; Koprowski, H.: Mechanisms of activation of human natural killer cells against tumor and virus-infected cells. Immunol. Rev. *44:* 125 (1979).

244 Santoni, A.; Herberman, R.B.; Holden, H.T.: Correlation between natural and antibody-dependent cell-mediated cytotoxicity against tumor targets in the mouse. I. Distribution of the reactivity. J. natn. Cancer Inst. *62:* 109 (1979).

245 Savary, C.; Lotzova, E.: Supression of natural killer cell cytotoxicity by splenocytes from *Corynebacterium parvum*-injected, bone marrow-tolerant, and infant mice. J. Immun. *120:* 239 (1978).

246 Schmidt, M.; Good, R.A.: Cancer xenografts in nude mice. Lancet *i:* 39 (1976).

247 Schwartz, R.S.: Another look at immunological surveillance. Lancet *293:* 181 (1975).

248 Seaman, W.E.; Blackman, M.A.; Gindhart, T.D.; Roubinian, J.R.; Loeb, J.M.; Talal, N.: β-Estradiol reduces natural killer cells in mice. J. Immun. *121:* 2193 (1978).

249 Seaman, W.E.; Gindhart, T.D.; Greenspan, J.S.; Blackman, M.A.; Talal, N.: Natural killer cells, bone and the bone marrow: studies in estrogen-treated mice and in congenitally osteopetrotic (mi/mi) mice. J. Immun. *122:* 2541 (1979).

250 Seeley, J.K.; Golub, S.A.: Studies on cytotoxicity generated in human mixed lymphocyte cultures. I. Time course and target spectrum of several distinct concomitant cytotoxic activities. J. Immun. *120:* 1415 (1978).

251 Seeley, J.K. Masucci. G.; Poros, A.; Klein, E.; Golub, S.: Studies on cytotoxicity generated in human mixed lymphocyte cultures. II. Anti-K562 effectors are distinct from allospecific CTL and can be generated from NK-depleted T cells. J. Immun. *123:* 1303 (1979).

252 Sendo, F.; Aoki, T.; Boyse, E.A.; Buofo, C.K.: Natural occurrence of lymphocytes showing cytotoxic activity to Balb/c radiation-induced leukemia RL♂ cells. J. natn. Cancer Inst. *55:* 603 (1975).

253 Shellam, G.R.; Hogg, N.: Gross-virus-induced lymphoma in the rat. IV. Cytotoxic cells in normal rats. Int. J. Cancer *19:* 212 (1977).

254 Shur, B.D.; Roth, S.: Cell surface glycosyl-transferases. Biochim. biophys. Acta *415:* 473 (1975).

255 Siliciano, R.F.; Henney, C.S.: Studies on the mechanism of lymphocyte-mediated cytolysis. X. Enucleated cells as targets for cytotoxic attack. J. Immun. *121:* 186 (1978).

256 Skov, C.B.; Holland, J.M.; Perkins, E.H.: Development of fewer tumor colonies in lungs of athymic nude mice after intravenous injection of tumor cells. J. natn. Cancer Inst. *56:* 193 (1976).
257 Smith, K.A.; Cornwell, G.G.; McIntyre, O.R.: The effect of poly I:C and interferon on lymphocyte DNA synthesis; in Lindahl, Kiessling, Osoba, Lymphocyte recognition and effector mechanisms, p. 101 (Academic Press, New York 0000).
258 Snell, G.D.: Histocompatibility genes of the mouse. II. Production and analysis of isogenic-resistant lines. J. natn. Cancer Inst. *21:* 843 (1958).
259 Snell, G.D.; Stevens, L.C.: Histocompatibility genes of mice. III. H-1 and H-4, two histocompatibility loci in the first linkage group. Immunology *4:* 366 (1961).
260 Snell, G.D.: Recognition structures determined by the H-2 complex. Transplant. Proc. *8:* 147 (1976).
261 Stein, K.E.; Schwarting, G.A.; Marcus, D.M.: Glycolipid markers of murine lymphocyte subpopulations. J. Immun. *120:* 676 (1978).
262 Strander, H.; Cantell, K.; Carlstrom, G.; Ingimarsson, S.; Jakobsson, P.; Nilsonne, U.: Acute infections in interferon treated osteosarcoma patients. J. infect. Dis. *133:* suppl., p. A-245 (1976).
263 Strom, T.B.; Bear, R.A.; Carpenter, C.B.: Insulin induced augmentation of lymphocyte-mediated cytotoxicity. Science *187:* 1206 (1975).
264 Stutman, O.; Paige, C.; Figarella, E.: Natural cytotoxic cells against tumors in mice. I. Strain and age distribution and target cell susceptibility. J. Immun. *121:* 1819 (1978).
265 Takasugi, M.;, Mickey, M.R.; Terasak, P.I.: Reactivity of lymphocytes from normal persons on cultured tumor cells. Cancer Res. *33:* 2898 (1973).
266 Takasugi, M.; Mickey, M.R.: Interaction analysis of selective and non-selective cell-mediated cytotoxicity. J. natn. Cancer Inst. *57:* 255 (1976).
267 Takasugi, M.; Akira, D.; Takasugi, J.; Mickey, M.: Specificities in human cell-mediated cytotoxicity. J. natn. Cancer Inst. *59:* 69 (1977).
268 Takasugi, M.; Koide, Y.; Akira, D.; Ramseyer, A.: Specificities in natural cell-mediated cytotoxicity by the cross-competition assay. Int. J. Cancer *19:* 291 (1977).
269 Talmadge, J.E.; Meyers, K.M.; Prieur, D.J.; Starkey, J.R.: Role of natural killer cells in tumor growth and metastasis in beige mice. Nature, Lond. *284:* 622 (1980).
270 Tan, Y.H.; Creagan, R.P.; Ruddle, F.H.: The somatic cell genetics of human interferon: assignment of human interferon locito chromosomes 2 and 5. Proc. natn. Acad. Sci. USA *71:* 2251 (1974).
271 Thoma, J.A.; Thoma, G.J.; Clark. W.: The efficiency and linearity of the radiochromium release assay for cell-mediated cytotoxicity. Cell. Immunol. *40:* 404 (1978).
272 Totterman, T.; Hayry, P.; Saksela, E.; Timonen, T.; Eklund, K.: Cytological and functional analysis of inflammatory infiltrates in human malignant tumors. Functional investigation with the infiltrating inflammatory cells. Eur. J. Immunol. *8:* 872 (1978).
273 Tovey, M.G.; Rochette-Egly, C.; Castagna, M.: Effect of interferon on concentrations of cyclic nucleotides in cultured cells. Proc. natn. Acad. Sci. USA *76:* 3890 (1979).
274 Trinchieri, G.; Santoli, D.; Knowles, B.: Tumor cell lines induce interferon in human lymphocytes. Nature, Lond. *270:* 611 (1977).
275 Trinchieri, G.; Santoli, D.; Koprowski, H.: Spontaneous cell-mediated cytotoxicity in humans. Role of interferon and immunoglobulins. J. Immun. *120:* 1849 (1978).
276 Trinchieri, G.; Santoli, D.; Dee, R.R.; Knowles, B.: Antiviral acitity induced by culturing lymphocytes with tumor-derived or virus-transformed cells. Identification of

the antiviral activity as interferon and characterization of the human effector lymphocyte subpopulation. J. exp. Med. *147:* 1299 (1978).

277 Trinchieri, G.; Santoli, D.: Antiviral activity induced by culturing lymphocytes with tumor-derived or virus-transformed cells. Enhancement of human natural killer cell activity by interferon and antagonistic inhibition of susceptibility of target cells to lysis. J. exp. Med. *147:* 1314 (1978).

278 Tsoukas, C.D.; Rosenau, W.; Baxter, J.D.: Cellular receptor for lymphotoxin: correlation of binding and cytotoxicity in sensitive and resistant target cells. J. Immun. *116:* 184 (1976).

279 Van Furth, R.; Hirsch, J.G.; Fedorko, E.: Morphology and peroxidase cytochemistry of mouse promonocytes, monocytes, and macrophages. J. exp. Med. *132:* 794 (1970).

280 Virelizier, J.L.; Allison, A.C.; Maeyer, E. de.: Production by mixed lymphocyte cultures of type II interferon able to protect macrophages against virus infections. Infect. Immunol. *17:* 282 (1977).

281 Virelizier, J.L.; Lipinski, M.; Tursz, T.; Griscelli, C.: Defects of immune interferon secretion and natural killer activity in patients with immunological disorders. Lancet *ii:* 696 (1979).

282 Vose, B.M.; Vanky, F.; Klein, E.: Lymphocyte cytotoxicity against autologous tumor biopsy cells in humans. Int. J. Cancer *20:* 512 (1977).

283 Vose, B.M.; Vanky, F.; Klein, E.: Human tumor lymphocyte interaction in vitro. V. Comparison of the reactivity of tumor-infiltrating blood and lymph-node lymphocytes with autologous tumor cells. Int. J. Cancer *20:* 895 (1977).

284 Vose, B.M.; Vanky, F.; Argov, S.; Klein, E.: Natural cytotoxicity in man: activity of lymph node and tumor-infiltrating lymphocytes. Eur. J. Immunol. *7:* 753 (1977).

285 Walberg, G.; Hiemstra, K.; Burge, J.J.; Singler, R.C.: Reversible inhibition of lymphocyte-mediated cytolysis by dimethyl sulfoxide (DMSO). J. Immun. *111:* 1435 (1973).

286 Warner, N.L.; Woodruff, M.F.; Burton, R.C.: Inhibition of the growth of lymphoid tumors in syngeneic athymic (nude) mice. Int. J. Cancer *20:* 146 (1977).

287 Weissmann, G.; Goldstein, I.; Hoffstein, S.; Tsung, P.K.: Reciprocal effects of cAMP and cGMP on micro-tubule-dependent release of lysosomal enzymes. Ann. N.Y. Acad. Sci. *253:* 750 (1975).

288 Welsh, R.; Zinkernagel, R.M.: Heterospecific cytotoxic cell activity induced during the first 3 days of acute lymphocytic choriomeningitis virus infection in mice. Nature, Lond. *268:* 646 (1977).

289 Welsh, R.: Cytotoxic cells induced during lymphocytic choriomeningitis virus infection of mice. I. Characterization of natural killer cell induction J. exp. Med. *148:* 161 (1978).

290 Welsh, R.; Kiessling, R.W.: Natural killer cell response to lymphocytic choriomeningitis virus in beige mice. Scand. J. Immunol. *11:* 363 (1980).

291 Welsh, R.; Kiessling, R.: Modification of target susceptibility to activated mouse NK cells by interferon and virus infections; in Herberman Natural cell-mediated cytotoxicity p. 963 (Academic Press, New York 1980).

292 Williams, A.F.: Differentiation antigens of the lymphocyte cell surface. Contemp. Top. mol. Immunol. *6:* 83 (1977).

293 Windhorst, D.; Padgett, G.: The Chediak-Higashi syndrome and the homologous trait in animals. J. invest. Derm. *60:* 529 (1973).

294 Winn, H.J.: Immune mechanisms in homotransplantation. J. Immun. *86:* 228 (1961).

295 Wolfe, S.A.; Tracey, D.E.; Henney, C.S.: Induction of 'natural killer' cells by BGG. Nature, Lond. *262:* 584 (1976).
296 Woodruff, M.A.; Klarner, N.L.: Effect of *Corynebacterium parvum* on tumor growth in normal and athymic (nude) mice. J. natn. Cancer Inst. *58:* 11 (1977).
297 Yefenof, E.; Klein, G.; Kvarnun, K.: Relationships between complement activation, complement binding, and EBV adsorption by human hematopoietic cell lines. Cell. Immunol. *31:* 225 (1977).
298 Young, W.; Hakomori, S.; Durdik, J.; Henney, C.S.: Identification of ganglio-N-tetraosylceramide as a new cell surface marker for murine natural killer (NK) cells. J. Immun. *124:* 199 (1980).
299 Zagury, D.; Bernard, J.; Thierness, N.; Feldmann, M.; Berke, G.: Isolation and characterization of individual functionaly reactive cytotoxic T lymphocytes: conjugation, killing and recycling at the single cell level. Eur. J. Immunol. *5:* 818 (1975).
300 Zarling, J.M.; Nowinsky, R.C.; Bach, F.H.: Lysis of leukemia cells or normal mice. Proc. natn. Acad. Sci. USA *72:* 2780 (1975).

Addendum

After the preparation of this review, several new and exciting developments have occurred. NK cells have been cloned [*Dennert; Nabel*] and will therefore be available as pure and homogeneous populations for studies of polyclonality, specificity, surface phenotype, and in vivo transfer. In other studies NC-and NK-mediated lysis of target cells was blocked by simple sugars which suggests the NK recognition structure may be lectin-like [*Stutman*]. The target structures appear to be present on immature cells but may be lost upon further differentiation [*Stern*].

References

Dennert, G.: Cloned lines of natural killer cells. Nature, Lond. *287:* 47 (1980).
Nabel, G.; Fresno, M.; Chessman, A.; Cantor, H.: Use of cloned populations of mouse lymphocytes to analyze cellular differentiation. Cell (in press).
Stern, P.; Gidlund, M.; Orn, A.; Wigzell, H.: Natural killer cells mediate lysis of embryonal carcinoma cells lacking MHC. Nature, Lond. *285:* 341 (1980).
Stutman, O.; Dien, P.; Wisun, R.; Lattime, E.: Natural cytotoxic cells against solid tumors in mice: Blocking of cytotoxicity by D-mannose. Proc. natn. Acad. Sci. USA *77:* 2895 (1980).

J.C. Roder, Department of Microbiology and Immunology, Queen's University, Kingston, Ont. K7L 3N6 (Canada)

Prog. Allergy, vol. 28, pp. 160–192 (Karger, Basel 1981)

Autologous and Syngeneic Mixed Lymphocyte Reactions and Their Immunological Significance[1]

Jack R. Battisto and Nicholas M. Ponzio

Department of Immunology, Research Division, Cleveland Clinic Foundation, Cleveland, Ohio, and Departments of Microbiology-Immunology and Pathology, and the Cancer Center, Northwestern University Medical and Dental Schools, Chicago, Ill., USA

Contents

[1] This work was supported by a grant from the National Institute of Allergy and Infectious Diseases, AI 12468 (to J.RB.) and by USPHS Grants CA22544 and AI 15247 (to N.M.P.). N.M.P. is also a recipient of Research Career Development Award CA00612 from the National Cancer Institute, NIH.

I. Introduction

Since the days of *Ehrlich*, [1] immunity to autologous antigens has been thought to be a highly abnormal, some time event. The clonal selection theory as proposed by *Burnet* [2,3] suggested that natural tolerance to self-antigens develops from the ontogenic elimination of potentially self-reactive lymphocytes. When an autoimmune reactivity was observed it was rationalized as being directed toward a self-antigen that was normally sequestered. In actuality, the immunological system has been found to display reactivity toward self-antigens through a number of isogeneic lymphocytic interactions (ILI) without apparent detriment to self [4–8]. One of the objectives of this presentation is to review each of the ILI systems, with a view to possibly gaining insights into the implications that such reactivities may have for the overall body economy, for the immunological system, as well as for certain specific entities such as autoimmune disease, immunological surveillance, and immunological responsiveness.

II. Murine Isogeneic Lymphocyte Interactions

A. Earliest Findings

Interaction of isogeneic cells have been described for both in vitro and in vivo systems. *Llacer and Uyeki* [4] reported that proliferative activity occurred in mixtures of autochthonous spleen and thymus cells from adult rats. That this phenomenon is not common to all species is apparent from the fact that such cells taken from adult mice do not show such replicative behavior. A further difference between these two species was pointed out by *Schwarz* [9] who has demonstrated that adult rat thymocytes will also respond in vitro to allogeneic cells while others have shown that those of the adult mouse will not [6].

In vivo reactivity between isogeneic cells of mice has been suggested by *Playfair and Krsiakova* [5] who described significant graft-versus-host reactions occurring in syngeneic recipients of lymphoid cells from 2-month-old thymus or 1-month-old spleen.

B. Syngeneic MLR (ILI Type 1)

1. Characterization

One of the more extensively studied instances of ILI in laboratory animals is that originally described by *Howe* et al. [6, 10]. In this system, thymus cells of neonatal CBA/J mice replicate when cultured in the presence of syngeneic, normal, adult spleen cells. The capability of thymic cells to divide when co-cultured with adult spleen cells was seen to be poor for embryonic thymic cells, to increase sharply for thymic cells taken shortly after birth, to begin to decline within the first week of life and to be lost by adulthood (fig. 1). The authors suggested ILI may represent an in vitro model for cognitory and discriminatory cellular events occurring routinely in vivo [6]. They called attention to the fact that the loss of ability to recognize self-antigens in the aging thymus could reflect an acquisition of immunological tolerance as *Burnet* had suggested and that the loss of responsiveness to foreign antigens by thymocytes could be due to relocation of the responsible cells in peripheral lymphoid organs. The marked increase in reactivity of thymic cells to self-antigens that occurs just after birth might be explained by the influence of thymic hormones upon immature cells, the antigenic effects of newly established intestinal flora, and/or the influences of maternal antigens or hormones present in colostrum.

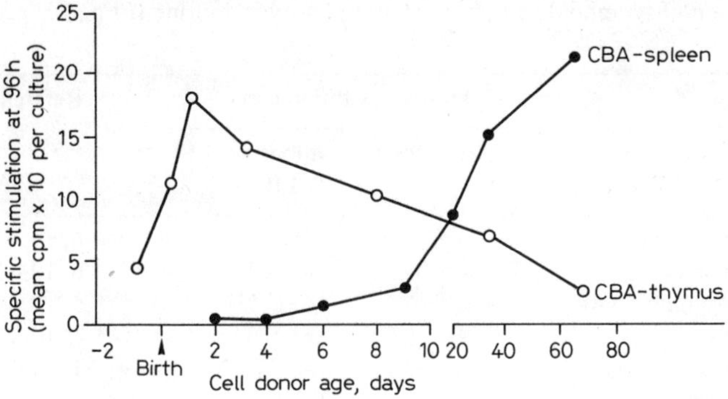

Fig. 1. Effect of cell donor age on the reactivity of CBA/J spleen and thymus cells in the ILI. 1×10^6 thymocytes from donors of various ages were cultured with 1×10^6 mitomycin-treated spleen cells from adult donors (○ – ○), 1×10^6 mitomycin-treated spleen cells from donors of various ages were cultured with 1×10^6 thymus cells from 2-day-old donors (● – ●). Each point represents the mean cpm of triplicate mixed cultures minus the mean cpm of appropriate controls. Reproduced by permission of the publisher, from *Howe* [11].

The stimulatory cells came under scrutiny as well, for these investigators showed that those of the spleen displayed this capacity best [6, 10]. In an extended study, *Howe* [11] showed that the stimulatory capacity of neonatal spleen cells is minimal but that it increases with increasing age of the cell donor (fig. 1). An evaluation of the capacity of various adult lymphoid tissues to stimulate in the ILI indicated that cells from the spleen and to a lesser extent cells from lymph nodes are capable of stimulation while those from the thymus and bone marrow are devoid of stimulatory activity [11–13]. *Howe* et al. [6, 10] speculated that the disparity in the stimulatory activity of spleen cells as compared to that of isogeneic cells from other sites may reflect quantitative or qualitative variations of the cellular antigens that elicit ILI. This will be discussed in section II.D. These workers also observed that spleen cells from thymectomized mice possessed reduced capability to evoke proliferation of the neonatal thymus cells [10, 14], suggesting that the thymus influenced expression of the requisite marker on the splenic cells.

2. Identity of the Stimulator Cells

Von Boehmer and Byrd [12, 13] were first to suggest that B cells are stimulatory in ILI. Indeed, their work indicated that whereas B cells are stimulatory in ILI, T as well as B cells are stimulatory for allogeneic

Table I. Comparison of lymphoid cells that are stimulatory in murine ILI and allogeneic MLR

Stimulator cell source	Ability to stimulate in		References
	ILI type 1	allogeneic MLR	
Unseparated neonatal spleen	–	ND	6, 10, 11
Unseparated adult spleen	+	+	6, 10–13
Unseparated adult lymph nodes	±	+	6, 10, 11
Adult bone marrow	–	±	6, 10–13
Adult B lymphocytes	+	+	12, 13
Adult T lymphocytes	–	+	16–18

ND = Not determined.

thymus cells from both neonatal and adult mice [15]. When 90% pure peripheral T cells were used as stimulating cells, only marginal stimulation of syngeneic neonatal thymus cells resulted, indicating that T cells alone are ineffective in stimulating syngeneic thymus cells (table I). However, since these T cells are stimulatory for allogeneic thymocytes, the authors suggested that stimuli other than those responsible for allogeneic stimulation induce proliferation of syngeneic thymus cells [16–18].

3. Genetics of Antigen(s) Responsible for ILI Type 1

Howe [11], through examination of the progeny of matings between CBA/J and A/J mice, called attention to data that indicated that capacity of spleen cells to stimulate in ILI was probably controlled by more than one gene. On the other hand, *Carter and Wegmann* [19] provided data showing the antigen to be regulated by one gene. They found that C3H/HeJ mice, while closely related to CBA/J mice, lack the spleen cell antigen which triggers ILI and by mating experiments between these mice showed that the antigen, which they termed Spl-1, segregates as a simple Mendelian trait.

The confusion that these two reports introduce relative to the heritance of the antigen that is stimulatory in ILI may be due to the fact that neonatal thymic cells were subsequently shown to be responding to two stimulatory antigens. This suggestion arises from the finding of *Ponzio* et al. [20], that adult murine lymph node cells respond blastogenically to a separate differentiation antigen on isologous and autologous splenic lymphocytes. This marker also contributes to the stimulation of neonatal thymic cells. Thus, the heritability of the major marker responsible for triggering neonatal thymic cells might well bear re-investigation. Furthermore, the marker responsible for stimulating adult lymph node T cell replication has not yet been studied from a genetic viewpoint.

C. ILI Type 2

The discovery of a second instance of murine ILI [20] broadened the idea that anti-self-clones do exist normally in the peripheral lymphoid system and heightened the necessity for adequate explanation.

1. Age Relatedness

This second ILI is best observed when both responder and stimulator cells are derived from donors greater that 10 weeks of age. Responsive cells appear restricted to lymph nodes since they could not be detected in adult spleen, bone marrow or thymus. Lymph node cells do not require the presence of spleen in order to differentiate into responder cells, since those derived from neonatally splenectomized mice are fully active. This suggests that responder cells either develop through inherent stimuli or as a consequence of exposure to environmental antigens cross-reactive with self-determinants. Furthermore, responder cells in this second ILI are known to be thymus-derived because treating lymph node cells with anti-θ-serum and complement deleted the capacity to react to isogeneic spleen cells. Since that initial finding, others have shown that when separated splenic T cells are used as responders, similar autologous or syngeneic mixed lymphocyte reaction (MLR) results [21, 22].

Stimulator cells in ILI type 2 bear Ig on their surfaces and appear in the spleen but not in the thymus, bone marrow or lymph nodes. The likelihood that Ig itself is the antigen being measured in either type of ILI seems remote since high densities of surface Ig appear early on bone marrow cells and the latter are not stimulatory in either type of ILI [20].

2. Similarities and Differences between Types 1 and 2 ILI

The similarities of ILI types 1 and 2 are that the responder cells in both instances are T cells (table II). Furthermore, the stimulator cells in both ILIs are non-T cells that more than likely are B cells, macrophages and/or dendritic cells. Experiments comparing the responsiveness of adult lymph node cells and that of neonatal thymic cells towards mitomycin C-treated lymphoid cells from a variety of sources suggest the presence of two isoantigens on B lymphocytes. Since both antigens were shown to be absent

Table II. Comparison of ILI type 1 and 2

Characteristic	ILI type 1	ILI type 2
Cell donor's age		
Responder	neonate (1–3 days)	>10 weeks
Stimulator	>2 weeks	>10 weeks
Responder cell		
Organ source	thymus	LN, spleen
Cell markers	Thy-1+	Thy-1+ Ly-1+, 2−
Stimulator cells		
Organ source	spleen LN	spleen, LN, BM, Peritoneal Exudate
Cell types	B	B, macrophage, dendritic cell
cell markers	Ig+, Ia+	Ig+ Ia+
Possible stimulating Antigen	Ia MDA-1 and 2	Ia MDA-2
Level of proliferative response	SI ≥ 20	SI = 2–6
Sequelae of response		
Cytotoxic potential	none	none
Suppressor function	ND	+
Helper function	ND	+

on bone marrow cells and develop with time, they have been classified as murine differentiation antigens (MDA) 1 and 2. Whereas both appear in the spleen, only one, MDA-1, is also detectable by ILI on lymph node cells. Both MDA-1 and 2 activate neonatal T cells, but MDA-2 triggers only adult lymph node T cells. One of the striking dissimilarities between the two ILI systems is the degree of replication that is detectable by the tritiated thymidine uptake method. Whereas the stimulation indices of ILI type 1 are generally in excess of 20, those for ILI type 2 rarely exceed 6. Other differences between the antigens that are stimulatory in the two ILI became apparent through maturational studies of their appearance on B cells [20].

D. Maturational Studies of Both Stimulatory Markers

1. Detection with Time

Since bone marrow cells are not stimulatory for either ILI they were used to reconstitute lethally X-irradiated mice and those cells that

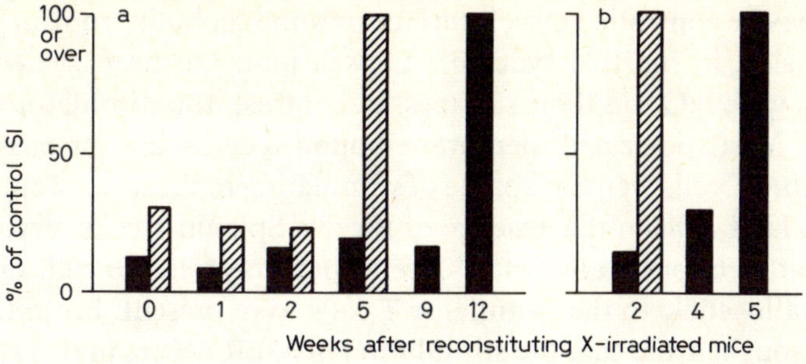

Fig. 2. Acquisition of murine differentiation antigens (MDA) by BM cells. Stimulation of Tn (■) by MDA-1 and of LN_a (▨) by MDA-2 on cells from spleens or X-irradiated, reconstituted mice. *a* Reconstituted with BM cells only: *b* reconstituted with BM and splenic T cells. Control stimulation of Tn by mitomycin-C-treated adult spleen cells at 72 h was 13,425 cpm and the sum of background cpm of responder and stimulator cells grown alone was 424. Thus, the stimulation index is 29.0. Adult LN cells stimulated by the same cells gave 1,012 cpm and the sum of the background cpm was 306; the resulting SI was 3.3. Reproduced by permission of the publisher, from *Battisto* [25].

migrated to the spleen were taken at various time intervals to be stimulators of both ILI. By this method MDA-2 became detectable on the splenic cells within 5–6 weeks, whereas MDA-1 became apparent only after 12 weeks [23, 24]. This observation suggested but did not establish that the markers might be separate entities (fig. 2).

2. Input of Splenic T Cells

Reconstituting X-irradiated mice with certain tissues and cells in addition to bone marrow cells accelerated the appearance of both MDA-1 [23] and MDA-2 [24, 25]. Splenic T cells, but not thymocytes, caused MDA-1 to be fully detectable within 5 weeks [23] and MDA-2 within 2 weeks [24]. Furthermore, bone marrow cells appeared to express MDA-1 partially when thymic supportive tissue was used. This information fits well with the earlier data of *Howe* et al. [10] who showed splenic cells from thymectomized mice were less stimulatory than normal spleen cells. Thus, the assistance of mature T cells appears to be required for MDA-1 expression on lymph nodal as well as splenic stimulators of ILI. What they provide to the maturing B cells is presently unknown.

3. Function of Surface Ig on Stimulator Cells

Studies in which subsets of B cells were serologically removed prior to using the remainder as stimulator cells suggested that isogeneic and allogeneic MLR may be controlled by cell surface immunoglobulin tropism [26]. Cellular division in the two types of ILI was found to be triggered by lymphocytes with IgG_1 on their surfaces. In contrast, the stimulator cells in allogeneic MLR possessed membrane-bound IgG 2A and/or possibly IgG 2B. Splenic T cells were incapable of stimulating replication of splenic or lymph nodal T cells in the absence of B cells. Splenic T cells served as weak stimulators of other allogeneic T cells, but only when B cells either isogeneic or allogeneic to the stimulator T cells were present. From these studies the thought arose that the stimulation in MLR occurs in two steps. First, immunoglobulins on cell surfaces appear to function to bring appropriate subsets of cells together. Next, antigenic recognition that results in blastogenesis occurs. Furthermore, the tropism or attraction that certain immunoglobulins have for some cell types may determine which subsets of cells participate in allogeneic MLR and which take part in isogeneic MLR [26].

E. How Widespread Is Murine ILI?

1. Which Strains of Mice Show ILI?

From the very beginning of the studies on ILI an important question has been whether the reaction is a general one. That is, do all strains of mice show the reaction and, if not, which are the ones that do? An extensive study conducted by *Howe* [11] indicated that most strains of mice exhibit weak ILI type 1, i.e. SI 1.8–2.9 in four strains. This result is quite unlike the response seen in CBA/J mice when the SI was 58. That study was done before the description of ILI type 2 and the question that naturally arises is whether the weak ILI responses are simply detecting MDA-2. The strong ILI type 1 response of CBA/J mice is attributable to stimulation of thymic cells by MDA-1 and 2 [23]. If this were true, one might infer that MDA-1 was lacking from the stimulatory B cells of mice from most strains (table III).

Smith and Pasternak [21] have recently studied the occurrence of syngeneic MLR (ILI type 2) in a number of mouse strains. Mice of all of the strains tested, except the NZB strain, exhibited the response. Thus, the replication of cells that experience antigen on other 'self' cells appears to be the norm. The absence of the response among NZB cells was suggested to be due to the pathogenetically related autoimmune disease that these mice display. This observation in mice is analogous to the defective autologous MLR observed in humans with systemic lupus erythematosus

Table III. Strain distribution of ILI types 1 and 2

Strain	H-2 haplotype	Demonstrated responsiveness	
		ILI type 1	ILI type 2
CBA/J	k	yes	yes
CBA/N[1]	k	no	yes
SJL/J	s	yes	yes
C57BL/6	b	yes	yes
BALB/c	d	yes	yes
A.TL	tl	ND	yes
A.TH	t2	ND	yes
NZB[1]	d	ND	no
MLR/1[1]	k	ND	no
BXSB/mp[1]	b	ND	no

[1]Various immune abnormalities have been demonstrated in these strains.

(SLE) to be described in a later section. Additionally, others [22, 27] have shown that there is a strong correlation between decreased syngeneic MLR and the genetically determined autoimmunity of mice. The decreased replication apparently resides in a defect of the responder T cell population and not in the stimulator cells.

The normally weak T cell proliferative response that is characteristic of ILI type 2 (relative to the response of ILI type 1) has been shown to be markedly enhanced by inclusion into the culture medium of the hydrophilic polymer, polyethylene glycol (PEG) [22, 28–30]. PEG is not a polyclonal activator of T or B lymphocytes, but is thought to function by promoting cellular interactions that normally occur between T cells, B cells and macrophages.

2. Special Case of CBA/N B Cells

An instance in which MDA-1 appears to be absent while MDA-2 is present has been reported by Dustoor et al. [31], who have shown that B cells from CBA/N mutant mice do not stimulate ILI type 1. Thymocytes from CBA/N neonatal mice were seen to be functional since they replicated when co-cultured with CBA/J and CBA/H spleen cells. Furthermore, thymocytes from $(CBA/N{\times}CBA/J)F_1$ and $(CBA/N{\times}DBA/2)F_1$ neonates were not stimulated by CBA/N spleen cells but were triggered by cells from the CBA/J parent. This indicated that CBA/N spleen cells probably lack the MDA-1-bearing subpopulation of B cells. While CBA/N spleen cells were shown to be inoperative in ILI type 1, they were fully stimulatory in ILI type 2. This observation suggests that MDA-1 and MDA-2 may be present on different subsets of B cells.

F. Identity of the Stimulatory Markers

1. Elimination of Viral Identity

Of some importance is knowledge of whether the MDAs represent self-antigens or alternatively are antigens acquired from the environment. Preliminary experiments of two sorts have indicated that the MDAs bear no identity with murine leukemia viruses (MuLV). The ILI are not eliminated or blocked with antibodies directed to MuLV glycosylated-protein 70 (gp 70). Furthermore, MuLV has not been recovered from ILI mixtures after replication has been allowed to progress to the maximum [*Del Villano* et al., unpublished data].

2. MDA-Specific Antibodies

While antibodies have not as yet been developed specifically for the MDAs, these would be invaluable in a number of different types of experiments. Attempts at producing such immunoreagents are in progress.

3. Are Both Markers on the Same B Cell?

Thus far, whenever MDA-1 has been shown to be present, MDA-2 has been detected as well. Thus, while MDA-2 can be seen to appear in the absence of MDA-1 there is uncertainty when both appear, whether they are present on the same set of B cells. The two instances where MDA-2 can be shown to be present in the absence of MDA-1 are on CBA/N splenic B cells and on spleen cells of X-irradiated CBA/J mice that have been reconstituted with BM cells for 5–6 weeks

Recently, it has been shown by *Steinman and Witmer* [32] and by *Ahmann* et al. [33] that non-T radiation-resistant splenic adherent cells are the predominant stimulators in murine allogeneic MLR while *Nussenzweig and Steinman* [34] have shown the same to be true for the syngeneic MLR. *Steinman and Witmer* [32] and *Nussenzweig and Steinman* [34] have characterized the potent stimulator of both types of MLR as a dendritic cell. No test of whether dendritic cells provoke syngeneic neonatal thymic cells to divide has yet been done. In syngeneic MLRs splenic adherent cells, thioglycollate-induced peritoneal exudate cells and normal peritoneal cells are all known to stimulate syngeneic lymph node T cells in the presence of PEG [28]. Thus, macrophages, B cells, and dendritic cells are all capable of acting as stimulators of syngeneic and autologous T cell replication. The question naturally arises: Are the antigens that control stimulation present in the same amount on all three sorts of cells?

4. Association of MDAs with Ia Antigens

Meo et al. [35] were first to show that anti-Ia sera were able to inhibit the murine *allogeneic* MLR. That the isogeneic MLRs were also inhibitable by anti-Ia serum was reported by *Finke and Battisto* [23]. This provided the suggestion that MDAs may be closely associated if not identical with Ia antigens. That Ia antigens expressed by the stimulator cells may also be involved in the induction of the T cell replication in the autologous and syngeneic MLRs has been suggested as well, by others [22, 28, 36]. Despite these findings, the possibility that the anti-Ia sera may have prevented proper cell-to-cell interaction through steric hinderance should not be overlooked.

G. Functions of the MDAs

1. Receptor for TRF?

Another suggestion that MDAs may be closely associated with Ia antigens has to do with studies in which a function for the MDAs was sought [24, 25, 31]. Data have been provided that show these differentiation markers on B cells may be involved with receipt of T cell helper signal. The helper factor was derived from an allogeneic supernatant and is antigen-non-specific, as well as non-Ia-bearing. Thus, it is akin to T cell-replacing factor (TRF). Bone marrow cells, known to lack both MDAs, were used to reconstitute X-irradiated CBA/J mice, and cells that homed to the spleen were examined by MLR at intervals to determine when the markers arose. In addition, purified splenic B cells from the reconstituted mice were exposed in vitro to sheep erythrocytes and TRF to determine their immunological capacity. Spleen cells obtained from mice 12 weeks after reconstitution with bone marrow cells were shown to have acquired MDA-1 and to have the capacity, following interaction with TRF, to produce the maximum number of cells synthesizing IgM and IgG antibodies to SRBC. Spleen cells examined 5–6 weeks following reconstitution expressed MDA-2 but were unresponsive to TRF.

In separate experiments, bone marrow-derived cells were found to mature earlier in the presence of splenic T cells, they expressed MDA-1 and were able to interact with TRF at 5–6 weeks. On the other hand, bone marrow-derived cells that acquired MDA-2 in the presence of splenic T cells at an earlier, 2-week interval, still remained unable to interact with TRF. These two correlations of marker appearances with cellular function suggest MDA-1 may be a receptor for TRF [24, 25, 31].

2. Corroboration of the Receptor Hypothesis by CBA/N B Cells

Corroborative evidence of another sort was derived from studies performed with B cells from CBA/N mutant mice. Immunological studies of these mice have shown that they are unresponsive to T cell-independent antigens [37] and apparently lack a late-maturing B cell population [38]. As has been indicated above, B cells from such mice fail to stimulate neonatal T cells in ILI type 1. During in vitro immunization with SRBC, CBA/N B cells were unable to interact with TRF [31]. The absence of these two functions in CBA/N mice thus lends further support to the hypothesis that MDA-1 may be a receptor for TRF.

Additional evidence for this function is derived from experiments in which TRF has been incorporated into ILIs of each type. TRF has been seen to cause inhibition of ILI type 1, but not type 2, by a factor of 5-fold or greater [*Dustoor* et al., unpublished observations]. Thus, MDA-1, but not MDA-2, appears to interact with a component(s) in TRF.

3. Helper Factors Derived from ILI

Supernatants from ILI type 1, when examined for their ability to serve as helper factor for in vitro cell-mediated lympholysis, were found to lack the second signal required for the maturation of pre-killer cells into fully fledged cytolytic T cells [*Dustoor* et al., unpublished observations]. However, when the autologous MLR was conducted in cultures containing PEG (an agent that enhances autologous MLR) plus heat-treated allogeneic stimulator spleen cells, an allo-specific cytotoxic response was generated. In the absence of the enhanced autologous MLR, heat-treated cells were unable to stimulate a cytotoxic response [39]. Interleukin-2 has since been shown to be produced in the syngeneic mixed lymphocyte reaction [40].

H. Knowledge of the T Cells that Replicate in the Two ILIs

1. Identity of the T Cells

While a great deal is known about the cells that act as stimulators in the ILIs, relatively little is known about the T cells that do the replicating. In early studies *Howe and Manziello* [14] showed that selective elimination of immature thymocytes by pretreatment of donor animals with cortisone acetate resulted in increased response of thymus cell suspensions in ILI type 1. This suggested that the more 'mature' T cell within the neonatal thymus is the cell that replicates. The same sort of result was observed by *Pasternak* et al. [41] in the syngeneic MLR (ILI type 2). The latter group showed further that the surface phenotype of responder cells in the syngeneic MLR is Thy 1^+, Ly 1^+, 2, 3^-, and this has been confirmed [22, 36]. Thus far, the phenotype of the neonatal thymic cells that respond in ILI type 1 has not been determined. A question that naturally arises is whether the T cells that replicate in the two ILIs are identical. Furthermore, since neonatal T cells are known to respond to both MDA-1 and 2, one wonders whether discrete subsets of T cells recognize each MDA separately.

III. The Human Autologous MLR

The murine ILI has a counterpart in the human, referred to as the autologous MLR. That human self-responsive cells exist has also been shown by: (a) lymphoid cells that bind to autologous antigens; (b) antibodies that interact with autologous cells, and (c) lymphocytes that are cytotoxic for target cells expressing autologous determinants. Relative to the autologous MLR, several groups have identified T cells that are capable of proliferating when co-cultured with mitomycin-treated or X-irradiated autologous lymphoblastoid cell lines [42–45], mitogen-induced blast cells [46–48] or normal peripheral blood, non-T cells [49–51]. Since, in all of these instances, the stimulator cells have not been purposely modified in any way, the proliferation of autologous responder cells may represent recognition of, and response to, unaltered self-determinants.

The autologous MLR compares favorably with allogeneic MLR with respect to characteristics that are expected of an immune response, such as specificity and memory [52, 53]. After primary stimulation, peak tritiated thymidine (^{3}H-TdR) incorporation generally is observed 1–2 days later in the autologous than in allogeneic MLR. However, restimulation of cells recovered 10–14 days after primary autologous MLR (a time at which baseline levels of ^{3}H-TdR uptake are observed) with fresh autologous stimulators results in an accelerated proliferative response (peak ^{3}H-TdR uptake at 2–3 days) that follows kinetics apparent in secondary allogeneic MLR. Furthermore, T cells primed with autologous, non-T stimulators in primary MLR give higher proliferative responses when challenged with autologous, as compared to allogeneic, non-T stimulators in secondary culture.

Given the known collaboration of T cells with both macrophages and B cells in the production of antibody [54–58], as well as the need for histocompatibility between sensitized cytotoxic T cells and relevant targets [reviewed in 59, 60], such self-recognition at the T cell level appears to be an integral component of normal immune responses in general. Therefore, the autologous MLR may be a reflection of such normal cellular interactions among lymphoid cells. What purposes such T cell proliferation to autologous, non-T cell stimulation can serve will be discussed in subsequent sections.

A. Nature of the Stimulator Cell in Autologous MLR

1. Lymphoblastoid Cell Lines

Several groups have reported on the ability of lymphoblastoid cell lines (initiated from peripheral blood lymphocytes of normal human donors) to stimulate fresh autologous T cells [42–45, 61, 62]. B lymphoblastoid cell lines stimulate vigorous proliferation of fresh T cells, whereas T lymphoblastoid cell lines are unable to induce significant responses (table IV). Since the majority of such continuously proliferating lymphoid cell lines carry the Epstein-Barr (E-B) viral genome, a viral-encoded product has been suggested as the stimulatory determinant [61]. However, E-B virus negative B lymphoblastoid cells have been observed to stimulate autologous T cells equally as well as E-B virus positive B lymphoblastoid cell lines [48].

2. Mitogen-Induced Blast Cells

Blast cells (derived from cultures where PBL and normal donors were stimulated with various mitogens) have also been shown to stimulate autologous T cell proliferation [46–48]. In these experiments, both conconavalin A (Con A) and phytohemagglutinin (PHA) – induced blasts were stimulatory. Since both of these lectins are generally considered T cell

Table IV. Nature of stimulator cells in the human autologous MLR

Cell type	Capacity to cause replication of autologous T cells	References
Lymphoblastoid cell lines		
T	–	42–45
B	+	61, 62
ConA-induced blasts	+	46–48
PHA-induced blasts	+	46–48
Peripheral blood mononuclear cells		
Monocytes	+	64, 66
B cells	+	50, 64, 70
Null cells	+	50

mitogens, at first glance this would appear to contradict the experimental results showing that B, but not T, lymphoblastoid cell lines are capable of stimulation in autologous MLR. Still, proliferation in both Con A and in PHA-induced cultures, although dependent upon the presence of T cells, involves B cells, as well [51]. B lymphoblasts are present in such mitogen-induced cultures and most probably represent the stimulatory population. Use of purified B lymphoblasts as stimulators in the human autologous MLR has not been reported. In the murine counterpart, however, mitomycin-treated lypopolysaccharide (LPS)-induced and gradient purified B lymphoblasts stimulate significant proliferation of autologous T cells [Ponzio, unpublished observations].

3. Normal Peripheral Blood Mononuclear Cells

Several groups of workers [49–51, 63, 64] have employed mononuclear cells from peripheral blood of normal donors to study the human autologous MLR (table IV). After separation into T and non-T fractions on the basis of E-rosette formation, optimal autologous MLR were obtained when T cells were co-cultured with autologous, X-irradiated, non-T stimulator cells. Since the stimulators were composed of B cells, null lymphocytes and monocytes, further separation procedures were performed. Removal of macrophages on the basis of phagocytosis of carbonyl iron or adherence to plastic failed to diminish, and in most cases resulted in enhanced stimulatory potential of the recovered cells [50, 63]. Macrophages were, thus, regarded as non-stimulatory in autologous MLR.

Gottlieb et al. [64] have shown that adherent macrophages from human peripheral blood are capable of stimulating significant allogeneic MLR, but are non-stimulatory in autologous MLR. Opposing results have been reported by *Beale* et al. [65], who separated peripheral blood monocytes on Sephadex G-10 columns. These investigators demonstrated that monocytes produce the highest degree of autologous T cell stimulation when compared to null or to B cell stimulators. Likewise, *Raff* et al. [66], using a monoclonal antibody (Mac-120) that reacts with human monocytes, demonstrated that monocytes are stimulatory in human autologous MLR. Moreover, their data suggest that only the subpopulation of monocytes that are Mac-120-positive (approximately 34% of peripheral blood monocytes) can stimulate autologous T cells.

Controversy also exists regarding the stimulatory potential of B lymphocytes and null cells. Various separation procedures have been employed to investigate this problem. *Kuntz* et al. [50] separated non-T cells on the basis of rosetting with mouse erythrocytes (B cell-enriched [67]) or with antibody-coated human erythrocytes (K cell-enriched [68]). These authors found that the null (K) cell fracion obtained by either technique stimulated far more ^{3}H-TdR incorporation by autologous T cells than did the B cell-enriched population. *Gottlieb* et al. [64], separated non-T cells into SIgM^{+} B cells and sIgM^{-} residual cell populations and found the sIgM^{-} population induced greater replication than sIgM^{+} stimulators. Nevertheless, significant proliferative responses were obtained using the purified sIgM^{+} B cell population as stimulators in the autologous MLR.

As has been noted earlier, the studies of *Steinman and Witmer* [32], and *Nussenzweig and Steinman* [34] suggest a primary role of dendritic cells as stimulators of allogeneic, as well as syngeneic MLR in the mouse. However, since purified B cells are capable of stimulating autologous T cells [64, 69], dendritic cells do not appear to be an absolute requirement for this response.

Thus, the methods employed to separate peripheral blood mononuclear cells and the purity of separated populations appear critical to the final interpretation regarding the nature of the stimulator cells in the human autologous MLR. Overall, the published data suggest that monocytes, as well as B lymphocytes and null cells, can all serve as stimulators in autologous MLR. Identification of a common stimulatory determinant on these various cells would help explain their ability to activate autologous T cells. Once characterized, the quantitative expression of such a determinant might be correlated with relative stimulatory potential in autologous MLR and serve as the basis for the variability noted by the several investigators.

Although responder cells in the human autologous MLR are generally thought to be T lymphocytes, relatively little has been published on the nature of the T cell subpopulation that proliferates. Recently, however, *Hausman* et al. [70] have described two subpopulations of T cells that respond in autologous MLR to different stimulator cells. One subpopulation of T cells responds to macrophage-depleted, B cell-enriched stimulators, while a distinct T cell subpopulation proliferates in response to HLA-DRw-bearing macrophages. Deletion of T cells responsive to autologous macrophages (by suicide with BUdr and light treatment or by treatment with a monoclonal antibody prepared against T cell blasts isolated from autologous MLR) resulted in decreased ability of the remaining cells to respond to autologous macrophages, while minimally affecting their ability to respond to autologous B cells. In addition, deletion of T cells that responded to autologous macrophages also inhibited the T cell-dependent reactivity to several conventional antigens, as measured by proliferation and lymphokine release. These data suggest, therefore, that T cells responsive to autologous macrophages are also involved in the T macrophage interactions required for reactivity to T-dependent antigens and lends functional relevance to the autologous MLR. To what degree these findings relate to the previously described T cell populations responding in the two forms of murine ILI remains to be clarified.

Thus, procedures used to separate T responder populations for human autologous MLR may possibly be a source of the variability noted when purified, autologous, stimulator populations have been used. The nature of the T lymphocyte(s) responsive in autologous MLR may be further defined in the future through use of the monoclonal reagents now available to discriminate human T cell subpopulations.

B. Nature of the Stimulatory Determinant

HLA-D and I region-coded products have been suggested as the stimulatory antigens in human and murine *allogeneic* MLR, respectively [63, 71–75]. Likewise, using similar methods of investigation, Ia-like determinants (HLA-D region products) appear also to mediate proliferation of T cells in the human autologous MLR [50, 62, 63, 76, 77]. Thus, rabbit anti-human Ia reagents have been used to block the proliferative response of T cells to autologous, normal human peripheral blood non-T cells and to autologous lymphoblastoid B cell lines. In addition, *Smith* [63] has employed sera from multiparous women to compare the nature of stimulatory determinants in autologous and allogeneic MLR. These sera were able to kill B, but not T cells in a complement-dependent assay. When the same B cell-enriched population was used to stimulate both allogeneic and autologous T cells, some multiparous sera were able to block both responses, while others inhibited stimulation of allogeneic, but not autologous T cells. Although the specificities of these antisera were not fully delineated, and effects on macrophages could not be totally ruled out, these data suggest that different cell surface antigens may stimulate autologous and allogeneic T cells.

That human Ia-like antigens may not be the only stimulatory determinants in human autologous MLR is also suggested by the observation that leukemic B cells of chronic lymphocytic leukemia patients are poor stimulators of autologous T cells [78], despite the presence of Ia antigens [79].

C. Sequelae of Autologous MLR

1. Generation of Cytotoxicity

In the allogeneic MLR, T cells are generated that cause destruction of target cells expressing antigens present on original stimulator cells. The autologous MLR, on the other hand, does not give rise to cytotoxic cells. Most investigators who have studied the human, as well as the murine autologous, MLR fail to detect cells that are cytotoxic when tested against autologous or syngeneic targets [21, 28, 39, 63, 80, 81]. However, *Miller and Kaplan* [82] have shown that T cells activated in autologous MLR can lyse LPS-induced, but not unstimulated autologous B cells. They have suggested that the target antigens on dividing B cells may be expressed in a form that renders the cells more susceptible to lysis. In comparison with the cytotoxicity generated in allogeneic MLR, *Miller and Kaplan* reported levels that were significantly lower, even though effector to target ratios were considerably higher (75 to 100:1). To what degree exogenous serum factors contributed to the observed cytotoxicity was not investigated; however, the autologous MLR-induced cytotoxic cells did exhibit specificity with regard to target cell lysis.

Cytotoxic effector cells derived from autologous MLR have also been described by *Tomonari and Aizawa* [83] and *Tomonari* [84]. These cytotoxic cells were able to lyse various autologous, allogeneic and even xenogeneic target cells, including lymphoblastoid and non-lymphoid cell lines and mitogen-induced blast cells. The cytotoxic cells recovered from these autologous MLR showed characteristics of natural killer (NK) cells, based on several criteria. These include virtually identical target cell range, sensitivity to ammonium chloride lysis, and cold target inhibition. However, effector cells taken from autologous MLR were capable of killing one NK-resistant target, suggesting activation of other cytotoxic cells as well. Since SRBC-rosetting peripheral blood lymphocytes were used as responders in these experiments, the data suggest that T cells activated in autologous MLR may directly perform cytotoxic functions. However, another possibility is that the responding T cells act as amplifier cells that produce soluble helper factors and recruit precursors of both the cytotoxic T and the NK lineage. The latter explanation is supported by the recent observations that in the murine autologous MLR, responding T cells express the Ly-1^+, $2,3^-$ phenotype [30]. This marker is characteristic of helper, not killer, cells. Furthermore, soluble helper factors for cytotoxic T cell precursors have been identified in supernatants of human autologous MLR [85]. Finally, *Hausman and Stobo* [86] have provided evidence that human T cells activated in the autologous MLR by macrophages are required for reactivity to soluble antigen and can provide influences necessary for maximal immunoglobulin synthesis.

2. Correlation of Autologous MLR and Suppressor Cell Activity

Several investigators have noted the involvement of suppressor cells in the autologous MLR [87–90]. *Smith and Knowlton* [89] reported that T cells activated in human autologous MLR exerted suppressive effects on both the proliferative and cytotoxic responses of fresh autologous T cells against allogeneic stimulators. This appears reminiscent of the development of spontaneous suppressor activity in cultures of normal murine spleen cells [91, 92]. However, in the mouse studies, purified T cells cultured alone did not develop suppressor potential, suggesting that stimulation by autologous, non-T cells (autologous MLR) was necessary.

Two other lines of evidence support the involvement of suppressor cells in the autologous MLR. Patients with SLE fail to demonstrate significant autologous MLR [87, 88, 90, 93, 94]. *Sakane* et al. [87, 88] have shown that active SLE patients have a defect in the ability of non-T cells to stimulate, as well as a defect in the ability of T cells to respond in autologous MLR. Similar observations have been made in the murine counterpart of the SLE patient, the NZB mouse. *Smith and Pasternak* [21], reporting on the strain distribution of autologous MLR in the mouse, have found that NZB mice exhibit little or no reactivity. The correlation of autologous MLR and suppressor cells in these two instances is strengthened, since SLE patients, as well as NZB mice have been shown to lack Con A-inducible suppressor cells [2, 87, 88, 95, 96].

In this regard, recently, *Moody* et al. [97] have demonstrated an impaired autologous MLR in patients with acute infectious mononucleosis (IM). This defect was localized to the inability of non-T cells from acute IM patients to stimulate autologous T cells. The deficiency was transient, since non-T cells from convalescent IM patients induced good autologous MLR. In contrast, T cells from IM patients responded well to both autologous and allogeneic non-T stimulators. Non-T cells from IM patients, however, were able to stimulate allogeneic T cell responses [*Weksler,* personal communication]. As these authors point out, given the evidence that suppressor cells are activated in autologous MLR, the biologic relevance of an impaired autologous MLR with respect to autoimmune phenomena becomes obvious. The appearance of antibodies in IM patients that react with autologous erythrocytes and lymphocytes, as well as the presence of anti-nuclear antibodies and rheumatoid factors, can all be explained by lack of suppressor cell-mediated control of humoral, anti-self-responses.

D. Effect of Corticosteroids on Autologous MLR

In an examination of various humoral factors that might influence the autologous MLR, *Ilfeld* et al. [98] demonstrated that the presence of hydrocortisone (in amounts equivalent to those found in human plasma) markedly suppressed the autologous MLR without any suppressive effect on allogeneic MLR. *Yu* et al. [99] also noted a similar diminution in autologous MLR performed with PBL taken from normal donors 4 h after receiving an i.v. injection of methylprednisolone. The studies by *Ilfeld* et al. [98] indicate that early cellular events are affected by hyrocortisone, since addition or removal of the drug early after initiation of culture resulted in the most dramatic effects. The mechanism of suppression achieved by hydrocortisone was suggested to be at the level of the self-responsive clone(s) of T cells. In the studies by *Yu* et al. [99], however, where preincubation of responder and stimulator populations was carried out, a 4-hour preincubation of stimulators, but not of responder cells, with the steroid resulted in suppression of autologous MLR. These authors did not, however, assess the responder or stimulator capacity of steroid-preincubated cells in allogeneic MLR.

E. Special Case of Tumor Cells as Stimulators in Autologous MLR

The ability of neoplastic lymphoid cells to stimulate proliferation of autologous lymphocytes from patients in remission or of lymphocytes from normal, healthy HLA-identical siblings in MLR has been attributed, in many cases, to tumor-specific antigens. The intent of this discussion is not to review the extensive work done in this area, yet several points relative to the autologous MLR appear worthy to note.

Ludgate and Gough [100] have demonstrated significant stimulation of PBL by autologous lymphoma cells (or cellular extracts) derived from spleens of patients diagnosed with various stages of Hodgkin's disease and other lymphomas. In contrast, spleen cells (or cell extracts) from normal individuals or non-tumor patients did not stimulate autologous PBL in MLR. These authors concluded that the determinants stimulating proliferation of PBL from lymphoma patients were the tumor-specific antigens. From the data, however, one cannot rule out the possibility that this reflects an instance of autologous MLR.

Further, *Opelz* et al. [101] observed that cells from only one third of patients with leukemia induced significant proliferation when co-cultured with responder cells derived from HLA-identical sibling donors. This again raises doubt as to whether such in vitro reactivity reflects stimulation by antigens that are tumor-specific or normal and autologous.

Obviously this is an important point to decide, for a variety of reasons. If in vitro reactivity to autologous tumor cells represents a true tumor-specific response, immunotherapy might logically be indicated. However, if such responses are examples of autologous MLR, immunotherapy would be unwarranted and might, in fact, be detrimental to host antitumor immune responses.

On the other hand, the stimulation induced by leukemic cells in HLA-identical responders has been used as a criterion to evaluate potential donors of bone marrow for patients following chemotherapy. Thus, a positive response might be interpreted as the ability of donor bone marrow to cause a graft-versus-host reaction following adoptive transfer to the leukemic patient. If, however, this response is in fact a normal autologous MLR, many such potential donors would be unnecessarily disqualified on the basis of a positive reaction.

Consequently, further analysis is required to determine the nature of in vitro proliferative responses when HLA-identical tumor cells are used as stimulators. Since autologous MLR do not generate detectable cytotoxic cells [21, 28, 39, 63, 80, 81], this may provide a distinguishing feature. The

Table V. Characteristics of the RCS-induced syngeneic proliferative response

Characteristics	References
Nature of responder cell	
Age of donor: neonate through adult	104, 105
Organ source: Thymus, lymph node, spleen	109, 110
Cell markers: Thy 1^+, Ly 1^+, 2^-	
Nature of stimulator RCS cell	
RCS donor: spontaneous, primary RCS mouse or recipient of transplanted RCS	
RCS markers: Thy 1^-, sIg^-, Ia^+, Lyb 2^+	105
Cell origin: B cell lineage; pre-B cell	
Simulating antigen: Ia ± tumor-associated marker	
Level of response: stimulation indes ≥20	104, 105
generally 2-fold > allogeneic stimulation	108, 110
Sequelae of response	
RCS-specific CML: none	108
Suppressor function: yes	108, 111, 112
Helper function: yes	111

presence in such cultures of cytotoxic cells capable of lysing relevant targets would strongly suggest a true tumor-specific response. Such an assessment of these patients would permit a more rational decision with respect to both prognosis and immunotherapeutic intervention.

Of some interest with respect to proliferative responses induced by tumor cells in autologous responders is a lymphoma that occurs in SJL/J mice. With advancing age, there is a greater than 90% incidence of spontaneous neoplasia in mice of this strain [102, 103]. The primary tumor, which is morphologically heterogeneous, generally arises in mesenteric lymph nodes and spleen. Although this lymphoma was first described as a reticulum cell sarcoma (RCS), more recent evidence indicates it is of a B cell lineage (table V) [104–107]. Several transplantable lines, developed from spontaneous primary tumors, have been carried for a number of years by serial in vivo passage in young (6 to 8-week-old) SJL recipients. After intravenous administration of 10^7 viable RCS cells, significant growth is noted within 7–10 days in all lymphoid tissues except the thymus. Growth in lymph nodes and spleen, measured as a percentage of body weight, can easily exceed 10% during this interval.

One pointed aspect of the host-tumor relationship is the ability of the tumor to induce marked proliferation in syngeneic, normal T cells derived from thymus, lymph nodes or spleen [104, 105]. Although the response has been extensively studied in vitro, similar responses have been demonstrated in vivo [108]. The responder cell has been typed as a Thy-1.2$^+$, Lyt-1$^+$, 2,3$^-$ T cell [109]. In comparison with allogeneic stimulation of the same responders, RCS cells induce proliferative responses that are 2-fold greater in magnitude. Furthermore, from BudR + light inactivation studies [110], there appears to be significant overlap in those T cells that respond to a given alloantigen and those that are responsive to RCS cells. In point of fact, RCS-responsive cells have been suggested to encompass all alloreactive SJL T cells [105, 110].

The determinant on RCS cells that stimulates proliferation appears to be coded for by genes within the I region of the H-2 complex. Relevant Ia antisera delete stimulatory cells when used with complement as well as inhibit the proliferative response when added directly to the mixed cell culture [105]. Such inhibition occurs at the stimulator cell level, since SJL T cells, in the presence of the same antisera, are unimpaired in their ability to respond to other alloantigens and to mitogens. There does not appear to be any involvement of viral or Mls encoded products in this proliferative response. Results of both allo- and hetero-immunization with RCS cells routinely produces antisera that are cytotoxic to RCS cells, but that will kill normal SJL lymphoid cells as well [*Ponzio,* unpublished observations]. Therefore, tumor-specific antigens do not appear to mediate the observed effects. That RCS cells may lack tumor-specific antigens (or else express weakly antigenic ones) is supported by the observation that despite the intense proliferative response that is induced by RCS cells, there is no development of detectable cytotoxic cells in such primary cultures in vitro, or following repeated immunization regimens in SJL recipients of X-irradiated RCS cells [108]. However, enhancement, as well as suppression, of both humoral [111, 112] and cell-mediated responses [108], either directly by RCS cells, or by products of the syngeneic RCS-induced proliferative response have been reported.

Therefore, the RCS-induced proliferative response compares favorably with the autologous MLR that is induced by normal non-T cell stimulators. The RCS-induced response may reflect an exaggerated normal autologous MLR. The heightened response may be due to the particular configuration and/or density of stimulatory Ia determinants alone or in combination with a new tumor-associated antigen. Thus, the RCS-induced proliferation of syngeneic T cells may provide a useful comparative model for further dissection of the proliferative response of T cells to autologous, non-T stimulators, in the autologous MLR.

Summary

The autologous and syngeneic mixed lymphocyte reactions (MLR) in both mice and humans have been reviewed. In the MLR of both species, the replicating cell is a T cell and the stimulator can be a B cell, macrophage or a dendritic cell. While two quite separate types of isogeneic lymphocyte interactions (ILI) have been described in mice only one has, thus far, been delineated in humans.

For the murine system, ILI type 1 refers to the replication that neonatal thymic cells display when co-cultured with syngeneic adult spleen cells. In ILI type 2, adult lymph node cells respond blastogenically in the presence of autologous or syngeneic adult spleen cells. In the syngeneic MLR of the mouse (ILI type 2) the replicating T cell is a Ly $1^{+}2,3^{-}$ bearing cell, while that which responds in ILI type 1 has not yet been examined.

Two separate antigens stimulatory in the ILIs may be present on different B cell subsets. Termed murine differentiation antigens 1 and 2 (MDA-1, 2), they are absent from bone marrow cells, but are acquired with time as the cells mature. Presence of splenic T cells hastens the appearance of these markers among maturing bone marrow cells. Thus far, all mouse strains tested demonstrate ILI type 2, attesting to the presence of MDA-2. Whether the phenotypically detected MDA-2s are identical, remains an unanswered question. On the other hand ILI type 1 is not detectable in all strains of mice, C3H and CBA/N mice being prime examples of those lacking MDA-1 that, together with MDA-2, is stimulatory in ILI type 1. Whether there is identity between either of the stimulatory antigens and Ia antigens remains to be determined. With regard to function of these antigens, several sets of data of a correlative nature have suggested that MDA-1 on B cells may act as a receptor of T cell helper factor.

The counterpart to murine ILI type 2 has been identified in humans and is termed autologous MLR. As yet there have been no reports that humans demonstrate ILI type 1.

The stimulator in human autologous MLR has been described as present on mitogen-induced blast cells and normal peripheral blood mononuclear cells that possess HLA-D region-coded products. B cells from chronic lymphocytic leukemia patients and from infectious mononucleosis patients have been shown to be poor stimulators of autologous T cells.

The human as well as the murine autologous MLR fail to generate cytotoxic cells specific for autologous or syngeneic targets. However, soluble helper factors for cytotoxic T cell precursors can be identified in supernatants of human autologous MLR, but not of murine ILI type 1.

Autologous MLR of both humans and mice have been noted to trigger suppressor cells. While the suppressor cell is not the cell directly involved in autologous MLRs, the finding suggests that the autologous MLR may function indirectly as an immuno-regulating event. Indeed, patients with systemic lupus erythematosus and NZB mice with autoimmune conditions both fail to demonstrate autologous MLR.

In instances where tumor cells (such as in Hodgkin's disease) have been shown to act as stimulators of autologous peripheral T cell replication, determination of whether the stimulation is attributable to tumor-specific antigens or simply reflects a normal or exaggerated autologous MLR may be most important to selection of proper therapy. A feature that may be useful to distinguish between the two quite separate events is the generation of cytotoxic cells, since the latter do not arise in autologous or syngeneic MLR. An illustration of tumor cells that stimulate strongly in syngeneic MLR but do not cause generation of killer T cells is the lymphoma from SJL/J mice. Originally described as a reticulum cell sarcoma, it is now thought to be derived from B cell precursors.

At this early state in the study of autologous cell interactions their potential significance is varied. The autologous MLR detected in vitro may actually reflect normal cellular interactions in vivo. Thus, tolerance to self-antigens may develop early in post-neonatal life as a consequence of terminal differentiation of T cells experiencing self-markers. Furthermore, the autologous MLR that is detectable among lymphoid cells of adults may reflect the maintenance of regulation of immune response against self-antigens. Whether the markers that are stimulatory for T cell replication have other functions, as well, has been suggested but remains to be definitively determined. In any event the importance of autologous and isologous cellular interactions to tumor immunology and to autoimmune diseases has already become apparent.

References

1 Ehrlich, P.: Über den jetzigen Stand der Karzinomforschung; in Himmelweit, The collected papers of Paul Ehrlich, vol. II, pp. 550–562 (Pergamon Press, Oxford 1957).

2 Burnet, F.M.: The clonal selection theory of acquired immunity, England (Cambridge University Press, Cambridge, 1959).

3 Burnet, F.M.: The concept of immunological surveilance. Prog. exp. Tumor Res., vol. 13, p. 1 (Karger, Basel 1970).

4 Llacer, V.; Uyeki, E.M.: Interaction of spleen and thymic cells in vitro. Int. Archs. Allergy appl. Immun. *35:* 88 (1969).

5 Playfair, J.H.L.; Krsiakova, M.: Syngenic splenomegaly in NZB and Balb/c mice. Transplantation *7:* 443 (1969).

6 Howe, M.L.; Goldstein, A.L.; Battisto, J.R.: Isogeneic lymphocyte interaction: recognition of self antigens by cells of the neonatal thymus. Proc. natn. Acad. Sci. USA *67:* 619 (1970).

7 Cohen, I.R.; Stavy, L.; Feldman, M.: Glucocorticoids and cellular immunity in vitro. Facilitation of the sensitization phase and inhibition of the effector phase of a lymphocyte anti-fibroblast reaction. J. exp. Med. *132:* 1055 (1970).

8 Weksler, M.E.; Birnbaum, G.: Lymphocyte transformation induced by autologous cells: stimulation by cultured lymphoblast lines. J. clin. Invest. *51:* 3124 (1972).

9 Schwarz, M.R.: Blastogenesis in cultures containing genetically dissimilar thymic cells. Immunology *10:* 281 (1966).

10 Howe, M.L.; Goldstein, A.L.; Battisto, J.R.: Detection of self recognition cells by in vitro isogeneic lymphocyte interaction. Proc. 5th Leukocyte Cult. Conf., 1970, p. 515.

11 Howe, M.L.: Isogeneic lymphocyte interaction: responsiveness of murine thymocytes to self antigens. J. Immun. *110:* 1090 (1973).

12 Boehmer, H. von; Byrd, W.J.: Nature of target cells in syngeneic and allogeneic mixed lymphocyte reaction in mice. Proc. 7th Leukocyte Cult. Conf., 1973, p. 643.

13 Boehmer, H. von; Byrd, W.J.: Responsiveness of thymus cells to syngeneic and allogeneic lymphoid cells. Nature new Biol. *235:* 50 (1972).

14 Howe, M.L.; Manziello, B.: Ontogenesis of the in vitro response of murine lymphoid cells to cellular antigens and phytomitogens. Proc. 6th Leukocyte Cult. Conf., 1971, p. 534.

15 Boehmer, H. von: Separation of T and B lymphocytes and their role in the mixed lymphocyte reaction. J. Immun. *112:* 70 (1974).

16 Boehmer, H. von; Shortman, K.; Adams, P.: Nature of the stimulating cell in the syngeneic and the allogeneic mixed lymphocyte reaction in mice. J. exp. Med. *136:* 1648 (1972).

17 Boehmer, H. von; Adams, P.B.: Syngeneic mixed lymphocyte reaction between thymocytes and peripheral lymphoid cells in mice: strain, specificity and nature of the target cell. J. Immun. *110:* 376 (1976).

18 Boehmer, H. von: Selective stimulation by B lymphocytes in the syngeneic mixed lymphocyte reaction. Eur. J. Immunol. *4:* 105 (1974).

19 Carter, J.; Wegmann, T.G.: Mendelian segregation of a tolerance-inducing self-antigen in the spleen. Cell. Immunol. *7:* 402 (1974).

20 Ponzio, N.M.; Finke, J.H.; Battisto, J.R.: Adult murine lymph node cells respond blas-

togenically to a new differentiation antigen on isologous and autologous B lymphocytes. J. Immun. *114:* 971 (1975).

21 Smith, J.B.; Pasternak, R.D.: Syngeneic mixed lymphocyte reaction in mice: strain distribution, kinetics, participating cells and absence in NZB mice. J. Immun. *121:* 1889 (1978).

22 Glimcher, L.H.; Steinberg, A.D.; House, S.B.; Green, I.: The autologous mixed lymphocyte reaction in strains of mice with autoimmune disease. J. Immun. *125:* 1832 (1980).

23 Finke, J.H.; Battisto, J.R.: Maturational requirements of murine differentiation antigens 1 and 2 that are operative in isogeneic lymphocyte interaction; in Battisto, Streilein, Immuno-aspects of the spleen, vol. 89 (Elsevier/North Holland/Biomedical Press, Amsterdam, 1976).

24 Battisto, J.R.; Finke, J.H.; Yen, B.: Correlation of B cell acquisition of differentiation antigens with capacity to interact with allogeneic affect factor (AEF); in Escobar, Friedman, Macrophages and lymphocytes – nature, functions and interaction. B. Proc. 8th Int. Congr. of the Reticuloendothelial Society, p. 229 (Plenum Press, New York: 1980).

25 Battisto, J.R.; Finke, J.H.; Yen, B.: Capacity to interact with T-cell replacing factor correlates with acquisition by B cells of murine differentiation antigen-1. Immunology *37:* 623 (1979).

26 Finke, J.H.; Ponzio, N.M.; Battisto, J.R.: Isogeneic and allogeneic lymphocyte interactions may be controlled by cell surface immunoglobulin tropism. Cell. Immunol. *26:* 284 (1976).

27 Hom, J.; Talal, N.: Decreased syngeneic mixed lymphocyte response in genetically susceptible autoimmune mice. Fed. Proc. *39:* 1129 (1980).

28 Ponzio, N.M.: Lymphocyte responses to syngeneic antigens. 1. Enhancement of the murine autologous mixed lymphocyte response by polyethylene glycol. Cell. Immunol. *49:* 266 (1980).

29 Katz, I.R.; Asofsky, R.; Thorbecke, G.J.: Suppression of spontaneous reticulum cell sarcoma development and of syngeneic stimulator cells by anti-μ treatment of SJL/J mice. J. Immun. *125:* 1355 (1980).

30 Katz, I.R.; Lerman, S.P.; Ponzio, N.M.; Shreffler, D.C.; Thorbecke, G.J.: Growth of SJL/J-derived transplantable reticulum cell sarcoma as related to its ability to induce T cell proliferation in the host. I. Dominant negative genetic influences of other parent haplotype in F-1 hybrids of SJL/J mice. J. exp. Med. *151:* 347 (1980).

31 Dustoor, M.M.; Tymoszczuk, W.; Yen, B.R.; Scher, I.; Battisto, J.R.: B cells from CBA/N mutant mice do not stimulate lymphocyte interactions. Proc. 13th Leukocyte Cult. Conf., 1979, p. 71.

32 Steinman, R.M.; Witmer, M.D.: Lymphoid dendritic cells are a potent stimulator of the primary mixed lymphocyte reaction in mice. Proc. natn. Acad. Sci. USA *75:* 5132 (1978).

33 Ahmann, G.B.; Nadler, P.I.; Birnkrant, A.; Hodes, R.J.: T cell recognition in the mixed lymphocyte response. I. Non-T, radiation-resistant splenic adherent cells are the predominant stimulators in the murine mixed lymphocyte reaction. J. Immun. *123:* 903 (1979).

34 Nussenzweig, M.C.; Steinman, R.M.: Contribution of dendritic cells to stimulation of the murine syngeneic mixed leukocyte reaction. J. exp. Med. *151:* 1196 (1980).

35 Meo, T.; David, C.S.; Rijnbeek, A.M.; Nabholz, M.; Miggiano, V.C.; Scheffler, D.C.: Inhibition of mouse MLR by anti-Ia sera. Transplant. Proc. *7:* 127 (1975).

36 Lattime, E.C.; Golub, S.H.; Stutman, O.: Lyt 1 cells respond to Ia bearing macrophages in the murine syngeneic mixed lymphocyte reaction. Eur. J. Immunol. *10:* 723 (1980).

37 Scher, I.; Ahmed, A.; Strong, D.M.; Steinberg, A.D.; Paul, W.E.: X-linked B-lymphocyte immune defect in CBA/N mice. I. Studies of the function and comparison of spleen cells. J. exp. Med. *141:* 788 (1975).

38 Kincade, P.W.; Paige, C.J.: in Cooper, Mosier, Scher, Vitetta, B-cell development in immunodeficient CBA/N mice in B lymphocytes in the immune response, p. 349 (Elsevier/North-Holland, New York: 1979).

39 Ponzio, N.M.; Matteoni, L.A.; Sarley, J.M.: Potentiation of allospecific cytotoxic T cell development by the autologous MLR; in Kaplan, The molecular basis of immune cell function p. 525 (Elsevier/North-Holland/Biomedical Press, New York: 1979).

40 Lattime, E.C.; Golub, S.H.; Gillis, S.; Pecoraro, G.A.; Stutman, O.: Ia antigen stimulated proliferation and IL-2 production in the syngeneic mixed lymphocyte reaction. Transplant. Proc. (in press, 1980).

41 Pasternak, R.D.; Bocchieri, M.H.; Smith, J.B.: Surface phenotype of responder cells in the syngeneic mixed lymphocyte reaction in mice. Cell. Immunol. *49:* 384 (1980).

42 Green, S.S.; Sell, K.W.: Mixed leukocyte stimulation of normal peripheral leukocytes by autologous lymphoblastoid cells. Science *170:* 989 (1970).

43 Weksler, M.E.; Birnbaum, G.: Lymphocyte transformation induced by autologous cells: stimulation by cultured lymphoblast lines. J. clin. Invest. *51:* 3124 (1972).

44 Steel, C.M.; Hardy, D.A.; Ling, N.R.; Dick, H.M.; MacKintosh, P.; Crichton, W.B.: The interaction of normal lymphocytes and cells from lymphoid cell lines. III. Studies on activation in an authochthonous system. Immunology *24:* 177 (1973).

45 Birnbaum, G.; Siskind, G.W.; Weksler, M.E.: Autologous and allogeneic stimulation of peripheral human leukocytes. Cell. Immunol. *3:* 44 (1972).

46 Weksler, M.E.: Lymphocyte transformation induced by autologous cells. II. Stimulation by mitogen-induced lymphoblast. J. exp. Med. *137:* 799 (1973).

47 Bluming, A.Z.; Lynch, M.J.; Kavanah, M.; Khiroya, R.: Transformation antigens on stimulated lymphocytes. J. Immun. *144:* 717 (1975).

48 Weksler, M.E.: Lymphocyte transformation induced by autologous cells. III. Lymphoblast-induced lymphocyte stimulation does not correlate with E-B viral antigen expression or immunity. J. Immun. *116:* 310 (1976).

49 Opelz, G.; Kiuchi, M.; Takasugi, M.; Terasaki, P.I.: Autologous stimulation of human lymphocyte subpopulation. J. exp. Med. *142:* 1327 (1975).

50 Kuntz, M.M.; Innes. J.B.; Weksler, M.E.: Lymphocyte transformation induced by autologous cell. IV. Human T lymphocyte proliferation induced by autologous or allogeneic, non-T cells. J. exp. Med. *143:* 1042 (1976).

51 Weksler, M.E.; Kuntz, M.M.: Synergy between human T and B lymphocytes in their response to phytohemagglutinin and pokeweed mitogen. Immunology *31:* 273 (1976).

52 Weksler, M.E.; Kozak, R.: Lymphocyte transformation induced by autologous cells. V. Generation of immunologic memory and specificity during the autologous mixed lymphocyte reaction. J. exp. Med. *146:* 1833 (1977).

53 Sheehy, M.J.; Sondel, P.M.; Bach, M.L.; Wank, R.; Bach, F.H.: HL-A LD (lymphocyte defined) typing: a rapid assay with primed lymphocytes. Science *188:* 1308 (1975).

54 Katz, D.H.; Hamaoka, T.; Dorf, M.E.; Benacerraf, B.: Cell interactions between

histoincompatible T and B lymphocytes. III. Demonstration that the H-2 gene complex determines succesful physiologic lymphocyte interactions. Proc. natn. Acad. Sci. USA *70:* 2624 (1973).

55 Rosenthal, A.A.; Shevach, E.M.: Functions of macrophages in antigen recognition by guinea pig T lymphocytes. I. Requirement for histocompatible macrophages and lymphocytes. J. exp. Med. *138:* 1194 (1973).

56 Ruhl, H.; Shevach, E.M.: The effect of alloantisera on antigen-induced T cell proliferation. J. Immun. *115:* 1493 (1975).

57 Erb, P.; Feldman, M.: The role of macrophages in generation of T helper cells. II. The genetic control of the macrophage T cell interaction for helper induction with soluble antigens. J. exp. Med. *142:* 460 (1975).

58 Munro, A.J.; Taussig, M.J.: Two genes in the major histocompatibility complex control immune response. Nature, Lond. *256:* 103 (1975).

59 Shearer, G.M.; Schmitt-Verhulst, A-M.; Major histocompatibility complex restricted cell-mediated immunity. Adv. Immunol. *25:* 55 (1977).

60 Zinkernagel, R.M.; Doherty, P.C.: MHC-restricted cytotoxic T cells: studies on the biological role of polymorphic major transplantation antigens determining T cell restriction-specificity, function and responsiveness. Adv. Immunol. *27:* 52 (1979).

61 Bausher, J.C.; Smith, R.T.: Studies of the Epstein-Barr virus-host relationship: autochthonous and allogeneic lymphycote stimulation by lymphoblast cell lines in mixed cell culture. Clin. Immunol. Immunopathol. *1:* 270 (1973).

62 Viallat, J.; Svedmyr, E.; Steinitz, M.; Klein, G.: Stimulation of peripheral human lymphocytes by autologous EBV genome-carrying lymphoblastoid cell lines. Cell. Immunol. *38:* 68 (1978).

63 Smith, J.B.: Stimulation of autologous and allogeneic human T cells by B cells occurs through separate B cell antigen systems. Cell. Immunol. *36:* 203 (1978).

64 Gottlieb, A.B.; Fu, S.M.; Yu, D.T.Y.; Wang, C.Y.; Halper, J.P.; Kunkel, H.G.: The nature of the stimulatory cell in human allogeneic and autologous MLC reaction; role of isolated IgM-bearing B cells. J. Immun. *123:* 1497 (1979).

65 Beale, M.G.; MacDermott, R.P.; Stacey, M.C.; Nash, G.S.; Hahn, B.H.; Seiden, M.V.; Jacobs Burkholder, S.L.; Loewenstein Pletscher, L.S.: Stimulating cell types in the autologous mixed leukocyte reactions in man. J. Immun. *124:* 227 (1980).

66 Raff, H.; Picker, L.; Stobo, J.: Macrophages required for T cell activation constitute a subpopulation of HLA-D positive cells. J. exp. Med. *152:* 581 (1980).

67 Gupta, S.; Grieco, M.H.: Rosette formation with mouse erythrocytes: probable marker for human B lymphocytes. Int. Archs. Allergy appl. Immun. *49:* 734 (1975).

68 Froland, S.S.; Natvig, J.B.: Identification of three different human lymphocyte populations by surface markers. Transplant. Rev. *16:* 114 (1973).

69 Horowitz, D.A.; Garrett, M.A.: Distinctive functional properties of human blood L lymphocytes: a comparison with T lymphocytes, B lymphocytes and monocytes. J. Immun. *118:* 1712 (1977).

70 Hausman, P.B.; Raff, H.V.; Gilbert, R.C.; Picker, L.J.; Stobo, J.D.: T cell and macrophages involved in the autologous mixed lymphocyte reaction are required for the response to conventional antigen. J. Immun. *125:* 1374 (1980).

71 Cresswell, P.; Geier, S.S.: Antisera to human B lymphocyte membrane glycoproteins block stimulation in mixed lymphocyte culture. Nature, Lond. *257:* 147 (1975).

72 Winchester, R.J.; Fu, S.M.; Wernet, P.; Kunkel, H.G.; Dupont, B.; Jersild, C.: Recog-

nition by pregnancy serum of non-HL-A alloantigens selectively expressed on B lymphocytes. J. exp. Med. *141:* 924 (1975).

73 Humphreys, R.E.; McCone, J.M.; Chess, L.; Herrman, H.C.; Malenka, D.J.; Mann, D.L.; Parham, P.; Schlossman, S.F.; Strominger, J.L.: Isolation and immunologic characterization of a human B lymphocyte-specific cell surface antigen, J. exp. Med. *144:* 98 (1976).

74 Geier, S.S.; Cresswell, P.: Rabbit antiserum to human B cell alloantigens. II. Mechanism of inhibition of stimulation in the mixed lymphocyte response. Cell. Immunol. *35:* 392 (1978).

75 Meo, T.; David, C.S.; Nabholz, M.; Miggiano, V.; Shreffler, D.C.: Demonstration by MLR test of a previously unsuspected intra-H-2 crossover in the B10.HTT strain: implications concerning location of MLR determinants in the Ir region. Transplant. Proc. *5:* 1507 (1973).

76 Hausman, P.B.; Stobo, J.D.: Specificity and function of a human autologous reactive T cell. J.exp. Med. *149:* 1537 (1979).

77 Bergholtz, B.; Albrechtsen, D.; Thorsby, E.: Stimulation of T lymphocytes by autologous, non-T lymphoid cells. Participation of HLA-D. Tissue Antigens *10:* 27 (1977).

78 Smith, J.B.; Knowlton, R.P.; Koons, L.S.: Immunologic studies in chronic lymphocytic leukemia; defective stimulation of T cell proliferation in autologous mixed lymphocyte culture. J. natn. Cancer Inst. *58:* 579 (1977).

79 Halper, J.P.; Fu, S.M.; Gottlieb, A.B.; Winchester, R.W.; Kunkel, H.G.: Poor mixed lymphocyte reaction stimulatory capacity of leukemic B cells of chronic lymphocytic leukemia patients despite the presence of Ia antigens. J. clin. Invest. *64:* 1141 (1979).

80 Greenblatt, H.C.; Howe, M.: Ontogenesis of proliferative and effector activity in the neonatal CBA/J mouse thymus. 9th Leukocyte Cult. Conf., 1974, Abstr. 95, p. 41.

81 Vande Stouwe, R.A.; Kunkel, H.G.; Halper, J.P.; Weksler, M.E.: Autologous mixed lymphocyte culture reactions and generation of cytotoxic T cells. J. exp. Med. *146:* 1809 (1977).

82 Miller, R.A.; Kaplan, H.S.: Generation of cytotoxic lymphocytes in the autologous mixed lymphocyte culture. J. Immun. *121:* 2165 (1978).

83 Tomonari, K.; Aizawa, M.: Induction of autoreactive cells by the preculture of human peripheral blood mononuclear cells with the autologous fresh plasma. J. Immun. *122:* 2478 (1979).

84 Tomonari, K.: Cytotoxic T cell generated in the autologous mixed lymphocyte reaction. I. Primary autologous mixed lymphocyte reaction. J. Immun. *124:* 1111 (1980).

85 Yu, D.T.Y.; Chiorazzi, N.; Kunkel, H.G.: Helper factors derived from autologous mixed lymphocyte cultures. Cell. Immunol. *51:* 305 (1980).

86 Hausman, P.; Stobo, J.: Autologous mixed lymphocyte reactivity: a model for immunologically relevant self recognition; in Krakauer, Immunoregulation and autoimmunity (Elsevier/North-Holland/ Biomedical Press, Amsterdam, in press, 1980).

87 Sakane, T.; Steinberg, A.D.; Green, I.: Failure of autologous mixed lymphocyte reactions between T and non-T cells in patients with systemic lupus erythematosus. Proc. natn. Acad. Sci. USA *75:* 3464 (1978).

88 Sakane, T.; Steinberg, A.D.; Arnett, F.C.; Reinertsen, J.L.; Green, I.: Studies of immune functions of patients with systemic lupus erythematosus. III. Characterization of lymphocyte subpopulations responsible for defective autologous mixed lymphocyte reactions. Arthritis Rheum. *22:* 770 (1979).

89 Smith, J.B.; Knowlton, R.P.: Activation of suppressor T cells in human autologous mixed lymphocyte culture. J. Immun. *123:* 419 (1979).

90 Sakane, T.; Green, I.: Specificity and suppressor function of human T cells responsive to autologous non-T cells. J. Immun. *123:* 584 (1979).

91 Burns, F.D.; Marrack, P.C.; Kappler, J.W.; Janeway, C.A.: Functional heterogeneity among the T-derived lymphocytes of the mouse. IV. Nature of the spontaneously induced suppressor cells. J. Immun. *114:* 1345 (1975).

92 Hodes, R.J.; Hathcock, K.S.: In vitro generation of suppressor cell activity: suppression of in vitro induction of cell-mediated cytotoxicity. J. Immun. *116:* 167 (1976).

93 Stobo, J.D.; Loehnen, C.P.: Proliferative reactivity of T cells to autologous cell associated antigens. Arthritis Rheum. *21:* suppl., p. 5210 (1978)

94 Kuntz, M.M.; Innes, J.B.; Weksler, M.E.: The cellular basis of the impaired autologous mixed lymphocyte reaction in patients with systemic lupus erythematosus. J. clin. Invest. *63:* 151 (1979).

95 Krakauer, R.S.; Waldman, T.A.; Strober, W.: Loss of suppressor T cells in adult NZB/NZW mice. J. exp. Med. *144:* 662 (1976).

96 Sakane, T.; Steinberg, A.D.; Green, I.: Studies of immune functions of patients with systemic lupus erythematosus. I. Dysfunction of suppressor T cell activity related to impaired generation of, rather than response to, suppressor cells. Arthritis Rheum. *21:* 657 (1978).

97 Moody, C.E.; Casazza, B.A.; Christenson, W.N.; Weksler, M.E.: Lymphocyte transformation induced by autologous cells. VIII. Impaired autologous mixed lymphocyte reactivity in patients with acute infectious mononucleosis. J. exp. Med. *150:*1488 (1979).

98 Ilfeld, D.N.; Krakauer, R.S.; Blaese, R.M.: Suppression of the human autologous mixed lymphocyte reaction by physiologic concentrations of hydrocortisone. J. Immun. *119:* 428 (1977).

99 Yu, D.T.Y.; Ramer, S.J.; Clements, P.J.: Effect of methylprednisolone on autologous mixed lymphocyte cultures. Transplantation *25:* 163 (1978).

100 Ludgate, M.E.; Gough, J.: Stimulation of autologous blood lymphocytes by malignant cells and homogenates. Br. J. Cancer *34:* 619 (1976).

101 Opelz, G.; Gale, R.P.; McClelland, J.D.: Relationship between leukemia antigens and stimulation in mixed leukocyte culture. J. natn. Cancer Inst. *59:* 95 (1977).

102 Siegler, R.; Rich, M.A.: Pathogenesis of reticulum-cell sarcoma in mice. J. natn. Cancer Inst. *41:* 135 (1968).

103 Carswell, E.A.; Wanebo, H.J.; Old, L.W.; Boyse, E.A.: Immunogenic properties of reticulum-cell sarcomas of SJL/ J mice. J. natn. Cancer Inst. *44:* 1281 (1970).

104 Lerman, S.P.; Chapman, J.M.; Carswell, E.A.; Thorbecke, G.J.: Properties of reticulum-cell sarcomas in SJL/J mice. I. Proliferative response to tumor cells of T-derived lymphoid cells from normal mice. Int. J. Cancer *14:* 808 (1974).

105 Ponzio, N.M.; David, C.S.; Shreffler, D.C.; Thorbecke, G.J.: Properties of reticulum-cell sarcomas in SJL/J mice. V. Nature of reticulum-cell sarcoma surface antigen which induces proliferation of normal SJL/J T cells. J. exp. Med. *146:* 132 (1977).

106 Carswell, E.A.; Lerman, S.P.; Thorbecke, G.J.: Properties of reticulum-cell sarcomas in SJL/J mice. II. Fate of labeled tumor cells in normal and irradiated SJL/J mice. Cell. Immunol. *23:* 99 (1976).

107 Taylor, C.R.: Immuno-histological observations upon the development of reticulum-cell sarcoma in the mouse. J. Path. *118:* 201 (1976).

108 Ponzio, N.M.; Lerman, S.P.; Chapman, J.M.; Thorbecke, G.J.: Properties of reticulum-cell sarcomas in SJL/J mice. IV. Minimal development of cytotoxic cells despite marked proliferation to syngeneic RCS in vivo and in vitro. Cell. Immunol. *32:* 10 (1977).
109 Lerman, S.P.; Chapman-Alexander, J.; Umetsu, D.; Thorbecke, G.J.: Properties of reticulum-cell sarcomas in SJL/J mice. VII. Nature of normal lymphoid cells proliferating in response to tumor cells. Cell. Immunol. *43:* 209 (1979).
110 Ponzio, N.M.; Chapman-Alexander, J.; Thorbecke, G.J.: Properties of reticulum-cell sarcomas in SJL/J mice. VI. Characterization of lymphoid cells that proliferate in response to RCS cells. Cell. Immunol. *41:* 157 (1978).
111 Hayama, T.; Umetsu, D.; Chapman-Alexander, J.; Thorbecke, G.J.: Enhancement and suppression of antibody formation in vitro by reticulum cell sarcoma (RCS). Fed. Proc. *39:* abstr. 1004, p. 458 (1980).
112 Mancino, D.; Lerman, S.P.; Watanabe, N.; Hirayama, N.; Caiazza, S.S.; Levinson, J.; Ovary, Z.: Inhibition of antibody production by irradiated reticulum cell sarcoma cells in SJL mice. Cell. Immunol. *50:* 451 (1980).

Nicholas M. Ponzio, PhD, Departments of Microbiology-Immunology and Pathology, and the Cancer Center, Northwestern University Medical and Dental School, Chicago, IL 60611 (USA)

Prog. Allergy, vol. 28, pp. 193–284 (Karger, Basel 1981)

Lymphocyte Locomotion and Migration

Delphine M.V. Parrott and P.C. Wilkinson

Department of Bacteriology and Immunology, Western Infirmary, Glasgow, Scotland

Contents

I. Introduction

Lymphocytes are intrinsically mobile cells and it is obvious that their mobility is a controlled process for they are not randomly dispersed throughout the body, but organized into discrete lymphoid organs with different structural components and different proportions of the various subpopulations of lymphocytes. Lymphocytes are constantly moving and constantly repositioning themselves; nevertheless, the lymphoid system functions as a single organ although separated into thymus, marrow, spleen, lymph nodes and gut-associated tissues, for these are connected to one another by the free-floating cells in blood vascular systems and the lymphatic system. The development of the immune system depends upon the movements of lymphocytes from their sites of origin in thymus, marrow or equivalent tissues to the other lymphoid organs and the tissues, and the ability to initiate an immune response depends upon the capacity of lymphocytes with appropriate receptors to move to the area where antigen is deposited or localized whether it be in the skin, the gut mucosa, the lymph nodes or spleen or elsewhere. In order to perform either of these functions lymphocytes must locomote actively out of the blood stream. When it is appreciated that the amount of time which lymphocytes spend as free-floating particles in blood or lymph is trivial in comparison with the time spent in moving through the tissues (e.g. a small T lymphocyte takes 18 h to recirculate of which less than 1 h is spent in blood or lymph) then it will be realized that not only is the study of lymphocyte migration and recirculation central to our understanding of immune competence but the study of lymphocyte locomotion is central to our understanding of lymphocyte migration.

It should be emphasized that, in this review, we use the words *locomotion* and *migration* to mean different things. Locomotion refers to the process by which individual lymphocytes translocate actively, either in vitro or in relation to other cell types in living tissue. We use the word 'migration' to refer to the major redistributions of lymphocyte populations between and within the organs and tissues of the body. Thus, lymphocytes 'locomote' across endothelium, and 'migrate' from the blood stream across a lymph node into lymph.

During the process of migration a lymphocyte must first adhere, then traverse endothelium, detach from endothelium and move over and amongst other cells whether in lymph node, spleen, lung, gut or skin. Each of these events whether involving adhesion or translocation is a locomotor process.

There are numerous sites in the body at which lymphocytes leave the blood and in so doing they assume the typical locomotor morphology familiar from in vitro studies. The site which immediately springs to the mind of all who are interested in lymphocyte traffic is the high-walled endothelial venules in lymph nodes and Peyer's patches. These have excited interest since they were discovered to be the routes by which lymphocytes cross from blood into lymph [87].

Experiments designed to investigate the migration of lymphocytes in vivo usually involve the injection of a bolus of isotopically labelled cells and then measurement of their arrival in selected tissues after an interval of time. Over the years terms such as lymph node-seeking and spleen-trapping have come into use. A favourite expression is 'homing', an emotive term that conjures up a picture of a lymphocyte with a 'ticket to ride' (scientific term: 'receptor') travelling in a train (blood stream, lymph) which it would leave only when it had reached the station (post-capillary venule with matching receptor) nearest to home. But such concepts and word pictures can be very misleading, for they imply that lymphocytes migrate in a purposeful way and that positive vectorial movement is involved.

In this review we examine the role of lymphocyte locomotion in events which are involved in the migration and recirculation of lymphocytes, the constant repositioning of lymphocytes within lymphoid tissues and the selective accumulations of lymphocytes in particular sites.

Lymphocyte recirculation from blood to lymph and back again is a complicated series of random locomotor episodes. For example, the times which any one lymphocyte takes for each 'round trip' depends on whether it is T or B, how many capillary beds it passes across and transiently adheres to, whether it migrates across the spleen, or only one lymph node, or several lymph nodes and a Peyer's patch and so on. Some subpopulations will be transiently retained by antigen in a lymph node or spleen others will pass through skin or gut. Estimates of modal transit times of 18–24 h should not be allowed to obscure the fact that some lymphocytes take as little as 1–2 h or as much as 72 h.

In order to examine the role of lymphocyte locomotion in vivo in any meaningful way it is necessary first of all to review our knowledge of lymphocyte locomotion in a more simple system than the whole animal. It is appropriate, therefore, that this review commences with a discussion of experiments which have studies lymphocyte locomotion in vitro and the allied topics of lymphocyte adhesion, to various substrata and other cells. Subsequently we go on to discuss lymphocyte migration placing particular emphasis on those topics which are especially relevant to lymphocyte locomotion, e.g. the time which subpopulations take to locomote across various tissues, the importance of the adhesiveness of lymphocytes to endothelium.

There are several recent reviews which give a more detailed and overall view of lymphocyte migration e.g. *Ford* et al. [67], *Sprent* [237], which consider the traffic of lymphocytes in relation to cellular interactions *de Sousa* [47], *Curtis* [36] and a very recent book of essays by *Trnka and Cahill* [244] on the anatomy and physiology of the lymphoid tissues to which the reader is referred for interest, information and more orthodox approaches to this problem.

II. Lymphocyte Locomotion in vitro

A. Mechanisms of Locomotion

1. Morphological Observations of Locomoting Lymphocytes. Early Observations

The earliest detailed studies of lymphocyte locomotion in vitro were made in the inter-war years. *Lewis and Webster* [140] watched lymphocytes moving out of explants of lymph nodes in plasma clot preparations in vitro and observed that the cells took irregular paths and showed considerable variation in speed, the mean speed being around 10–20 µm/min. They described the polarized morphology adopted by moving cells with the nucleus placed anteriorly and with an elongated tail containing cytoplasmic organelles. *McCutcheon* [148] watched the movement of lymphocytes taken directly from blood and cultured over 12 h. The number of moving cells increased during the period of culture and their speed gradually increased from 4 µm/min after 1 h to 15 µm/min after 9 h. *Lewis* [138] used time-lapse cinematography to study the movement of rat lymph node cells which had been cultured for 24 h in vitro and described their morphology in detail. The cells first put out a broad pseudopodium in the direction in which they were about to move. Behind the pseudopodium a 'constriction ring' remained in situ with respect to the substratum till it reached the tail. As the ring reached the tail, the tail lengthened giving rise to the 'hand-mirror' morphology of moving lymphocytes which has been observed many times since (fig. 1, 2). *de Bruyn* [41] made observations similar to those of *Lewis* [138] but suggested that constriction rings were caused by external factors determined by the substratum on which the cells were moving and were not an essential feature of locomotion. These early morphological descriptions have been confirmed by many workers since. Moving lymphocytes usually move with a pronounced tail and with the nucleus well forward immediately behind the leading edge, so that the front end of the cell is oval or rounded. However, considerable variations on this description are frequently seen especially if lymphocytes are squeezing through narrow spaces or between other cells, in which case the morphology may be determined by physical constraints (fig. 3). The typical hand-mirror morphology is that seen in unimpeded cells locomoting on two-dimensional surfaces (fig. 1). *Lewis* [138, 139] postulated that the shape changes seen in locomoting lymphocytes were due to gelling and exertion of contractile tension within the cytoplasm, a postulate that

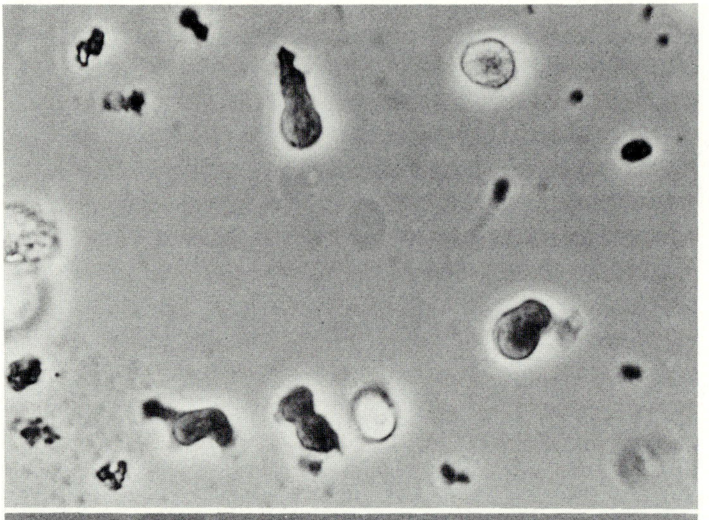

a

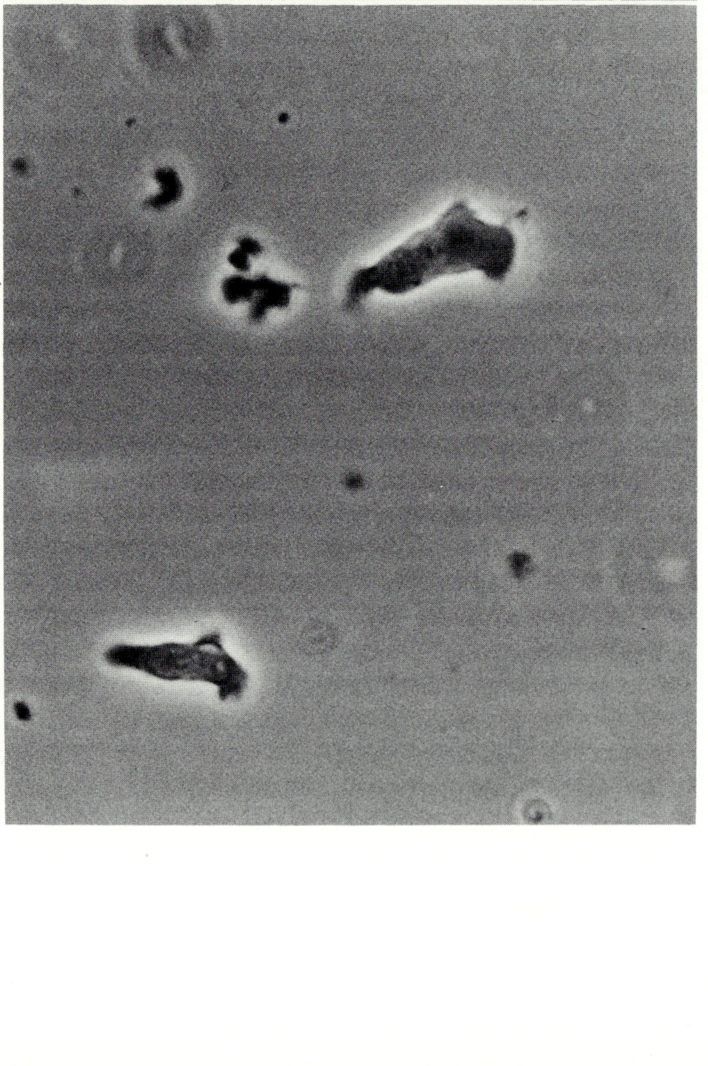

b

is consistent with more recent discoveries of the role of networks of microfilaments of polymerized actin and related proteins in cell locomotion.

2. Effect of Culture in vitro on Locomotor Capacity of Lymphocyte Populations

Many authors have emphasized the increasing locomotor capacity of lymphocytes with time in culture in vitro first observed by *McCutcheon* [148]. *Marshall and Roberts* [160] made cinematographic studies of the locomotion of human blood lymphocytes cultured with phytohaemagglutinin (PHA), which suggested that, on mitogenic transformation and cell division, a population of daughter cells with locomotor capacity was formed. Both large and small cells were motile after culture and the locomotor morphology was similar to that described above with large blunt pseudopodia being thrown forward, sometimes in pairs to form Y-shaped cells, and with prominent tails which retained a remarkably constant orientation in relation to the direction of movement. Lymphocytes in pairs were sometimes seen, attached to one another by their tails. Often, after cell division, the tails were formed from the central constriction between the two daughters which remained adherent to one another. *Biberfeld* [15] cultured human lymphocytes in the presence of PHA and showed a gradual increase in the percentage of cells in locomotor morphology from 4% at 1 h of culture to 20% at 50 h. *Rydgren* et al. [213] reported similar findings on the effect of culture in PHA on motility of human lymphocytes in plasma clots tracing cell-paths by time-lapse cinematography. *Schreiner* et al. [222] reported that 5% of unstimulated mouse spleen B cells were in locomotor morphology, whereas after 24–48 h culture in the presence of trypsin or lipopolysaccharide 40–90% of the same population were in locomotor morphology. We have carried out a time-lapse cinematographic study and analysed the movement of human peripheral blood lymphocytes taken freshly from blood and after a 48-hour period in culture with PHA [262]. Only 6% of the lymphocytes direct from blood were motile (whereas 74% of monocytes in the same preparation were motile). After 48 h culture in vitro, 44% of the lymphocytes showed translocations

---

Fig. 1. a Human blood lymphocytes locomoting on glass. The cells were cultured for 48 h in minimal Eagles' medium + 10% fetal calf serum, then filmed locomoting in 50% human plasma. A high proportion of the population was motile. Note hand mirror morphology with the prominent anterior nucleus [from ref. 262]. *b* Lymphoblasts from the same population showing locomotor morphology.

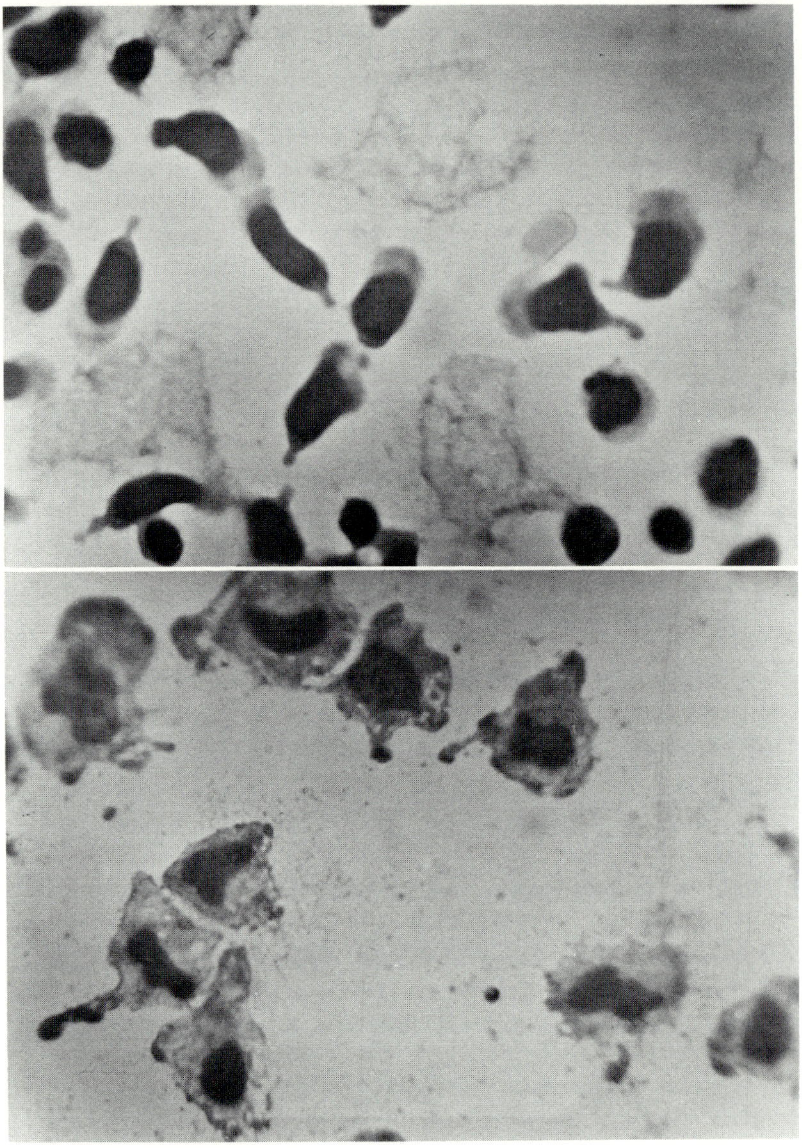

a

b

Fig. 2. Rat thoracic duct lymphocytes migrating under agarose [courtesy of Dr. *A.O. Anderson*]. Fixed Giemsa-stained preparations. *a* Fresh thoracic duct lymphocytes. *b* T lymphoblasts after 72 h culture in presence of Con A. Note polarized morphology and prominent uropods in both populations.

and the moving cells included lymphoblasts and small lymphocytes (table I).

Lymphocytes activated in vitro in other ways also show an increase in locomotor capacity. *McFarland and Heilman* [150] used human mixed lymphocyte cultures from two unrelated donors. They saw little locomotor activity during the first 24 h of culture but thereafter locomotor forms were seen. These cells were immunologically stimulated by the mixed lymphocyte reaction and blast cell transformation was prominent.

Other work emphasizes that addition of an activating agent (other that serum) is not essential to induce locomotor capacity in lymphocytes in vitro, and that a period of culture in vitro without such agents but in the presence of serum is sufficient. *Berman* [13] observed hand-mirror morphologies and locomotion around the surfaces of macrophages (peripolesis) in lymphocytes cultivated for 24 h in fetal calf serum (FCS) without added activating agents. FCS itself is of course known to have a direct activating effect on lymphocytes and may be mitogenic. *O'Neill and Parrott* [177] found that human blood T cells did not move into micropore filters when fresh from blood but would do so after a period of culture. Non-T cells including B cells were motile immediately after removal from blood, but became more so on culture. Cells cultured in concanavalin A (Con A) showed an increase in locomotion over a 72-hour period, as did cells cultured in the absence of a mitogen. On the other hand, *Schmalsteig* et al. [218] found T lymphocytes to be motile directly after separation from human blood. Stimulation in vivo as well as in vitro induces the formation of lymphocyte populations with locomotor capacity. In studies using filter assays, the locomotion of lymphocytes from unstimulated mouse lymph nodes was poor, but 3–9 days after contact sensitization with oxazolone, a population of highly motile lymphocytes and lymphoblasts could be obtained from the draining lymph node [212]. *Pinkston* et al. [196] using an agarose assay saw increases of locomotion of both human T and B lymphocytes during the first few hours of culture in the presence of human serum.

It is reasonable to conclude from the above studies that small lymphocytes in blood or in unstimulated lymphoid tissues are not highly motile, though a minority of locomotor forms can usually be found. Lymphoblasts, on the other hand, are highly motile cells and various procedures which induce blast transformation will also enhance motility. However, in culture, an enhancement of motility is seen even without blast transformation, and post-blast cells which have reverted to small lymphocyte

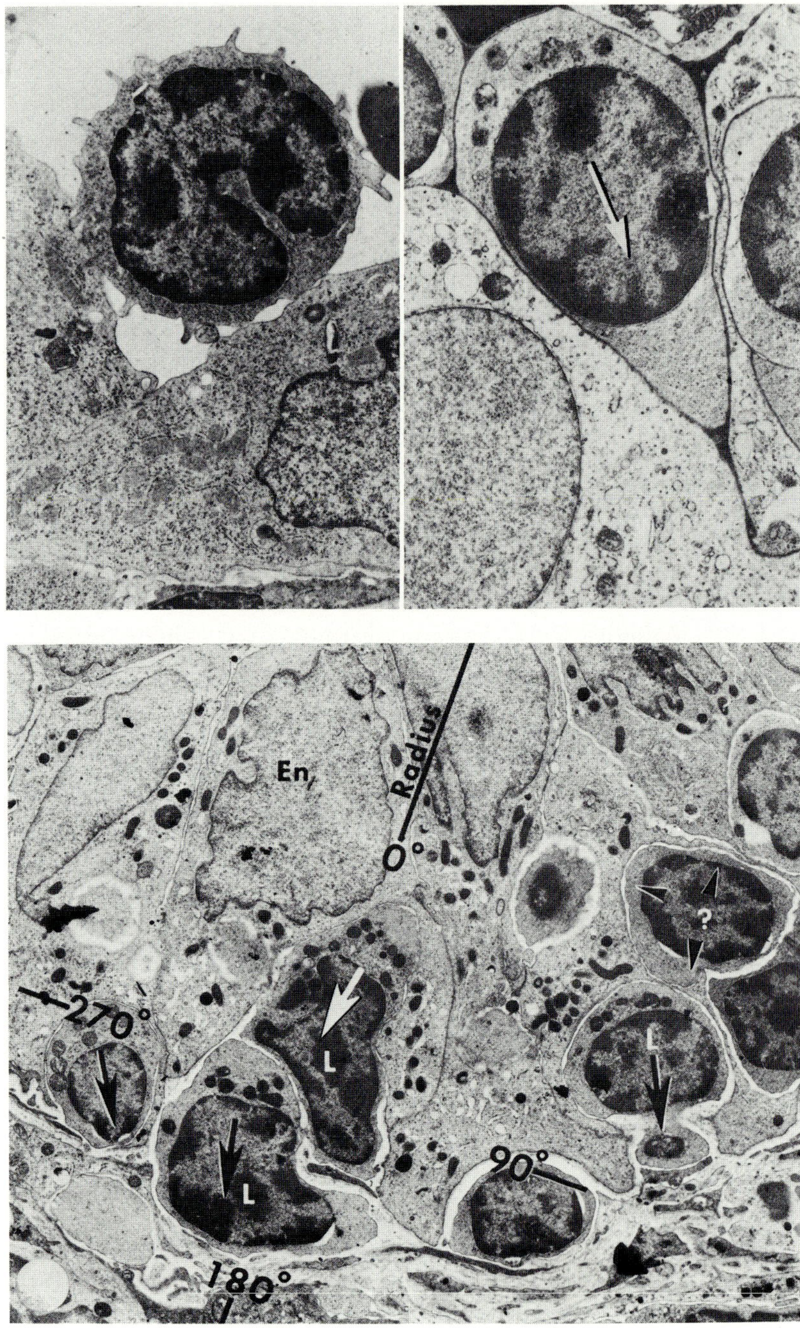

a b

c

Table I. Locomotion of human blood lymphocytes and monocytes on glass in 50% plasma + casein 1 mg/ml (chemokinetic medium)

	Lymphocytes direct from blood	Lymphocytes after 48 h in Eagles + 10% fetal calf serum + PHA	Monocytes direct from blood
Number of cell paths analysed	34	57	46
Number of cells moving	2 (6%)	25 (44%)	34 (74%)
Mean speed (only moving cells included), μm/min	8.0	5.9	8.8
Percent turning angles at			
<20°		46	33
<40°		68	58

The data were obtained by tracing the paths of cells from a time-lapse film. Note that the proportion of motile lymphocytes increases after culture in vitro. The turning behaviour of lymphocytes and monocytes is similar. Both cell types show a tendency to make most of their turns at narrow angles.

morphology may also show good motility. These differences in locomotor behaviour of lymphocytes at different stages of differentiation may be very pertinent to the behaviour of lymphocyte populations migrating into tissues in the living animal.

3. The Uropod

The tail of the moving lymphocyte is often called the 'uropod'. This structure has been a source of considerable interest. As mentioned, it is a constant feature of locomoting lymphocytes and lymphoblasts in culture (fig. 1, 2) more so than the tails of locomoting neutrophils or monocytes,

---

Fig. 3. Interactions of rat lymphocytes with high endothelial venules [courtesy of Dr. *A.O. Anderson* and taken from ref. 4]. *a* A lymphocyte adhering to the luminal surface of the high endothelial venule by villous and membranous contact points. *b* A lymphocyte moving between endothelial cells. The direction of locomotion is indicated by the arrow. Note organelles posterior to the nucleus. Note also that morphology is constrained by the surrounding cells. Lymphocytes frequently do not show typical hand-mirror morphologies under such conditions. *c* Lymphocytes moving through HEV. The luminal surface is at the top. The lymphocytes are oriented with the nucleus leading and the organelles posteriorly and are moving (downwards) into the node. The angle formed by a line drawn through the polar axis of each lymphocyte (arrows) was measured using a 360° navigational compass which was aligned with the radius of the vessel.

which are evanescent structures, appearing and disappearing frequently during movement. The uropod has been implicated not only in movement but also in interactions of lymphocytes with other cells or materials in their environment. *Sharp and Burwell* [225] showed that in mixed cultures of primed lymphocytes with macrophages, the lymphocytes tended to migrate on and around the surfaces of macrophages as if exploring them, a phenomenon they named *peripolesis. McFarland and Heilman* [150] and *McFarland* et al. [152] observed that the uropod was especially involved in this process, in studies using mixed lymphocyte cultures of human blood. In time-lapse films taken during several days following the commencement of culture, lymphocytes approached macrophages, contacted the macrophage surface with their anterior end and manoeuvred into a position where the uropod could contact the macrophage. They then appeared to explore various points on the macrophage surface with their uropods. Clusters of lymphocytes were frequently found in association with macrophages and these macrophage-associated lymphocytes often transformed into lymphoblasts, as confirmed by marked incorporation of ^{3}H-thymidine by such clustered lymphocytes. Interactions were seen between blast forms in these clusters and small lymphocytes, the two cell types being attached by their uropods. Similar findings, implicating the uropod in the interaction of immunologically reactive lymphocytes with macrophages and lymphoblasts, have been reported by *Werdelin* et. al. [259] and *Nielsen* et al. [171]. Ultrastructural studies revealed that the uropod contained the Golgi apparatus and centrioles and was rich in microtubules [15]. Mitochondria were also found there (fig. 3) and sometimes rough endoplasmic reticulum. The uropod was observed to be surrounded by 'microspikes' or 'microvilli' [151] whereas these structures were absent from the front end of the cell. They may be identical to retraction fibres seen at the tails of moving cells of many types, which form as the tail is pulled forward by the forward movement of the front end of the cell. This may cause posteriorly placed microfilaments adherent to the substratum to be pulled together into narrow bundles surrounded by plasma membrane rich in substratum-adhesive molecules. Microvilli have been found to be enriched in surface immunoglobulin [43]. Such terminal retraction fibres are probably not identical to the microvilli often seen distributed over the whole surface of lymphocytes.

Biberfeld et al. [16] have implicated uropod-forming cells in antibody-mediated cytotoxicity, since human Chang cells in the presence of anti-Chang antibody are killed by cells which infiltrate uropods into their substance. *Glauert and Sanderson* [82] also reported that in K cell-mediated cytotoxicity the K cells insert projections into the substance of the target cell and that cytotoxic T cells [215, 216] insert similar uropod-sized projections into target cells. The identity of these structures with uropods may be questioned since, although they look similar in electron photomicrographs, time-lapse films of cytotoxic T cells and K cells [216] show very rapid shape changes and appearance and disappearance of these projections, probably within seconds, whereas the locomotor uropod is a semi-permanent feature of the polarized cell.

Morphological observations of lymphocytes exploring other cells such as macrophages, or even acellular debris [152] with their uropods have suggested that the uropod is a specialized structure involved in immunological recognition [209]. In moving lymphocytes, spontaneous accumulation of cell-surface recognition molecules such as membrane immunoglobulin and Fc receptors is seen at the tail of the moving cell [19, 222]. The membrane protein concentration at this end of the cell is probably high and this concentration of membrane protein is accompanied by submembranous accumulation of actin-myosin filaments under the cap and in retraction fibres [19, 79]. The tails not only of lymphocytes but of other leucocytes such as neutrophils are the most adherent part of the cell and frequently become tethered to the substratum. Thus, the postulate that the uropod adheres to, and acts as, a recognition unit for structures on other cells seems reasonable. However, the uropods of cells in movement and of those engaged in the contact phenomena described above may not be identical structures, the evidence that they are being mainly based on their microscopical similarities alone. Caution should be exerted in interpreting these structures as identical on present evidence.

4. Locomotion and Membrane Movements

Lymphocyte locomotion is probably basically similar to that of other tissue cells that have been studied, involving as it does adhesion to a substratum, morphological polarization characterized by formation of a microfilament-rich leading edge, and forward movement generated by contraction of actin filament networks anchored at a point or points in the leading edge of the cell. Blood cells such as neutrophils and monocytes move rapidly (ca 20 μm/min) compared to fibroblasts which locomote at ca 20 μm/h. Lymphocytes move on glass more slowly than neutrophils, but when in contact with the surfaces of other cells their speed is comparable to that of neutrophils (see below). The arrangement of the cytoplasmic contents in the moving lymphocyte is different from that of the neutrophil, since, as stated above, in lymphocytes the nucleus is placed anterior to the centrioles and Golgi region, whereas in neutrophils the nucleus lies behind these structures.

The role of microfilaments in locomotion has been studied best in fibroblasts. There has been a great deal of interest in the distribution and activity of actin and myosin in lymphocytes, but research has concentrated on the role of microfilaments (and microtubules) in redistribution of membrane structures – patching and capping – and little is known about the details of their role in locomotion. There is a huge literature on *membrane* movements in lymphocytes, much of which is not directly helpful in understanding the mechanism of locomotion itself. This literature has been comprehensively reviewed by *Schreiner and Unanue* [221] and *Loor* [143].

Membrane movements obviously take place when cells move and these movements can be observed, for example, in fibroblasts, when particles are carried backwards from the leading edge of the moving cell towards the nucleus [107] or in the capping of cross-linked molecules in locomoting cells. Various explanations have been put forward, including backward membrane flow, or the active transport of aggregates or particulates on the cell surface by direct linkage to submembranous microfilaments. Cells form networks or bundles of microfilaments at the point of contact with a substratum or with a large object such as a phagocytosable particle [203] and a cluster of membrane proteins might induce microfilament organization in the membrane immediately beneath it by the same mechanism, and the membrane proteins might then be moved as the submembranous microfilaments move.

Certain lymphocyte surface components, such as membrane immunoglobulin, Fc receptors and the Thy leukaemia antigen, cap spontaneously, even though uncross linked by extrinsic ligands as the cell begins to locomote [19, 222] and myosin, actin and tubulin have been found concentrated under such caps by immunofluorescence [79,223]. Other membrane components such as Thy 1 and H-2 antigens do not, when uncross-linked, cap as the lymphocyte assumes a locomotor morphology, but do cap slowly when cross-linked by antibody. These latter caps are not associated with actin or myosin or with locomotor morphology. *Braun* et al. [19] suggested that, in the first type of capping, membrane components are attached directly to submembranous microfilaments and are swept back actively by them to the tail as the cell changes shape. In the second type of capping there was no relation to microfilaments, rather the cap formed because the cross-linked complex formed an area of restricted lateral diffusibility relative to the surrounding membrane. The spontaneous capping of membrane Ig, Fc receptors or thymus leukaemia antigen was a phenomenon that followed locomotion. However, if the same membrane components were cross-linked with antibody, all three types of surface molecule capped rapidly, myosin redistributed with the cap, and the addition of antibody itself stimulated locomotion. The lateral mobility of surface Ig and Thy 1 in lymphocytes has been measured by fluorescence photobleaching [52]. These components showed high mobility in the non-cross linked form, but addition of anti-Ig immobilized surface Ig. Thy 1 (which is monovalent) was not immobilized by one layer of antibody, but could be with a double layer .

Addition of anti-Ig to surface immunoglobulin-bearing B lymphocytes has been shown to trigger mobilization of intracellular calcium [20]. This calcium release required cross-linking of the bound receptors and did not occur when other common surface macromolecules were cross linked with antibody. Since the surface Ig-anti-Ig reaction has been shown to act as a locomotor stimulus, this report suggests a role for cytoplasmic calcium in lymphocyte motility. There is much evidence now for such a role for calcium in neutrophil chemotaxis [194], but calcium dependence has been little studied in relation to lymphocyte locomotion.

Most of the evidence for a linkage between cell surface components and contractile proteins is based on indirect techniques such as immunofluorescence, but direct biochemical evidence has also been obtained that intracellular actin is directly attached to cross-linked membrane Ig in lymphocytes [63], and to H-2 antigens in mastocytoma cells [132].

5. Locomotion of Lymphocytes of Different Types

From the studies quoted above, it is clear that lymphocyte locomotion is strongly influenced by extrinsic agents. Lymphocytes stimulated with mitogens or with antigen, as in mixed lymphocyte cultures, or even following culture in the absence of added stimuli, locomoted better than lymphocytes from unstimulated or uncultured populations, and locomotion was frequently observed in lymphoblasts. However, there have been few studies in which the type of the locomoting cell was defined. Separated T and B cell populations purified from blood are both able to locomote after a period of culture [177]. T lymphoblasts taken from the lymph nodes of mice 3 days after contact sensitization with oxazolone show high

Table II. Locomotion of established lymphoid cell lines to chemoattractants in filters

Line	Distance (μm) migrated into filter towards			Checkerboard evidence[1] for (where done)	
	Neg. control	endotoxin-activated plasma (10%)	casein (1 mg/ml)	chemokinesis	chemotaxis
1. Human B lymphoblasts					
ANA2	45	83	N/T		
BEC11	37	69	53		
BR18	43	66	53		
MICH	26	24	30		
SMI	55	49	76		
SMI_4	29	98	75	+	weak
Tay	38	53	48		
RPMI1788	24	80	91	+	+
BAR_1	50	102	N/T	+	–
F89	21	78	24	+	+
FAL_1	43	72	38		
FLE_1	36	44	N/T		
OR 1_1	61	93	N/T	+	+
JIM_1	23	63	N/T	+	+
2. Human T lymphoblasts					
Ac. lymphatic leukaemia line from Sloan-Kettering Institute	23	34	53	+ (see table IV)	+
3. Burkitt's lymphoma					
EB4	13	16	16		
BJAB	40	37	35		
Daudi	39	45	35		
Namaliva	48	55	46		
Raji	16	29	31		
U698	38	59	48		
8A (Daudi/Raji/hybrid)	25	31	37		
4. Mouse plasmacytoma					
MOPC 70A	N/T	56	47		
315 34λ	35	39	51		
AB8	21	24	25		
PI2	37	33	64		
H38	13	13	13		
H17+	22	36	34		
H21	27	34	20		
5563T	38	47	24		

N/T = not tested.

[1] See table IV for full explanation of the checkerboard assay.

motility in the absence of chemoattractants [212]. Studies of established B lymphoblast lines confirm that lymphoblasts are actively motile. These cells have been observed to locomote on or under the surface of other tissue cells [197], and to migrate into micropore filters in response to chemoattractants [212]. *Nilsson and Ponten* [173] observed that while cultured lymphoblasts were highly motile, cultured lymphoma and myeloma cells were much less so. In our laboratory a similar pattern of locomotion of the same cell types to chemotactic factors was noted (table II). These findings may be of interest because the less motile cells (lymphomas, myelomas) form solid tumours in vivo, whereas lymphoblasts, when neoplastic (e.g. from leukaemias), do not. Little is known about the locomotor capacities of lymphocyte subsets, though cytotoxic T cells form an obviously motile population as observed in time-lapse films [215, 216].

The locomotor capacity of lymphocytes is influenced considerably by the surface on which movement is studied. For example, as mentioned above, B lymphoblasts locomote into micropore filters and on or under the surfaces of other tissue cells but, when tested in identical media, they do not locomote on glass or protein-coated glass [unpublished observations]. This is almost certainly due to differences in the ability of cells to adhere on different surfaces, and brings us to an extended consideration of the influence of the substratum and of cell adhesion on lymphocyte locomotion.

B. Lymphocyte Locomotion and Adhesion

1. Artificial Substrata

Lymphocytes resemble other tissue cells in locomoting by crawling on substrata, and successful locomotion demands balanced attachments and detachments of parts of the moving cell which are contiguous with the substratum. A cell which does not adhere well is unlikely to move well and the relatively low adhesiveness of lymphocytes, taken directly from blood or lymph nodes, to artificial substrata may explain the poor motility of these cells on such substrata. Cell adhesion to uncoated glass or plastic is non-energy-dependent and independent of cell viability and so is perhaps not pertinent to the present discussion. If such surfaces are coated with biological materials, such as proteins or materials released by cells previously coating the substratum, the adhesion is of quite a different type which requires viable cells and is energy and divalent cation-dependent [89]. Lymphocytes separated directly from blood adhere poorly to serum-coated glass or tissue culture plastic compared to granulocytes or macrophages, a property which is frequently used to purify the latter cells from mixed cell populations. Nevertheless, there is always a small proportion of lymphocytes which does adhere to serum-coated glass, and B lymphocytes are more adherent to artificial surfaces than T lymphocytes, the well-known nylon wool-separation procedure being based on this difference in adhesiveness. One reason for poor lymphocyte adhesiveness on two-dimensional surfaces may be their relatively poor ability to spread. Neutrophils and monocytes which attach to surfaces extend veil-like processes across the surface before they show orientation and locomotion [262, 276], whereas small lymphocytes remain rounded. Thus, only a small proportion of the lymphocyte surface is in contact with the substratum and this may not provide sufficient anchorage for adhesion. If this is the case, lymphocytes would be expected to attach better and locomote better on three-dimensional surfaces, where the proportion of cell surface in contact with the substratum would be higher. This does appear to be the case, as comparisons of lymphocyte locomotion in three-dimensional filters with their locomotion on flat surfaces, discussed below, suggest.

It is possible that the enhancement of locomotion seen in lymphocytes which have been cultured, or in lymphoblasts, could be due to an increase in adhesiveness. Small lymphocytes may have insufficient cytoplasm to spread well, whereas lymphoblasts, which have more cytoplasm, show better spreading. Direct observations lend some support to this view, since locomoting lymphoblasts may show more prominent anterior lamellipodia than small lymphocytes; however, the cell body in both cell types in locomotion remains compact and unspread. Time-lapse cinematographic observations in our laboratory [unpublished] have shown better adhesion and locomotion of human blood lymphocytes and lymphoblasts when taken from 48- to 72-hour PHA cultures, on serum-coated glass, than immediately after removal from blood. Cell surface-bound PHA could itself attach to the glass and help to anchor the cells. Such observations require to be put on a firm quantitative basis, though quantitative assays of cell adhesion have not yet been applied to the study of lymphocytes from various sources, or cultured in different ways.

Lymphocyte adhesion to substrata can be enhanced using ligands with both a high affinity for lymphocyte surfaces and a capacity for binding to substrata. Such ligands could act as anchors, binding the lymphocyte to the substratum. *Mori* et al. [168] studied the effect of Con A on the behaviour of human lymphoma cells and non-lymphocytic cell types on flat surfaces.

In media containing various concentrations of Con A in FCS, lymphocyte adherence after 30 min increased from 20% of cells adherent in 5 μg/ml Con A to 95% adherent in 25–50 μg/ml Con A. The adherent cells flattened out and showed locomotor morphologies, but observations of cell translocations were not made. After prolonged culture (9 h), the cells reverted to a spherical form and became detached. We filmed lymphocytes bound to glass coverslips coated with Con A at concentrations down to 1 μg/ml [unpublished observations]. Although the lymphocytes spread and showed elongated morphologies they were too tethered to move, suggesting that Con A forms too tight an adhesion to the substratum to allow translocation. *Wong* et al. [265] carried out a similar study of chicken lymphocytes. These cells stuck to glass or plastic when washed free of serum but became detached if protein was added. Studies were made of the attachment of chick thymocytes or bursacytes in Hanks chicken serum + lectins. The cells became adherent in 200 μg/ml Con A with 25% protein, or in PHA (Difco PHA-P) at a 1 in 100 dilution with 1% added protein, but did not adhere in pokeweed mitogen. Adherence was energy-dependent and was reversed when the medium was replaced by a lectin-free solution. Another ligand which allowed adherence of chicken lymphocytes was alloantibody against chicken blood group antigens which were represented on the surfaces of the lymphocytes under study. Once again, no studies of cell locomotion were made.

Alexander and Henkart [1] have reported a detailed study of the adherence of human Fc receptor-bearing lymphocytes to substrata coated with immune complexes. They used complexes made with rabbit anti-DNP or TNP and the appropriate antigen, which they bound to plastic substrata and then allowed human blood lymphocytes to adhere. 5–30% of the blood cells became adherent, representing the Fc receptor-bearing population. Over 90% of the adherent cells flattened on the surface and 65% became elongated with uropods. Scanning electron microscopy showed loss of microvilli and the appearance of ruffles and long microextensions (characteristic of locomoting cells). Time-lapse cinematography showed that 30–50% of the adherent cells were able to translocate with a mean speed of 4.7 μm/min. They also examined cells on surfaces coated with high doses (1 mg/ml) of Con A, PHA, wheat germ agglutinin (WGA) and ricin. 5–50% of the cells became attached. Of these, only 9% on Con A showed locomotor morphology, 1% on WGA and 16% on PHA.

Haston et al. [unpublished] have recently studied the effect of fibronectin purified from FCS on adhesion of lymphocytes to glass surfaces.

Fibronectins might be considered good candidates for such a study since they are active as mediators of adhesion of fibroblasts and other cell types. However, the fibronectin preparation used failed to increase lymphocyte adhesion. This finding is perhaps hardly surprising since there is a large quantity of fibronectin bound to the lymph node stroma [240b]. If lymphocytes were to become bound to fibronectin, it would be likely to impede their locomotion in vivo. It is likely that the molecular interactions which mediate adhesion of fibroblasts to substrata are different from those which mediate adhesion of lymphocytes and of other invasive cells. Indeed, failure to adhere to fibronectin could be a prerequisite for the penetration of motile cells through connective and other tissues.

We have used a human B lymphoblast line, RPMI1788, for comparative studies of lymphocyte locomotion on glass surfaces and in micropore filters [*Shields and Wilkinson,* unpublished]. These cells migrated into filters (8 or 12 μm pore size cellulose nitrate filters, Sartorius, Göttingen, Germany) in media containing serum albumin (1 mg/ml) but no other protein, and this cell line can thus be used for studies with chemotactic factors in filter assays. However, when placed on glass, they failed to adhere or to locomote. Similarly chemotactic factors such as casein, to which they responded in filters, caused them to detach from glass. These cells would, however, adhere and polarize on Con A-coated glass. One possible difference between the glass slide and the filter is that the former presents a flat two-dimensional surface, and the latter a tight three-dimensional lattice for the cell to move on. A cell of low adhesiveness, such as lymphocytes are, may not get sufficient of its surface area into apposition with a two-dimensional surface to gain purchase whereas, in the three-dimensional mesh of the filter, a greater proportion of the surface area of the cell may be contact with the substratum, and thus the cell may gain sufficient adhesion to locomote. The filter mesh may also provide a constraint for cells that detach, so that they can still make progress even when their attachment is poor, whereas under the same conditions on a glass slide they can be seen to skid and float at the mercy of chance currents in the fluid medium surrounding them.

Native serum albumin binds appreciably to micropore filters or to glass [260] and albumin-coated filters can readily be prepared by soaking in serum albumin followed by washing. Figure 4 shows that murine lymph node lymphocytes show an increase in locomotion into such filters which is dependent on the concentration of albumin originally used to coat the filter. A similar dose-response curve was obtained in studies of human

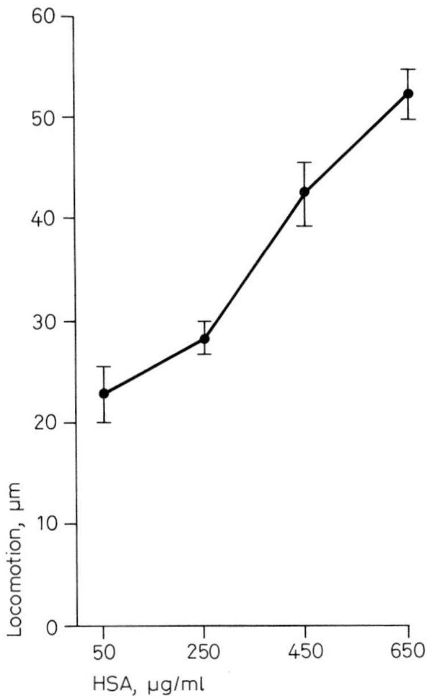

Fig. 4. The locomotion of mouse lymph node lymphocytes into filters (Sartorius; 8 µm pore size) coated with HSA (abscissa original HSA concentration used to coat the filter). Note the marked dose-dependent chemokinetic effect of the HSA. Similar dose-response curves have been obtained using human blood lymphocytes studied after 48 h culture in PHA [unpublished observations of the authors].

lymphoblasts on albumin-coated filters. Thus, albumin behaves as a chemokinetic factor for lymphocytes and lymphoblasts as it does for neutrophils, probably because albumin-coated filters provide a three-dimensional substratum that favours locomotion. We have been unable to show of the same chemokinetic effect for lymphocytes moving on a two-dimensional glass slide coated with albumin.

2. Lymphocyte Locomotion and Adhesion on the Surfaces of Other Cells

Human B lymphoblasts in mixed cultures with fibroblasts or with normal human glial cells on a solid plastic substratum adhered to and locomoted in contact with the feeder cells in preference to the plastic [197]. They migrated on the surfaces of these cells and could be seen to underlap the feeder cells so that they were confined to a groove between the ventral surface of the cell and the substratum. In this position they were moving on a three-dimensional rather than a two-dimensional surface. In such cultures lymphoblasts were observed to locomote at speeds of around 30 μm/min, which is comparable to the fastest speeds reported for neutrophils migrating on serum-coated glass [149] and considerably faster than the expected locomotion of lymphocytes on cell-free surfaces. *Ponten* [197] did not observe any contact inhibition of locomotion of lymphoblasts on contact with glia or fibroblasts, nor conversely of the glia or fibroblasts on contact with lymphoblasts. In the same studies, lymphoblasts did not show contact inhibition on contact with each other, nor did they form stable associations with each other while on solid surfaces. Nevertheless, the same cells, in suspension in culture, aggregate and normally grow as clusters unattached to the container. *Haemmerli* et al. [95] observed that rat leukaemia cells (L522), in culture with normal human fibroblasts or rabbit carcinoma cells, flattened and locomoted into the space between the tissue cell and the glass. When they left this space they rounded up. Reflection-contrast microscopy [94] showed that the lymphocytes made strong adhesions (dark areas) under the tissue cells but, if they moved out from under the cell, their adhesion quickly became weaker judged by loss of the dark interference pattern. The human lymphoblast line RPMI 1788 adhered and migrated very poorly on glass or plastic but migrated more vigorously on (or under) BHK fibroblasts and tended to remain in contact with such fibroblasts rather than migrating onto the plastic [*Shields*, unpublished observation], *Chang* et al. [30] have recently reported the locomotion of mouse splenic lymphocytes activated by mixed lymphocyte culture, Con A or lipopolysaccharide (LPS) in contact with murine 3T3 fibroblasts or L cells. These lymphocytes migrated under, preferably to over, the fibroblasts at speeds of up to 20 μm/min. Most of the locomoting cells from mixed lymphocyte cultures were T cells. The ability of lymphocytes to locomote on fibroblasts was not related to the H-2 haptotype of either the lymphocytes or the fibroblasts.

Haston [110] has reported experiments in which mouse lymph node lymphocytes, or thymocytes, were placed in mixed cultures with lymph node reticular cells, a population of large, flat, fibroblast-like cells which are normally present as stromal cells in lymphoid tissue. These are much larger than lymphocytes (up to 400 μm in length), bear Fc receptors and are phagocytic. Thymocytes and lymph node lymphocytes showed a strongly selective attachment to the reticular cells rather than to the artificial substratum and, after several days in culture, the appearance was one

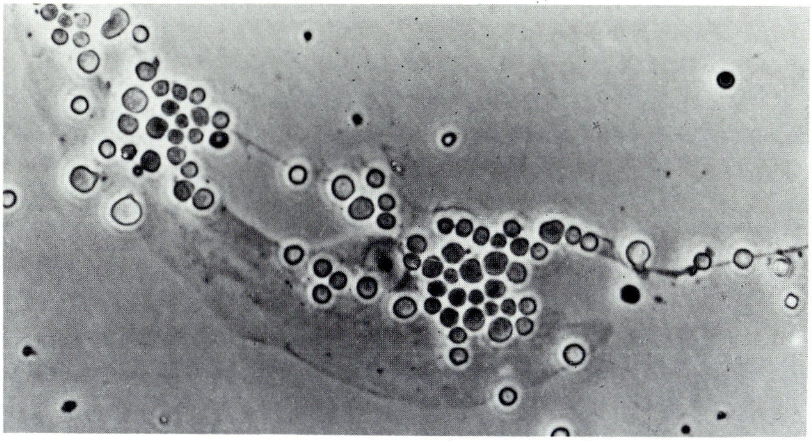

Fig. 5. Mouse lymphocytes taken from lymph nodes were added to a culture of lymph-node reticular cells grown for 5 days on Falcon tissue culture dishes in RPMI-FCS and 5% CO_2. The lymphocytes and reticular cells were cultured for a further 24 h. The picture shows large numbers of lymphocytes bound to a reticular cell and very few bound to the non-cellular substratum. The dark grey lymphocytes were underlapping the reticular cells and time-lapse cinematography showed that these were locomotor cells. The light refractile cells were overlapping the reticular cells and showed much less movement [picture courtesy of Dr. *Wendy Haston* from ref. 110].

of clusters of lymphoid cells associated with the reticular cells with the surrounding substratum relatively cell-free (fig. 5). Thymocytes were rather poorly motile. However, lymph node lymphocytes were highly motile, showing a strong preference for locomotion under the reticular cell layer, rather than on the bare plastic substratum. 78% of lymphocytes showed translocations under these cells with a mean speed of 23 $\pm$ 12 μm/min. It was almost always lymphocytes which underlapped the reticular cells, rather than those which overlapped them, which were the most motile. Their direction of locomotion was sometimes observed to be determined by the contours of the reticular cell layer, especially on confluent layers of reticular cells, i.e. they showed a form of contact guidance. *Haston and Shields* [unpublished] filmed the movement of RPMI 1788 on monolayers of the same reticular cells and these human lymphoblasts locomoted vigorously in contact with such monolayers. Thus, various lymphocyte types show an impressive capacity to locomote on or under the surfaces of various types of tissue cell, and one must conclude that when one is studying lymphocytes on artificial substrata one is studying them under conditions which are far from optimal for locomotion. The experiments with reticular cells are especially interesting since this is a surface with which lymphocytes normally have considerable contact in vivo, and this may provide a useful in vitro model for the movement of lymphocytes in lymphoid tissue.

Several of the above studies indicated that lymphocytes or blasts which migrated between the reticular cell (or fibroblast) layer and the substratum, or between two layers of cells, migrated better and showed stronger adhessions than lymphocytes on the top surface of such cells. It is possible that such cellular layers show a different pattern of membrane components on their top and their bottom surfaces, and that lymphocytes prefer the latter. A simpler explanation might be that the lower surface provides a three-dimensional substratum, and the top surface a two-dimensional one for the lymphocytes to move on.

The tendency for lymphocytes to show peripolesis and to cluster around macrophages has already been mentioned in relation to the role of the lymphocyte uropod in this process. Thymocytes [141, 144, 214, 227], blood lymphocytes [13, 150] and lymphocytes from lymph nodes and other lymphoid tissues [141, 142, 144, 214, 227, 259] all showed adhesion to macrophages. Adherence to peritoneal macrophages was stronger than to alveolar macrophages, neutrophils, eosinophils or fibroblasts [141, 227]. Adherence was stronger in the presence of an immune reaction. *Salvin* et al. [214] observed that antigen-sensitized guinea-pig lymphocytes showed locomotion both in the presence and the absence of antigen and that they adhered to macrophages in the presence of antigen while unsensitized lymphocytes did not. *Lipsky and Rosenthal* [141, 142] studied both antigen-independent and antigen-dependent binding of guinea-pig thymocytes and lymphocytes to syngeneic macrophages. The antigen-independent binding of thymocytes occurred even if the thymocytes were dead, but antigen-dependent binding required living syngeneic lymphocytes and was followed by DNA synthesis in the lymphocytes which were physically associated with antigen-bearing macrophages. These observations are similar to those of *McFarland* and colleagues [150–152] of human blood cells in mixed leucocyte cultures in which, similarly, thymidine incorporation was seen in the macrophage-associated lymphocytes. *Werdelin* et al. [259] and *Nielsen* et al. [171] reported that the lymphocytes involved in such reactions in sensitized guinea-pigs were mainly T lymphocytes by morphological criteria. Adhesive interactions of immune lymphocytes and macrophages are thus a rather specialized form of intercellular adhesion which could play such roles in immune reactions as, for example, to allow the cell-cell contact necessary for macrophage-T helper cell interactions.

Adhesion of lymphocytes to endothelium is of especial interest, not only in relation to migration across high endothelial venules (HEV) in lymphoid tissue (fig. 3), but also in relation to migration into inflammatory sites. Much emphasis has been placed on the interactions between lymphocytes and HEV and much less on lymphocyte-endothelium interactions in other sites, though lymphocytes migrate into many tissues where the vascular endothelium is not specialized, including sites of inflammation. In the latter sites, the interaction between lymphocyte and endothelium may resemble that between neutrophils and endothelium which has been well reviewed by *Lackie and Smith* [134]. In any event the physical and chemical events involved in lymphocyte adhesion to HEV on the one hand and to non-specialized endothelia on the other may be quite different, and different populations of lymphocytes are probably involved in both. *de Bono* [40] studied the adhesion of blood, lymph node, thymic or bone marrow lymphocytes from non-immune pigs to allogeneic or syngeneic aortic endothelium in vitro. All cells adhered well and showed motile forms. They adhered poorly to human umbilical cord endothelium and to non-endothelial pig cells. They adhered much better to cultured aortic endothelial cells than to fresh aortic endothelium, and attachment to cultured cells decreased with time after addition of the lymphocytes. The reasons for these variations are not clear. Adhesion required viable lymphocytes but the lymphocytes adhered to both living and formalinized endothelial cells. This is in contrast to adhesion of non-immune lymphocytes to macrophages where, as discussed above, the macrophage but not the lymphocyte required to be viable. Viability is not of course a necessary requirement for receptor interactions between cells but would be required if the cytoskeleton were involved in the attachment. The more specialized interaction between lymphocytes and the HEV has been the subject of many in vivo studies but it has not been so easy to devise in vitro models for it. *Stamper and Woodruff* [239, 240a] and *Woodruff* et al. [270] have used an assay of rat lymphocyte adhesion to HEV in which section of lymph nodes were fixed with glutaraldehyde and adhesion of lymphocytes to the sections was studied at 7°C. Rat lymphocytes from thoracic duct, spleen or lymph nodes, but not cells from thymus or bone marrow, showed selective adhesion to HEV but not to other structures in these sections. The authors interpreted this to mean that recirculating cells but not other lymphoid cells selectively recognized the HEV. The lymphocytes required to be alive, metabolizing and with an intact microfilament system, and in the presence of calcium. Pretreatment of the lymphocytes with trypsin, but not neuraminidase, inhibited adhesion. Adhesion to HEV was not species-restricted and therefore unlikely to be major histocompatibility complex-related. As in the experiments of *de Bono,* [40] the endothelium appeared to play a passive and the lymphocytes an active role. Since glutaraldehyde cross-links membrane structures, it would be reassuring if a similar selectivity could be shown for the interaction of lymphocytes with unfixed HEV. *Woodruff and Kuttner* [267] did show adherence to unfixed HEV, but not as strongly as to HEV fixed in glutaraldehyde.

C. Chemokinesis and Chemotaxis of Lymphocytes and Lymphoblasts

In the preceding section, we have presented evidence that lymphocytes are highly motile cells capable of speeds comparable to those of neutrophils or monocytes in environments that are favourable to locomotion, and provided that a population of cells with a good locomotor capacity is used. In this section, we wish to consider how the environment controls locomotion invitro and, in particular, the effects of chemical substances in this control. Environmental factors may determine the speed at which cells move without affecting their direction; thus regulating *random* locomotion. They may also determine the *direction* of locomotion. It should be evident from the preceding discussion that lymphocyte locomotion is dependent on the nature of the adhesions between cells and substratum (on flat two-dimensional substrata at least); thus, contact phenomena are likely to be important determinants of locomotion. Lymphocytes prefer to migrate in contact with the surfaces of other cells rather than on artificial substrata and their locomotion may be guided by the contours of these cells. This would be a form of *contact guidance* which might give directionality to cell locomotion, without being a chemotactic phonomenon. Lymphocytes moving randomly could also be trapped in an area of high adhesiveness, thus accumulating there: once again no chemotactic effect would be involved. Contact inhibition of locomotion could lead to a non-random distribution of cells by a similar mechanism. However, this phenomenon has not been demonstrated in lymphocytes.

Soluble chemical substances can act as chemokinetic factors by determining the rate at which cells move (orthokinesis) or the frequency of turning of the cells (klinokinesis). They can also act as chemotactic factors when cells exposed to chemical gradients show morphological orientation and directional locomotion in the gradient [129]. They may also act to generate more than one of the above reactions simultaneously in moving cells. The chemokinetic and chemotactic reactions of lymphocytes are discussed in some detail below, although they are only two of many possible locomotor reactions which lymphocytes may show to environmental agents. It merely happens that methods are available for evaluating chemotaxis and chemokinesis, and that other reactions are not so easy to observe or measure.

Numerous attempts have been made over many years to demonstrate chemotaxis of lymphocytes but, until very recently, these attempts were unsuccessful [108, 109]. Moreover, since chemokinesis of blood cells has only recently been accepted as a phenomenon, it is true to say that, until a few years ago, no effect of any sort of soluble environmental chemical substance on lymphocyte locomotion had been documented. This was an anomalous situation since, clearly, lymphocytes migrated from the blood stream into inflammatory sites in vivo, showing a similar time course and distribution to macrophages, cells in which chemotactic reactions had been demonstrated to be of importance. In 1971, *Ward* et al. [256] reported that lymphocytes would move into filters in response to culture supernates of lymphoid cells from immune animals to which antigen was added after culture. *Higuchi* et al. [116] reported that thoracic duct lymphocytes migrated into filters towards immunoglobulin which had been chemically modified by treatment with a protease. The assay methods used in these studies did not permit the distinction to be made between chemokinesis and chemotaxis. Various methods are now available which allow this distinction to be made; these are: (a) visual observation of directional locomotion of cells towards sources of chemotactic gradients usually by time-lapse filming [2, 275]; (b) morphological orientation towards gradient sources in specially designed chambers [273], and (c) use of the checkerboard assay for measuring chemotaxis into filters. The latter assay was described by *Zigmond and Hirsch* [275] and allows an assessment of the effect of both the absolute concentration of an attractant on the distance locomoted into a filter (chemokinesis) and of the concentration gradient (chemotaxis). A positive effect of positive gradients together with a negative effect of negative gradients on the distance moved by neutrophil leucocytes through filters has been accepted as evidence for a chemotactic effect of the factor in question. This is based on assumptions detailed by *Zigmond and Hirsch* [275], one of which is that the cells under study behave as a single population, which is usually true of neutrophils, but may not be so of lymphocytes. Nevertheless, this is the only convincing method for demonstrating chemotactic effects using assays which sample the locomotion of a population of cells (filter and agarose assays).

So far, lymphocyte chemotaxis has only been documented using the checkerboard assay. We and others have been unable to observe directional locomotion of lymphocytes on glass or plastic, but there are obviously problems with adhesion of lymphocytes to these materials, and artificial two-dimensional substrata may be inappropriate for these studies. *Anderson and Warren* [7] have made tracks of rat thoracic duct lymphocytes moving under agarose towards sources of various chemotactic factors. These studies have not been published in detail yet, but an abstract has appeared. Lymphocytes showed displacement suggestive of chemotaxis in gradients of various factors including endotoxin-activated serum, C3a and leucocyte extracts. Some analyses of tracks of lymphocytes in these experiments are reproduced in figure 6. Such studies clearly require to be extended and documented in full. As things stand at present chemotaxis of lymphocytes is much less securely established as a phenomenon than it is in neutrophils. In contrast, chemokinetic reactions are easy to demonstrate in lymphocytes, and the speed of their locomotion can readily be altered using various proteins and other materials discussed below.

D. Responses of Various Lymphocytes to Chemoattractants in Filters

The first report that the locomotion of lymphoid cells was influenced by gradients of attractants, judged by checkerboard assays, was that of *Russell* et al. [212]. In these studies, human B lymphoblasts from a number of established in vitro cell lines were exposed to various gradients of protein chemotactic factors such as endotoxin-activated plasma, casein and alkali-denatured serum albumin. The lymphoblasts migrated vigorously into filters (12 μm pore size) in response to such agents and showed a chemokinetic reaction, since migration increased as the attractant concentration was increased. Many of the cell lines also showed a chemotactic response to gradients of the same factors as judged by checkerboard assays (table III). Thus, when exposed to a given attractant, lymphocytes can show a complex locomotor response which is both kinetic and tactic, as is also true of neutrophils. In the same series of experiments, we examined the responses of mouse lymph node lymphoblasts obtained between 3 and 9 days after contact sensitization with oxazolone. These cell populations (chiefly T cells) showed considerable unstimulated locomotion in the

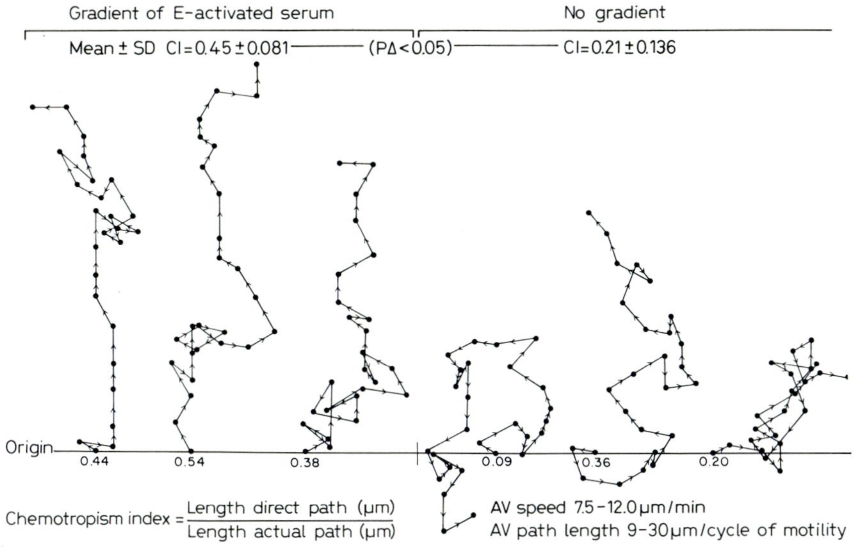

Fig. 6. Locomotion pathways of rat thoracic duct lymphocytes moving under agarose. Cells tracked during a period of 60 min by video-cinematography. The chemotropism index (chemotactic index) is derived from *McCutcheon* [149]. The nearer it approaches to unity, the greater the directional response to the gradient. Note that the cells tracked in gradients of endotoxin-activated serum showed higher chemotactic indices than those tracked in the absence of a gradient [unpublished data of Dr. *A.O. Anderson*].

absence of an attractant (fig. 7) and a chemokinetic response to attractants, but no evidence of chemotaxis. This suggested that lymphoblasts from different sources showed different types of locomotor response to attractants and raised the possibility that B lymphoblasts had a chemotactic capacity but T lymphoblasts not. However, we were later able to test an established human acute lymphatic leukaemia T lymphoblast line which did show evidence of chemotaxis in a checkerboard assay. This result has not previously been published and is therefore shown in table IV.

A series of established cell lines of human B lymphoblasts, human lymphoma cells and mouse plasmacytoma cells has been compared in our laboratory [*Shields,* unpublished observations] for their ability to respond to chemoattractants. Table II shows that the majority of the lymphoblast lines studied responded to chemoattractants, and several of these showed evidence of both chemotaxis and chemokinesis in checkerboard assays. The lymphoma lines and mouse plasmacytomas, on the other hand, while

Table III. Response of human lymphocytes to chemotactic factors in checkerboard filter assays compared with neutrophils and monocytes

	Percent chemotactic increment in positive gradients (first figure of pair) and negative gradients (second figure of pair) of the following factors				
	endotoxin-activated plasma/serum	casein	f Met Phe	f Met Leu Phe	Alk-den HSA
Lymphocytes					
B cell line F89	+33/–24				
B cell line J1M	+20/–13				
B cell line BAR	+1/–4				
B cell line RPMI 1788		+13/–9	+25/–16, +/–13	0/–1, +5/–1, +8/–1	+18/+4
T cell line		+38/–6			
Peripheral blood cells					
Cultured in PHA	+12/–13	+22/–11, +24/+14			
Cultured in Con A					
Cultured: No mitogen		+33/–11			
Neutrophils	+37/–17	+31/–33	+27/–9	+23/–42	+23/–16
		+29/–24			+28/–10
		+12/–25			
Monocytes		+22/–2			

The chemotactic increment is calculated from the experimental and calculated values in checkerboard assays (see table IV for detailed explanation) as follows:

$$\frac{\text{Observed locomotion } (\mu m) - \text{calculated locomotion}}{\text{Calculated locomotion}} \times 100.$$

A positive increment in positive gradients together with a negative increment in negative gradients indicates a chemotactic response.

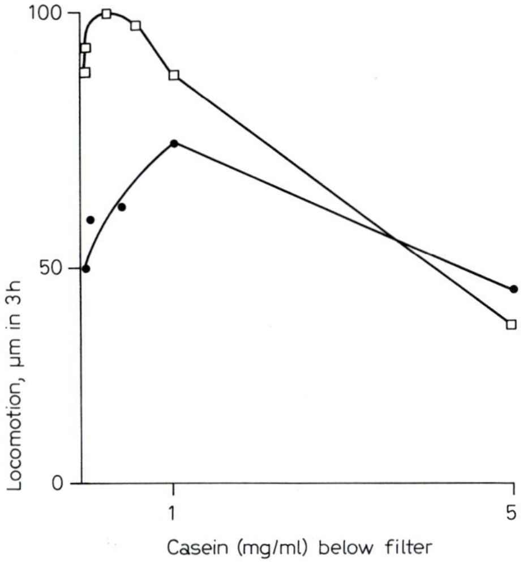

Fig. 7. The locomotion towards casein in filters of normal mouse auricular lymph node lymphocytes (●) compared with lymphoblasts and lymphocytes from auricular lymph nodes 72 h after local sensitization with oxazolone (□). Note the very high locomotion of the contact-sensitized cells even in the absence of casein [from ref. 212].

being motile, showed poor responses to attractants. This observation is reminiscent of the finding of *Nilsson and Ponten* [173] of a similar pattern of locomotor behaviour of lymphoblasts and lymphoma cells migrating in contact with the surfaces of various tissue cells in visual assays.

When human lymphocytes were studied directly after removal and separation from blood, a small proportion migrated into filters, but this proportion could be increased following a period of culture in vitro. Blood lymphocytes were cultured for 48–72 h in Eagle's medium with FCS and a mitogenic dose of PHA. Such cultured lymphocytes were very motile and showed chemokinetic and chemotactic reactions in endotoxin-activated plasma, casein and other protein factors. Many of the moving cells were lymphoblasts, but small lymphocytes also migrated vigorously towards attractants [264]. Interestingly, low concentrations (a 100-fold to a 1,000-fold lower than the mitogenic dose) of PHA induced locomotor reactions in such cells, and checkerboard assays suggested that these reactions, which were never more than moderately strong, were both chemokinetic and chemotactic. As stated earlier, lectins may modify lymphocyte adhesion to substrata and it remains to be evaluated to what extent the locomotor response to PHA is due to alteration, by the lectin, of the

Table IV. Effect of varying the absolute concentration and the concentration gradient of casein on the locomotion of a human T cell (acute lymphocytic leukaemia) line into filters

		Locomotion into a 12-μm pore size filter (3 h at 37°C). Mean of 5 fields in 2 filters ± SEM: concentration of casein (μg/ml) below filter				
		0	50	300	550	800
Concentration of	0	*17 ± 1.2*				
casein (μg/ml)	50		*21* ± 1.1	22 ± 1.4(21)	31 ± 1.6(22)	39 ± 2.0(22)
above filter	300		28 ± 2.2(26)	*26* ± 2.7		38 ± 2.5(29)
	550		25 ± 1.6(34)		*37* ± 2.9	
	800		32 ± 2.4(43)	37 ±2.3(45)		*50* ± 2.8

The cell line was obtained from Dr. *John Hanson,* Sloan-Kettering Institute, N.Y. Cells were cultured in RPMI +20% FCS and used at a concentration of 5×10^5 per ml in Gey's solution, pH 7.2. The figures without parentheses in the above table indicate the distances migrated by cells in a series of chambers each containing a different concentration of casein above and below the filter. Down the diagonal from upper left to lower right, the casein concentration is the same above and below the filter, i.e. there is no gradient and the cells are responding to the absolute concentration. Note that under these conditions, locomotion distance varies as the absolute concentration of casein varies, i.e. the cells show chemokinesis. Above the diagonal the cells were moving in a positive gradient, below it in a negative gradient. The figures in parentheses are estimates, calculated after the equation in ref. 275, of locomotion expected in any chamber assuming the cells responded only to absolute concentration but not to gradients. Note that in fact cells in positive gradients have locomoted farther than expected, and in negative gradients less than expected, i.e. there is evidence of a chemotactic reaction.

substratum on which the lymphocytes move. Following this paper, *O'Neill and Parrott* [177] showed that addition of a mitogen to lymphocyte cultures was not necessary to induce locomotor responses in cultured human blood lymphocytes and that a period of culture in serum-supplemented medium alone (for 24–72 h) was sufficient. When T cells were separated on nylon wool from non-T cells, both populations were able to respond by migration into filters towards casein or endotoxin-activated serum. Checkerboard assays did not give clear cut evidence for chemotaxis, since a positive effect of positive gradients, but not a negative effect of negative gradients, could be demonstrated. This is the kind of result one might expect with a heterogeneous population of cells, one population showing chemotaxis and another chemokinesis but no chemotaxis; however, other explanations can also be envisaged. *Parrott* et al. [190] later studied locomotion of human T cell subsets. Tγ showed little ability to respond to attractants, whereas Tμ moved very well towards casein. Since these subsets are functionally different, suppressor cells carrying the Tγ marker, and helper cells, the Tμ marker, this observation is of potential importance in understanding the migration patterns of functionally different T subsets in vivo.

Following the observation [246, 247] that cross-linking of surface Ig on mouse lymphocytes with anti-Ig could induce locomotion, *Schreiner and Unanue* [220] studied the effect of anti-Ig on the locomotion of lymphocytes into filters. When using high doses of anti-Ig (10–100 μg/ml) they observed chemokinesis but no chemotaxis. In a later study using rat lymphocytes and much lower doses of anti-Ig (100 ng/ml), *Ward* et al. [257] were able to demonstrate a chemotactic reaction of B cells in gradients of anti-Ig. This ability to respond chemotactically to low, but not high, doses of anti-Ig is reminiscent of our finding of chemotactic locomotion of human lymphocytes to submitogenic, but not mitogenic doses of PHA cited above. Since anti-Ig and PHA are both cross-linking ligands, it is possible that chemotaxis is only induced by minimal concentrations of such ligands. One requirement for chemotaxis is that cells should be able to detect concentration differences across their own length [272]. A high concentration of a cross-linking agent which induced clustering and capping might impair this discrimination by segregating and reducing the functional number of receptors, and if so the cell would detect gradients of such ligands most efficiently under conditions where membrane perturbation was minimal.

Since polyvalent ligands such as lectins and anti-Ig can induce locomotor and chemotactic reactions in lymphocytes, it is possible that other polyvalent ligands could behave similarly. In this context, and since an ability of lymphocytes to respond chemotactically in sites of immune reactions may be useful, it was obviously of interest to examine the effects of antigen on lymphocyte locomotion. Mice were immunized with serum albumin in *Bordetella pertussis* adjuvant and challenged with the same antigen 21 days later [263]. Following challenge, lymphocytes were taken from the lymph nodes draining the challenge site. In primed mice, these lymphocytes, studied by a checkerboard filter assay, showed chemotactic reactions to the antigen, whereas those from control unprimed mice, given only a challenge dose of antigen, showed only chemokinesis. The response of immune lymphocytes to antigen was studied over a period from 3 to 9 days after challenge. Cells responded during all of this time. At 3 days many of the locomoting cells were lymphoblasts. Later the number of blast forms declined and by 9 days the locomoting population was predominantly made up of small lymphocytes. In retrospect, it was unfortunate that serum albumin was chosen as the antigen for these studies, since it is such a good chemokinetic agent for normal lymphocytes, thus making the chemotactic reactions more difficult to interpret; however, when mice were immunized in the same way with ovalbumin, which was not chemokinetic, the lymphocytes from the primed group, but not from the unprimed group, responded chemotactically. In further unpublished studies using the same experimental protocol and mice immunized with DNP-IgG, chemotactic reactions to DNP-IgG and to DNP-HSA were seen, but not to unmodified IgG or HSA. In the serum albumin experiments, about 20% of the lymphocytes in immunized lymph nodes were induced to locomote and about the same proportion were observed to bind fluorescein-labelled antigen, many of these being lymphoblasts. This proportion seems too high for all the cells to have been derived from the primed population. Many may have been responding by non-antigen-specific chemokinesis, or possibly some of the unprimed cells may have had soluble antibody bound to Fc receptors and thus been able to bind antigen although not primed to it.

The few studies of lymphocyte chemotactic factors carried out up to the present time, and summarized above, suggest that responsive populations of lymphocytes can migrate towards many of the factors (e.g. endotoxin-activated serum or plasma, casein, denatured proteins), which are known attractants for neutrophils or mononuclear phagocytes. Other neutrophil attractants, such as formyl peptides, seem to be rather poor attractants for lymphocytes. The studies with anti-Ig and with antigen suggest that lymphocytes may also show specialized forms of chemically stimulated locomotion based on their possession of intrinsic membrane immunoglobulin. Polyclonal ligands such as lectins can also induce locomotor reactions. Interestingly, studies with membrane-modifying agents suggest that a different class of membrane-binding sites may mediate locomotion to cross-linking agents on the one hand and to conventional (phagocyte) chemotactic factors on the other. The response to PHA was readily abolished by treatment of the cells with proteases (trypsin, pronase) but not much affected by lipid-specific agents. On the other hand, responses to casein, denatured albumin or to activated plasma were diminished in cells pretreated with *Clostridium perfringens* phospholipase C or staphylococcal sphingomyelinase C, but only minimally so by proteases or glycosidases [264]. Thus, the glycoprotein-binding sites for lectins are presumably clear of the bilayer and accessible to proteolytic attack, whereas those for the other chemotactic factors may be fairly closely integrated with the lipid bilayer and disorganized by agents which damage membrane lipids, but relatively inaccessible to proteolytic attack.

Chemokinetic reactions are easily demonstrable in lymphocytes using filter assays. Many factors which are chemotactic for neutrophils and which were shown also to be chemotactic for defined lymphocyte populations, e.g. B lymphoblasts, only induce chemokinetic reactions in other lymphocyte populations, e.g. oxazolone-induced mouse lymph node cells. It is possible that lymphocytes only show chemotactic reactions under rather specialized conditions, and that the locomotion shown by lymphocyte populations in vivo is, for the most part, chemokinetic. Chemotaxis implies accumulation of cells at a gradient source and retention of the cells once they arrive at the source, since any cell which moves out will be attracted back in by the gradient. Such a mechanism would be an obvious disadvantage to a recirculating cell, but might be advantageous to a similar cell moving into a site of inflammation. As will be made apparent in the discussion of lymphocyte locomotion in vivo, small lymphocytes may show selective binding to high endothelial venules, whereas lymphoblasts are much less selective. Lymphoblasts bind to endothelia in many sites and move into foci of inflammation i.e. their behaviour is closer to that of the blood phagocytes than that of the small lymphocytes.

The term chemokinesis covers effects of environmental chemicals on the rate of locomotion or frequency of turning of cells, and obviously cells can be accelerated or decelerated by numerous mechanisms, e.g. metabolic activation or inhibition, changes in level and activity of contractile proteins in the cell cytoplasm or changes in the character of the adhesive forces binding the cell to the substratum. The important chemokinetic effect of serum albumin, the protein present in highest concentration in the cells' environment, is almost certainly mediated by substratum-bound albumin which provides a suitable surface for locomotion. Albumin is a pure chemokinetic agent for both lymphocytes [263] and neutrophils [130, 261], and has no chemotactic activity. Presumably other environmental agents with chemokinetic properties will be discovered. We have recently studied the effects of the soluble factors [interaction modulation factors; 37, 38], released by T and B lymphocytes, on locomotion of lymphocytes into filters. The factor released by T cells was earlier shown to reduce cell-cell adhesion of B cells but not of T cells, and conversely, the B cell factor reduced cell-cell adhesion of T cells but not of B cells. These factors were found to accelerate locomotion of the opposite cell-type into filters but not of the cell type which produced them. This effect was chemokinetic, not chemotactic [*Wilkinson and Curtis,* unpublished], thus, under conditions where adhesion was moderately reduced, locomotion of heterotypic lymphocytes was enhanced by such factors.

Some preliminary observations of the effects of drugs may throw some light on the mechanism of chemokinetic locomotion of lymphocytes. As would be anticipated, cytochalasin B, which paralyses microfilaments, suppressed lymphocyte locomotion [29]. *Schreiner and Unanue* [220] reported that drugs that elevate intracellular cGMP levels (acetylcholine, carbamylcholine and cGMP itself) increased spontaneous motility of lymphocytes, especially B lymphocytes. Drugs which elevate cAMP levels (dibutyryl cAMP, theophylline, cholera toxin, isoproterenol) inhibited the locomotion induced by anti-Ig, except in cells pretreated with colchicine. Human B lymphoblasts migrating in response to activated plasma moved faster in the presence of colchicine or vinblastine than untreated cells [212], and rat spleen lymphocytes similarly moved faster in the presence of colchicine, but not of lumicolchicine [29]. Chemotactic effects in checkerboard filter assays were lost in colchicine-treated cells [212]; however, in visual assays using neutrophils, it is evident that colchicine does not abolish the ability of cells to orient towards gradient sources [2], but does impair the coordination of directional movement. The role of microtubules in locomotion is still not clear.

There has been considerable recent interest in low molecular weight chemotactic factors for neutrophils, especially formyl peptides and certain classes of lipids. It has been reported briefly, without documentation, that lymphocytes lacked receptors for formyl-methionyl tripeptides [12]. We have examined the locomotor responses of cultured human blood lymphocytes and B lymphoblast lines to these peptides. We were unable to show any chemotactic response to formyl-methionyl-leucyl-phenylalanine (f-Met-Leu-Phe); however, these cells usually showed some chemokinetic reaction to this tripeptide. In two checkerboard assays, the dipeptide, f-Met-Phe was both chemotactic and chemokinetic [226].We do not understand why there should be a difference between these peptides, but although the reactions of lymphocytes to formyl peptides are not impressive, the results suggest that the evidence for a binding site on lymphocytes for these peptides should be re-examined. One difficulty with studying chemotactic reactions of lymphocytes to these peptides and other low molecular weight factors may be that gradients of such small molecules under the usual conditions of assay may be rather short-lived [274] and by the end of the 3-hour assay period required for lymphocyte locomotion into filters the peptide concentration may not be greatly different on the two sides of the filter. Another neutrophil chemotactic factor which has been shown to be chemokinetic, but not chemotactic for lymphocytes, is the arachidonic acid derivate, HHT (12-L-hydroxy-5,8,10-heptadecatrienoic acid) [146]. Arachidonic acid itself also had chemokinetic activity, as did α-thioglycerol, a reducing agent which has been shown previously to stimulate various other lymphocyte functions. HHT is derived from arachidonic acid by cycloxygenase action, and indomethacin, which is a cyclogenase inhibitor and anti-inflammatory agent, was shown to suppress arachidonic acid-induced lymphocyte locomotion. Such observations may prove to be of some interest since recent studies [125, 147] suggest that prostaglandins, particularly PGE_2 may markedly reduce lymphocyte output in efferent lymph from sheep lymph nodes, and that such prostaglandins are released in lymph nodes after antigenic challenge. It is possible that the presence of such materials may retain lymphocytes in lymph nodes by chemotactic or negative chemokinetic mechanisms.

III. Lymphocyte Migration and Locomotion in vivo

As emphasized in section I it is not the intention here to give an overview, yet alone an extensive review of the literature on lymphocyte migration and recirculation but rather to pinpoint those areas in which understanding of lymphocyte locomotion is particularly relevant.

Few lymphocytes freshly isolated from blood or lymph show active locomotion in vitro although lymphoblasts taken from efferent lymph from stimulated lymph nodes [104], from colostrum [229] or from the gut mucosa [21] may do so. In histologic and ultrastructural studies, however, lymphocytes may be caught during passage across endothelium and epithelium and in many cases the morphology which is assumed is the same (the 'hand mirror form') as that seen in in vitro studies. Lymphocytes have been observed with this morphology when crossing the post-capillary venules into lymph nodes (fig. 2) [4, 60, 174], into Peyer's patches [217] the marginal zone of a Malpighian body in rat spleen [174], the sinusoids of bone marrow [42], crossing vascular endothelium into sites of experimental allergic neuritis [11], during rejection of a kidney allograft [193] and during an inflammatory response [88] in immune arthritis.

Locomoting lymphocytes have also been seen within the epithelium of mammary alveoli [224] and the epithelium covering intestinal villi [162, 243], even in transit across the dome epithelium covering Peyer's patches to or from the intestinal lumen [*Bruce and Parrott,*, unpublished].

The process of active locomotion is thus most easily seen when lymphocytes are moving out of a lumen (capillary venule or sinusoid) and into a tissue or out of a tissue (mammary gland, gut mucosa) into a lumen (alveoli, intestinal lumen), although lymphocytes are continually or frequently in locomotion whenever they are within tissues. The assumption of locomotor morphology by lymphocytes is not confined to their entry into the organized lymphoid tissues. In the following sections of the review we define the recirculatory properties of different populations of lymphocytes and attempt to relate those different properties to their locomotor characteristics and consider whether the changes which antigen induces in recirculation of lymphocytes can be related changes in locomotion.

The migration of lymphocytes through non-lymphoid tissue, taking the skin and the intestinal mucosa as our principal examples, is then described. There are substantial differences between the concentration of different populations of lymphocytes in the skin and the blood stream. Different again are the concentrations and populations of lymphocytes in the intestinal mucosa and the blood stream. It is obvious that selection processes must occur to achieve these different concentrations and populations of lymphocytes and at first thought it would most likely be that selection occurs at entry by selectively stimulating the locomotion of limited numbers and type of lymphocytes. But with mobile populations selection can also occur within tissues if one cell type moves faster than others or even at exit points. We, therefore, consider what role locomotion, whether directed or random, plays in the selection of lymphocytes in non-lymphoid tissues.

A. The Traffic of Lymphocytes from Blood to Lymph. Transit Times

The pioneer work of *Gowans* [85, 86] and of *Hall and Morris* [99, 100] established that the majority of small lymphocytes circulate continuously from blood to lymph. Subsequent work demonstrated that although their portals of entry may be similar, T and B lymphocytes follow different pathways within the lymphoid tissues [184], and indeed take very different times to fulfil similar journeys.

The capacity for some lymphocytes to recirculate is already present in the fetus [192] of the sheep by 75 days of gestation. The ovine placenta is epitheliochorial and prevents the exposure of the fetus to exogenous antigen. One must assume, therefore, that recirculation of lymphocytes is not initiated in response to exogenous antigen, and small T lymphocytes, at least, recirculate very vigorously in the sheep fetus [24]. Much emphasis has been placed on the capacity of lymphocytes to recirculate and on the similarity of modal transit times [245] of lymphocytes from blood to lymph in the sheep, mouse and rat. Most measurements in the sheep are around 18 h [26, 106] and the modal transit time of the majority of lymphocytes in mice and rats is also around 18 h. In fact these modal transit times only apply to young adult animals; they are longer in older adult sheep [78] and shorter in fetal sheep [24].

Furthermore, most measurements of modal transit times are only applicable to small T lymphocytes. It is known that B lymphocytes move more slowly than T cells in rats [65, 120] and measurement of a definite model transit time for B cells is often not practicable [120]. Thus, after intravenous injection in the rat of normal (i.e. mostly T) lymphocytes, 33% are recovered in the thoracic duct over a period of 48 h, while 19.5% of separated B cells are recovered over the same time [120]. In the mouse, the difference is more striking: 39–65% of normal unseparated lymphocytes are recovered over 48 h, while only 1–2% of B cells are recovered over the same time [235]. There is indirect evidence that B lymphocytes in the sheep also move slower than T lymphocytes [77] and comparatively few B lymphocytes recirculate in the sheep fetus [24]. *Strober* [241] has shown that B cells in the rat only acquire the capacity to recirculate when they become memory cells whilst T cells lose the capacity to recirculate when they become memory cells [96].

Other populations of lymphocytes also recirculate at different rates or do not recirculate at all. The differential distribution of the Lyt subsets in the mouse, and of T cell subsets within the lymphoid tissues in man, suggests that suppressor cells recirculate less than helper cells [14, 51, 99]. Most B lymphoblasts destined for the gut circulate only once since they are destined to become sessile plasma cells [102, 124] but the few B lymphoblasts that do recirculate do so more quickly than B cells [120]. Only a small proportion of T lymphoblasts and effector T cells have the capacity to recirculate [236]. Yet both T and B lymphoblasts are highly motile cells in vitro, small B lymphocytes are less motile in vitro and small T lymphocytes when freshly isolated from blood have little or no motility at all. It is difficult, therefore, to equate the capacity for lymphocytes to recirculate with their capacity for in vitro motility, and we now develop the theme that the different populations or lymphocytes recirculate differently because they have different locomotor properties, and while some intrinsic motility may be a necessary prerequisite for recirculation, highly stimulated motility may be inimical to recirculation.

B. Destination of Injected Lymphocytes

1. Time Course Studies

A technique which has provided much evidence relevant to the present thesis is of that which involves injecting a bolus of labelled cells and sampling the changing patterns of distribution with time after intravenous injection (fig. 8). This technique is unphysiological but it nevertheless does give evidence of routes and transit times across individual tissues which cannot be gained from measuring modal transit times from blood to lymph. A variety of isotopes have been used for this purpose and their suitability for the task varies [200]. Among the most commonly used are ^{51}Cr for quantitation of small lymphocytes, ^{125}I-UdR for quantitation of large lymphocytes and lymphoblasts, ^{3}H-adenosine, ^{3}H-uridine, ^{3}H-thymidine for quantitation and for autoradiographic analysis of the location of lymphocytes within tissues. A typical time course study following the injection of ^{51}Cr-labelled lymph node cells is shown in figure 8 [69] and this is similar to studies by others [200]. Following their infusion, labelled lymphocytes are normally rapidly removed from circu-

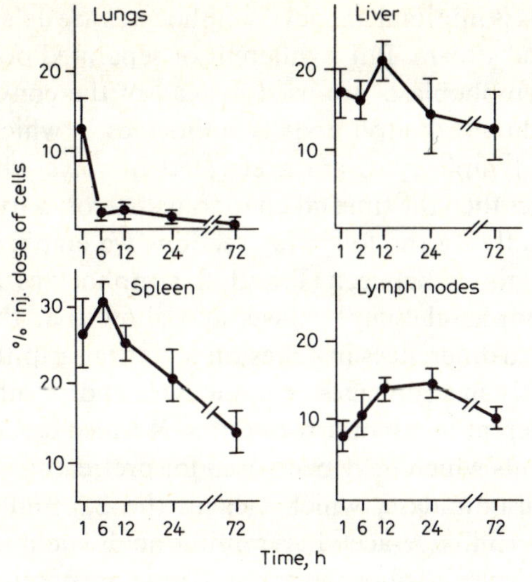

Fig. 8. Time course of the migration of ^{51}Cr-labelled small lymphocytes through lungs, liver, spleen and lymph nodes of CBA mice [data from ref. 69].

lation (within 1 h), as shown by the fast drop in radioactivity from the blood. Immediately after injection lymphocytes enter the lungs and at early times after injection the concentration of radioactivity is high but normally this falls sharply to about 2% by 6 h and is constant thereafter. The rise in radioactivity in the liver occurs immediately after the lungs, of is around 15–20% of the labelled cells injected, and stays within that range over 3 days. The number of cells in the spleen climbs during the first 2 h to 20–30% of injected activity, remains relatively constant for 8–12 h and then falls gradually. A small number of labelled cells arrives in lymph nodes within minutes and this number climbs relatively slowly, reflecting the release of cells from the larger capillary beds of lungs, liver and spleen and does not peak until 12–24 h after injection.

The attention of most investigators has been focussed on this arrival of lymphocytes in lymph nodes and their interaction with the HEV of lymph nodes. But the work of *Freitas and de Sousa* [70–73], *Andrews* et al. [8], and *Ford* et al. [66, 67] has emphasized that the events in the lungs, liver and spleen are not trivial and must also be taken into account when analysing the processes of lymphocyte migration. It is true that in most non-lymphoid tissues the interaction of small circulating lymphocytes with the walls of blood vessels is minimal and few lymphocytes are delayed and fewer still leave the blood stream. But if different or separated populations of lymphocytes or lymphoblasts are used instead of the conventional mixtures of thoracic duct or lymph node lymphocytes in which T cells predominate or if the lymphocytes are pretreated in ways which change their surface properties then the time taken to transit across various tissues will change dramatically. In the following sections we emphasize the difference in migration times between T and B lymphocytes and between mesenteric and peripheral lymphoblasts and show that these differences can be attributed to differences in adhesion to various capillary beds and other cells, especially macrophages or monocytes and/or differences in their inherently different locomotor properties. We also describe the effects of some of the agents which have been used for pretreatment of lymphocytes, including neuraminidase, which cleaves the 2,3 and 2,6 glycosidic linkages between terminal *N*-acetylneuraminic acids and mucopolysaccharide, the enzymes, trypsin, phospholipase A and phospholipase C, which all decrease cell adhesion, plant lectins, Con A and PHA, which have an affinity for sugar residues on cell surfaces, and sulphated polysaccharides such as dextran sulphate, which change the net surface charge and adhesion of lymphocytes [8].

2. The Migration of Lymphocytes and Lymphoblasts through the Lungs (fig. 9)

The first capillary bed which all intravenously injected lymphocytes encounter is the lungs and it fortunately provides us with good experimental examples of many of the points we wish to emphasize. It has been suggested that the initial hold-up of injected labelled lymphocytes within the lungs is due to the retention of in vitro damaged cells [106, 187], but since labelled lymphocytes are also held up at early times after injection when given by an intra-arterial route [258] it seems reasonable to deduce that the lungs are normally and for physiological reasons a slow capillary bed [200]. Intravenously injected lymphoid cells must pass through the capillary bed of the lungs but the time which they take depends upon the experimental animal used and the population, subset or state of differentiation of the lymphoid cell. All lymphoid cells move slowly through the lungs of sheep [78] although they move much faster in the fetus than adult sheep [24]; indeed, the rapid clearance of the lung capillary bed probably accounts for greater speed of recirculation in the sheep fetus [24]. Immature thymocytes move very slowly through the lungs [84, 183] in

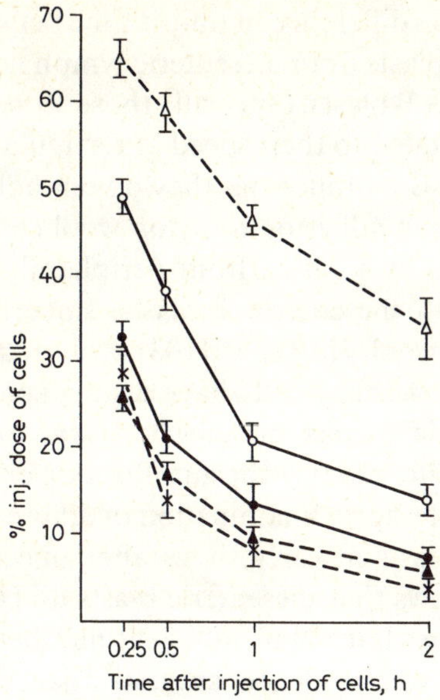

Fig. 9. Time course of the migration of ^{51}Cr-labelled small lymphocytes and (^{125}I) UdR-labelled lymphoblasts across the lungs of CBA mice [*Shields and Parrott,* unpublished]. ● = ^{51}Cr-labelled lymphocytes from mesenteric lymph nodes; ▲ = ^{51}Cr-labelled lymphocytes from peripheral lymph nodes; × = ^{51}Cr-labelled nylon wood purified (T) from peripheral lymph nodes; ○ = (^{125}I) UdR-labelled mesenteric lymphoblasts; △ = (^{125}I) UdR-labelled peripheral lymphoblasts.

comparison with small lymphocytes. Small T lymphocytes move more rapidly across the lungs than small B lymphocytes [70] but when small lymphocytes are activated into lymphoblasts these then move more slowly through the lungs than small lymphocytes (fig. 9), and it depends upon which lymph nodes they are derived from whether they move slowly or very slowly. Cells activated within the gut-associated lymphoid tissues, whether taken from the mesenteric lymph nodes of mouse (fig. 9) [106], thoracic duct lymph of mice [74, 236], the mesenteric lymph nodes or thoracic duct lymph of rat [232] or the intestinal lymph of the sheep [103] move much more rapidly through the lungs than lymphoblasts activated in the periphery, whether taken from mouse peripheral lymph nodes [207, 208], rat peripheral lymph nodes [232] or peripheral efferent lymph in the sheep (fig. 9) [103].

It is interesting that B blasts taken from mesenteric sources apparently move through the lungs as rapidly as T blasts from mesenteric sources as evidenced by the fact that there was little difference in transit time between a mixture of B and T blasts and pure T blasts from mesenteric lymph nodes [206]. This rapid transit of thoracic duct B blasts, especially those secreting immunoglobulin, would seem to be related to their speed of recirculation [120] and to the fact that unlike small B lymphocytes they do not adhere to nylon wool [25, 74]. But differences in adhesion to nylon wool cannot explain the difference in transit times of T blasts from peripheral sites through the lungs in comparison with mesenteric T blasts, since both sources of T blasts pass through nylon wool. *Hall* et al. [103] proferred the explanation that peripheral lymph node blast cells have a lung/spleen-somatic lymph node-'homing' potential whilst mesenteric lymph node blasts are destined for their 'homing' site, namely the gut. But this explanation would not appear to be the whole story. Examination of differences between peripheral and mesenteric blasts at early times after injection [*Shields and Parrott,* unpublished] shows that mesenteric blasts do begin to enter the small intestine at early times but not in sufficient numbers to account for the discrepancy. Moreover, evidence suggests that most peripheral lymphoblasts are inflammatory or activated T blasts and their 'home' if anywhere is an inflammatory site in the skin, peritoneal cavity or gut [10, 207, 208]. Certainly when there is one or more sites of inflammation then the level of radioactivity in spleen, liver and lungs is reduced [208]. In the absence of inflammatory sites peripheral blasts may have no alternative but to traverse and retraverse the lungs, liver and spleen until they die or redifferentiate into small lymphocytes. There is also evidence, however, that peripherally activated lymphocytes can leave the circulation in the lungs. The level of radioactivity in the blood drops very rapidly after the injection of peripheral lymphoblasts [*Shields and Parrott,* unpublished], suggesting that margination can occur. Moreover, when the retention of peripheral lymphoblasts in the lungs is exacerbated in mice by injecting reserpine which induces severe hypotension [205], sometimes as many as 50% of injected lymphoblasts are still retained in the lungs at 24 h. There are also increased numbers in blood, liver and spleen, but decreased numbers in peripheral lymph nodes. Autoradiographs of lungs show that many blast cells are within the alveolar walls [*Rose and Parrott,* unpublished]. Subsequently, when the effects of reserpine wears off the cells resume migration normally. Reserpine treatment of recipients has some effect on the transit time of mesenteric lymphoblasts but it is less than that on peripheral lymphoblasts and less still on small lymphocytes. *Rouse and Wagner* [210] showed that in vitro activated cytotoxic T cells also move very slowly through the lungs and demonstrated by means of autoradiography that the cells are present in the alveolar walls.

3. The Effect of Pretreatment with Lectins and Enzymes on Lung Transit Times of Lymphocytes

Pretreatment of small lymphocytes with various substances so as to change their surface properties also alters the time of migration through the lungs. For example, pretreatment with Con A causes a temporary hold-up of small lymphocytes both T and B in the lungs [75] accompanied by a deficit of cells in the blood [233]. Autoradiographic examination [74] showed subsequently that the cells were mainly in the alveolar walls but sometimes in small groups around alveolar macrophages. Pretreatment of lymphocytes with the sulphated polysaccharide dextran sulphate, however, causes retention both in the blood and the lungs, and autoradiographic analysis shows again that many lymphocytes were present in the alveolar walls [73]. Moreover, treatment of recipient mice with dextran sulphate has a similar effect to pretreatment of injected lymphocytes [73]. Pretreatment of lymphocytes with various enzymes also alters the progress of lymphocytes through the lungs. Neuraminidase decreases transit time; trypsin causes increased retention in the blood stream but variable changes in the lungs [66, 71] and phospholipase A (PL-A) [71] causes an increase in transit time with a decrease in cells in the blood.

It is extremely unlikely that one simple explanation can be found to account for all the alterations in transit times to lymphocytes and lymphoblasts which have just been described. Small B lymphocytes probably move more slowly than T cells because they adhere or marginate temporarily to the lungs' capillaries. A similar explanation could account for the effect of Con A treatment of lymphocytes. Trypsin-treated cells are unlikely to adhere, their prolonged retention in the lungs is probably a consequence of their being unable to leave the blood stream elsewhere whilst dextran sulphate probably causes both greater adherence to the lung capillary bed and inhibits entry to spleen and lymph nodes. The effects of PL-A are different yet again. Like trypsin the main effect of PL-A is to decrease adhesiveness but unlike trypsin, PL-A treated lymphocytes do leave the blood stream. *Freitas and de Sousa* [71], therefore, speculated that PL-A, which is known to stimulate random and chemokinetic locomotion of lymphocytes [264], would have caused the treated cells to leave the blood stream in the lungs. Peripheral lymphoblasts already have high random motility without stimulation or treatment (see above) and it is a reasonable deduction from the data presented that the considerable time which these lymphoblasts take to clear the lung capillary bed is a consequence of extravasation, though whether these cells ultimately return to the blood stream in the lungs or exit through lymphatic drainage has yet to be tested. Further evidence in support of our theory that highly stimulated motility is incompatible with rapid recirculation can be obtained by studying the transit of cells across the liver.

4. Transit of Lymphocytes and Lymphoblasts across the Liver

After leaving the heart, lymphocytes are presented with the opportunity of entering the various lymphoid tissues, the bone marrow or non-lymphoid tissues, including the gut and the liver. They are delivered in proportion to the blood flow and enter the tissue according to the concentration of cells in the delivering blood, so within half an hour of injection some lymphocytes appear in all lymphoid and non-lymphoid tissues. The build-up in lymph nodes is slow because at early times large numbers of lymphocytes are transiently removed from circulation by the larger tissues, liver, bone marrow and spleen. The liver has a large capillary bed and contains large numbers at early times after injection of labelled cells (fig. 8), although the time of peak accumulation will vary according to the rate of release of cells from the lungs, the lymphocyte or lymphoblast population studied and whether or not the cells have been pretreated with enzymes or lectins. As with the lungs B lymphocytes cross the liver more slowly than T lymphocytes, peripheral lymphoblasts more slowly than mesenteric lymphoblasts. Pretreatment with Con A delays transit [70, 74, 233]. Pretreatment with dextran sulphate [73], PL-A [71] or trypsin [66, 71, 72], had little effect on the progress of lymphocytes through the liver. Pretreatment with the enzyme neuraminidase [71, 80] and with the viruses NDV and influenza, because they contain neuraminidase [268, 269] also delays lymphocytes, particularly T lymphocytes within the liver. Such treatment does not kill the cells for they recover after 24 h and move on [71, 72].

It is often assumed that lymphocytes only accumulate in the liver if they are dead. The liver does indeed remove dead and effete lymphocytes from circulation, but it is obvious that many lymphocytes do leave the blood stream in the liver, cross the liver and leave, still viable, via the lymphatics. *Smith* et al. [230] and *Miller and Adams* [163] found substantial numbers of lymphocytes, especially T lymphocytes [163], in the afferent lymph to the hepatic lymph node in the sheep. *Smith* et al. [232] showed that peripheral T lymphoblasts also leave the liver via the lymphatic drainage in the rat. There are several lines of evidence which indicate that it is primarily cytotoxic or suppressor T lymphocytes which are held up or leave the blood stream in the liver. The peripheral blasts used by *Smith* et al. [232] were alloreactive; *Rouse and Wagner* [210] showed that in vitro-activated cytotoxic T cells crossed the liver slowly and *de Sousa* et al. [51] have shown that more Lyt 23 (suppressor/cytotoxic) lymphocytes are retained in the liver often in association with Kupffer cells than Lyt 1 (helper) cells. *Smith* et al. [232] have speculated that the liver has a specific function of culling cytotoxic cells and those that appear in the lymph have presumably escaped destruction. We prefer the notion that they are present in the liver and subsequently leave via the lymph as a consequence of their greater locomotor and/or adhesive qualities.

5. Transit across the Spleen

The spleen is a major lymphoid organ which all lymphocyte types and subsets can enter with ease and in which they are concentrated from the blood; as a consequence removal of the spleen has a greater or lesser effect on the migratory times of all lymphocytes. The arrival time of small lymphocytes in the lymph nodes is speeded up but at the same time their retention is increased in the lungs, liver and bone marrow [18, 200]. Splenectomy affects the distribution of peripheral blasts to a similar extent as that of small lymphocytes but has comparatively little effect upon mesenteric lymphoblasts [*Rose and Parrott,* unpublished].

Accumulated data from a variety of sources including investigations of the traffic of thymocytes in mice [189] and rats [84], thoracic lymphocytes in the rat from blood or lymph [84] across isolated spleen [64] all show that the transit time of T lymphocytes across the spleen is around 5–6 h while B cells may take more than 24 h [70, 172]. The transit time for mesenteric and TDL lymphoblasts across the spleen is substantially shorter (4–5 h) than that for peripheral lymphoblasts (8–10 h) [232]. The difference in transit times for B cells versus T cells and peripheral versus mesenteric blasts is partly a consequence of their different locomotor and adhesive properties but it may also be a consequence of their taking different migration pathways through the spleen.

6. The Segregation of Lymphocytes within the Spleen. The Importance of the Marginal Zone

The spleen, like other organized lymphoid tissues has clearly defined compartments and micro-environments. The lymphoid tissue, or white pulp, is separated from the red pulp by a clearly defined marginal zone. The white pulp is further subdivided into a periarteriolar sheath (T or T-dependent area [189]) and lymphatic nodular area (T-independent or B area [184]). The majority of small lymphocytes enter the white pulp as do peripheral lymphoblasts but mesenteric lymphoblasts are found in both red and white pulp and some may bypass the white pulp altogether [191, 232]. Some immature thymocytes are unable to enter the white pulp [55, 84, 189]. They appear to be held up in the marginal zone area of the spleen which has the type of microvasculature which indicates slow blood flow [114]. Intravenously injected cells and antigenic material are found here shortly after injection, and labelled cells may remain for some time before moving on. The marginal zone of the spleen fulfils a similar function to the HEV in lymph nodes of concentrating lymphocytes from the blood as well as providing a site of interaction of lymphocytes with antigen, and it is significant that *van Ewijk* et al. [250] observed under scanning electron microscopy that lymphocytes exhibit numerous microvilli on their surface which make contact with underlying reticular cells when in the marginal zone but withdraw them once in the white pulp. In the mouse both T and B lymphocytes enter by the marginal zone [184]. Subsequently, T cells move on to the TDA (periarteriolar sheath) and then leave the spleen while B cells remain in perifollicular (B) areas. The route across the rat spleen [172] is somewhat different because T and B areas are arranged slightly differently. As in the mouse both T and B lymphocytes enter the white pulp via the marginal zone, the T cells then move to the TDA which they begin to leave at 2–3 h and most exit by 5–6 h. B cells leave the marginal zone by 3–6 h; they form a band at the junction between the TDA and lymphocyte corona and by 24 h B cells are only present in the corona of the germinal centre and around this time they start to leave the spleen. The re-entry of cells to the blood from the white pulp is not fully understood but it may take place through connecting bridges [165] within the marginal zone.

As might be expected pretreatments of lymphocytes with lectins and enzymes alters their rate of migration across the spleen. Con A-treated small lymphocytes are held up first in the red pulp, the marginal zone and then subsequently in the periphery of the follicle [74], and autoradiographs of dextran sulphate-treated lymphocytes present a similar appearance, findings which are compatible with slow traffic times through the spleen.

Neuraminidase-treated small lymphocytes enter the spleen in smaller numbers than normal because of their low concentration in the blood [66,

71] but those that do so are distributed normally in the spleen. However, trypsin-treated lymphocytes are held up in the red pulp and perifollicular area [69] as indeed are PL-A_2-treated cells. But whereas with the lectin-treated cells one is confident that the lymphocytes are slower to move through the spleen than normal, a similar explanation is less likely for trypsin and PL-A-treated cells, as evidenced by the differential effects of splenectomy on the subsequent progress of lymphocytes. Lectin-treated lymphocytes arrive in the lymph nodes in splenectomized mice but trypsin or PL-A-treated cells are unable to enter lymph nodes in normal or splenectomized mice [70–73].

It would appear that the marginal zone of the spleen presents a critical labyrinth for lymphocytes to negotiate. If they are too sticky or too active they cannot get out of the labyrinth; if they are not sticky enough they cannot get in.

C. The Entry of Lymphocytes into Lymph Nodes

1. Transit Times of Lymphocytes and Lymphoblasts across Lymph Nodes

There are two routes of entry into lymph nodes, through the afferent lymphatics and marginal sinus and from the blood stream through the post-capillary venules. The proportion and numbers of lymphocytes which enter by each route varies according to the location of the lymph node. Very few lymphocytes enter peripheral lymph nodes via the afferent lymphatics unless they are draining the site of antigenic stimulation [100]. Lymph nodes further alon the lymphatic chain receive efferent lymph containing large numbers of lymphocytes from other nodes and the mesenteric lymph node receives cells both from Peyer's patches and the gut mucosa [87]. If labelled normal small lymphocytes are infused into the afferent lymph of a peripheral lymph node in sheep they traverse the lymph node fairly rapidly to appear in the efferent lymph [25, 59, 78] but there are marked differences in the speed of transit according to cell type. Small T (^{51}Cr-labelled) lymphocytes move quickly with a modal transit time varying between 6 and 12 h; sIg^+ (B cells) move much more slowly with no clearly defined modal transit time (fig. 10) [25]. Interestingly antigenically activated lymphoblasts taken from the efferent lymph, and labelled with ^{3}H-thymidine before infusion into the afferent lymph of another lymph node, take much longer than normal to traverse the lymph node from afferent to efferent lymph. Such activated cells which are highly

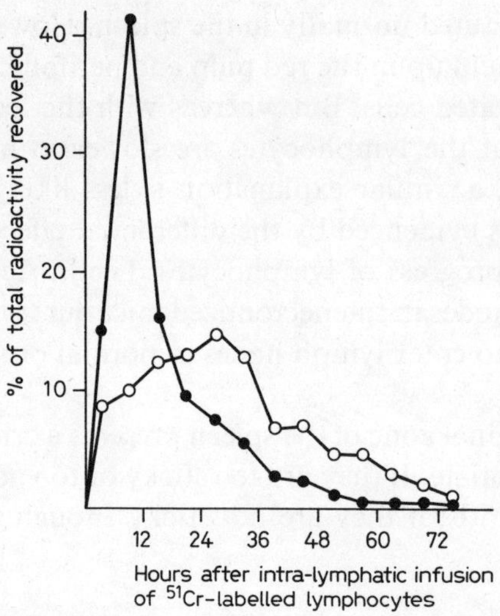

Fig. 10. Time course of the migration of ^{51}Cr-labelled sIg–(●) and sIg+(○), i.e. T and B lymphocytes, from afferent to efferent lymph across the popliteal lymph node of a sheep [reproduced by kind permission of Dr. *R.N.P. Cahill,* ref. 25].

motile in vitro [104], rapidly enter the stroma of the lymph node, but leave the node much more slowly than normal [59]. The obvious interpretation is that the lymphoblasts released a lymphokine which caused a temporary 'shut down', a well-documented occurrence early after antigen stimulation (see below), but it is worth contemplating that the lymphoblasts are so sticky (or so active in locomotion) that they are reluctant to leave the lymph node.

2. Transit Times of Lymphocytes and Lymphoblasts across Lymph Nodes after Entering through HEV

There would appear to be general agreement that the normal transit times for small T lymphocytes from blood to lymph is around 15–18 h whilst B lymphocytes take 24–48 h [26, 65, 235]. At first sight it would appear that lymphocytes entering from the blood take longer to cross lymph nodes than those entering by the afferent lymphatics. If, however, account is taken of the time which elapses before significant numbers of lymphocytes reach the lymph nodes after an intravenous injection, i.e. around 6–12 h (fig. 10) [78], then it will be seen that small lymphocytes take very similar times to cross lymph nodes whether they enter through afferent lymphatics or across HEV.

Mesenteric and peripheral lymphoblasts take varying lengths of time to cross from blood to lymph according to the lymph node studied [232], but although mesenteric lymphoblasts usually take less time than peripheral blasts the difference is almost certainly a consequence of the preliminary hold-up of peripheral blasts in the lungs and liver.

Pretreatment of small lymphocytes with lectins increases the transit time across lymph nodes, as with all other tissues [70, 74, 233], and affects T lymphocytes more than B lymphocytes [70].

3. The Route of Lymphocytes across HEV

Lymphocytes entering lymph nodes from the blood stream do so by crossing post-capillary venules rather than the capillary bed [87]. Why they should choose this route has engaged the attention of many workers in the field of lymphocyte traffic and indeed continues to do so.

Marchesi and Gowans [157] suggested that lymphocyte emigration was qualitatively different from other leucocytes because their regular route across the HEV was by emperipolesis, i.e. they moved intracellularly rather than intercellularly. This suggestion is now more or less discounted following extensive serial sectioning of lymph nodes by *Schoefl* [217], *van Ewijk* et al. [250] and *Anderson and Anderson* [4] and the demonstration that lymphocytes are present in channels between endothelial cells and the walls of the channels are stainable by intralymphatic injections with horseradish peroxidase [4]. Typically, the lymphocyte appears to make contact with the endothelial cells by means of its pseudopod and macrovilli, and then to adhere by numerous points of contact and to flatten by moving across the endothelial wall. There does not appear to be any part of this process of diapedesis which is unique and different from that which occurs elsewhere when lymphocytes pass across endothelium or epithelium [249], but the fact that as many as one in four lymphocytes (mostly small T lymphocytes) in the blood stream will adhere to and traverse an HEV into a lymph node [23, 111] demonstrates that this process is nevertheless extremely efficient. Many attempts have been made to account for this efficiency including the postulate that carbohydrates of the lymphocyte membrane formed receptors specific for HEV and a whole series of experiments were designed to test the particular hypothesis [80, 266]. But a succession of careful experiments, first by *Freitas and de Sousa* [70–73] and then by *Ford* et al. and *Smith* et al. [232], disposed of this particular theory and the arguments clearly set out by *de Sousa* [46] and *Ford* et al. [67] will not be repeated here.

In the process of traversing the HEV the lymphocyte must achieve a nice balance between too little and too much adhesion similar to that which is involved in negotiating the marginal zone of the spleen. When perfusing rat mesenteric lymph nodes *Anderson and Anderson* [4] demonstrated that it is necessary to add trypsin to the perfusate in order to dislodge lymphocytes adhering to HEV. In complementary experiments *Ford* et al. [66], using ingenious isolated rat lymph node preparations, demonstrated that trypsin-treated lymphocytes could not adhere to cross the HEV into the substance of the node although neuraminidase-treated cells did so with ease. On the other hand, Con A-treated lymphocytes [74] appeared in autoradiographic studies to stick in the lumen of the HEV whilst dextran-treated cells, although able to cross the HEV, seemed unable to detach themselves and move away into the lymph node cortex [73].

4. The Significance of HEV Morphology

Much emphasis has been placed on the pseudo-columnar form of HEV but the typical morphology is not a permanent feature. It does not appear until a few days after birth and after lymphocyte traffic has commenced [164]. HEV are thin in neonatally thymectomized mice [189] and rats [84] and in nu/nu mice [50], but can be variable in height according to location [4, 50]. The passage of lymphocytes across the venule is not dependent upon the height of the endothelial walls since it takes place when the venules are flat [84, 189]. It has been claimed that lymphocyte traffic as such restores venule height in neonatally thymectomized rats [84], but it seems more likely that it is also associated with the restoration of immune responsiveness [49]. Many investigators have noted that the height of HEV increases when lymph nodes are in receipt of a strong antigenic stimulus [5, 22, 183] and the endothelial cells may be stimulated to cell division. As would be expected the height of HEV diminishes when deprived of antigenic stimulus by severing of afferent lymphatics to a lymph node [119]. In some pathological situations venules in sites other than lymph nodes take on the appearance of the pseudo-columnar endothelium of HEV, as in chronic granuloma in sheep [231] and in immune synovitis [88], although in these situations the endothelial cells are said to bulge or swell rather than increase in height. Similar appearances also occur in venules in rejecting skin allografts [56, 254] and kidney allografts [193, 198]. In all these situations there is a perivascular infiltrate of lymphocytes and blast cells associated with 'activated endothelial cells' and resemblance to HEV is often remarkable [88].

Several attempts have been made to define some metabolic specialization of the high endothelial walled cells. Studies by electron microscopy have emphasized their 'activated' appearance with large, notched nuclei; the cytoplasm contains plentiful ribosomes and rough endoplasmic reticulum [6, 217]. *Ford* et al. [67] have been pursuing this line recently and showed by autoradiography that ^{35}S-sodium sulphate was taken up by HEV but not by other endothelia [67]. More recently, *Andrews* et al. [8] have been attempting to purify the substance which incorporated $^{35}SO_4$ and they are now examining a compound termed fraction 5. They have shown that brief incubation of lymphocytes in fraction 5 produced similar morphological changes to those which occur when lymphocytes attach to HEV, namely retraction of microvilli [250]. Brief incubation with trypsin and then fraction 5 made the surface of lymphocytes almost smooth, explaining why trypsin-treated lymphocytes are unable to enter lymph nodes [66].

Preliminary experiments [8] indicated that intradermal injection of fraction 5 increased the localization of ^{51}Cr-labelled small lymphocytes over and above that induced by simple inflammation, but the low amounts of radioactivity present make confident quantitation difficult. *Andrews* et al. [8] favour the idea that the sulphated molecule is secreted by HEVs into the blood stream where it acts as a signal to lymphocytes to adhere and then retract their microvilli. This idea still assumes that lymphocytes and not other leucocytes are uniquely responsive to this material, and offers no explanation for the selective process which occurs in the marginal sinus of the spleen nor how the flat endothelium characteristic of sheep post-capillary venules [59] achieve their purpose. It, nevertheless, offers one of the most promising approaches to the 'HEV' problem.

5. The Changes in Lymphocyte Kinetics following Injection of Antigen.

The subcutaneous injection of most antigens causes an increase in migration of lymphocytes to the draining lymph node [53, 101, 271] and after intravenous and intraperitoneal injections an increase in migration of lymphocytes to the spleen [271]. These increases in the draining lymphoid tissue are accompanied by a concomitant decrease in non-draining tissues. Two mechanisms have been proposed to account for the increase lymphocyte flux: (a) a change in the structure of lymph node or spleen so as to increase the mechanical trapping of circulating lymphocytes and facilitate their interaction with antigen-trapping mechanisms [53, 271], and (b) that antigens in some way change the locomotion of lymphocytes persuading a greater proportion to leave the circulating blood and to move towards the source of antigen, i.e. to move towards a gradient source. It remains possible that both these mechanisms do exist and act upon a minority of lymphocytes but we now know they are not the main mechanisms by which antigen-induced lymphocyte migration is achieved. In 1976 *Cahill* et al. [23] measured the transit of ^{51}Cr-labelled lymphocytes through the popliteal lymph nodes of sheep following injection of antigen. They found that after an initial brief drop or shut-down [101] in output from the stimulated node there was a large increase in output which matched the increased input or recruitment without altering transit time. There was thus no evidence of trapping or delayed exit of lymphocytes after antigen stimulation. At the same time *Herman* et al. [115] observed that after the injection of antigen there was an increase in blood flow to the responding lymph node of rabbits. Subsequently *Hay and Hobbs* [111], who noted that the increased lymphocyte output in the efferent lymph of sheep occurred at the same time as the regional blood flow to the node was increased, proposed that vascular changes and increased lymphocyte traffic may be related phenomena. More recently, *Ottaway and Parrott* [178], utilizing an isotope technique whereby lymphocyte traffic and blood flow are measured simultaneously, showed that the migration of small (^{51}Cr-labelled) lymphocytes to lymph nodes is directly related to blood flow (fig. 11). There is, therefore, no need to seek for further explanation of the means by which the major increase in lymphocyte traffic to stimulated lymphoid tissues is brought about. It has been suggested [59] that in lymph nodes, such as the mesenteric which receives a substantial input of cells from the afferent lymphatic, the input of lymphocytes from the blood stream is reduced. In theory, therefore, this should disturb the correlation between blood flow and small lymphocyte traffic to mesenteric lymph nodes but preliminary observations in the mouse do not support this. It should be borne in mind, however, that the correlation between blood flow and lymphocyte traffic to lymph nodes only applies to small lymphocytes, and not lymphoblasts, and it has not been tested as yet whether there is any difference between T and B cells.

The mechanism by which antigen brings about changes in blood flow is not known though it is probably due to the release of vasoactive amines.

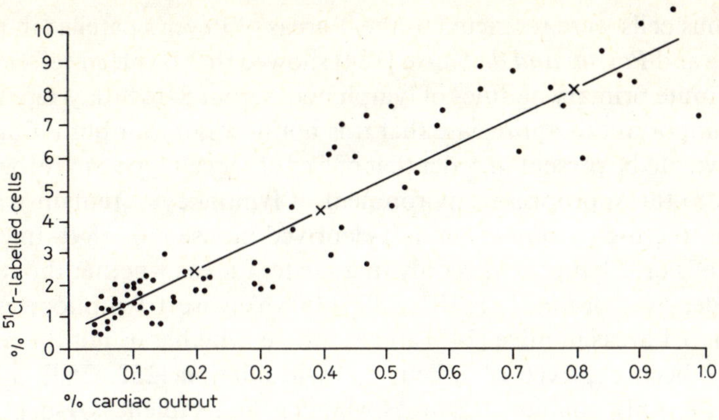

Fig. 11. The relationship between the migration of ^{51}Cr-labelled small lymphocytes and blood flow in mouse lymph nodes [Data from ref. 178].

Hay et al. [113] have reproduced the phenomena of shut-down and recruitment of lymphocytes to the lymph node in sheep by means of cell-free lymph in the absence of administered antigen, and have been attempting to identify the relevant factors. Preliminary evidence has shown that E-type prostaglandin is present but is almost certainly not the only factor involved.

D. The Sorting Out of Lymphocytes within the Organized Lymphoid Tissues.

Lymphocytes utilize common routes of entry into the organized lymphoid tissues but once inside they segregate from one another into clearly defined T and B areas [184], whether in spleen, lymph node, Peyer's patch or appendix [182, 252]. This separation process was termed ecotaxis by *de Sousa* [45] who postulated that lymphocytes acquired these migratory characteristics during differentiation, along with surface antigens and phenotypes. Ecotaxis occurs within hours of injection of labelled cells and has been demonstrated in mouse, rat, rabbits, sheep and man. T and B areas can be recognized by lymphocytes early in development as evidenced by the fact that *Pearson* et al. [192] could discern depleted T areas in the spleen and Peyer's patches of thymectomized fetal lambs in utero, and the discrimination by T or B cells of separate areas is possible within a few days of birth in mice. *Waksman* [255] demonstrated that thymus cells were restricted to the T areas of Peyer's patches in newborn mice and *Parrott and de Sousa* [184] showed that B spleen cells localized in minute primary nodules of lymph nodes from 5- to 8-day-old mice.

It is important to appreciate that it is not necessary for both T and B lymphocytes to be present in order that one or other cell type can recognize or ecotax to the appropriate environment. B lymphocytes remain within their own micro-environment in a T-deprived mouse [50, 189]. Injected thymus cells or T lymphocytes only migrate to T areas whether the recipient is T-deprived or not [121, 181, 189, 235]; likewise T lymphocytes are restricted to T areas in mice [249] and in rabbits whether or not the recipients have been deprived of B cells by X-irradiation [252, 253], and B lymphocytes only migrate to B areas whether the recipient is T-deprived or not.

The postulate that a lymphocyte acquires the capacity to ecotax along with cell differentiation and the acquisition of the surface components which define phenotype was tested experimentally by *de Sousa* et al. [51] who compared the migration of Lyt 1 and Lyt 23 T cells. They showed by autoradiography that fewer isotopically labelled Lyt 23 migrated to lymph nodes than Lyt 1 cells; most Lyt 23 cells were recovered in spleen and liver. These observations correspond well with those of *Beverley* [14] who showed that fewer Lyt 23 cells recirculate than Lyt 1 cells. But of greater interest was the observations revealed by autoradiography that Lyt 1 and Lyt 23 cells appeared to segregate from one another in the lymphoid tissues, for Lyt 23 cells were mostly found in B cell areas, i.e. the periphery of the spleen follicle and the outer cortex and the intermediate zone between cortex and medulla of lymph nodes, whereas the Lyt 1 were almost exclusively found in T cell areas. Another interesting feature of the distribution of Lyt 23 cells was their frequent association with phagocytic cells both in the liver with Kupffer cells and in the medullary cords of lymph nodes with polymorphs.

What determines the environmental preferences of lymphocytes? Looking at autoradiographs of lymph nodes and spleens taken at different times of the injection of labelled cells, it is so tempting to envisage the cells as moving in a purposeful way to their allotted location, but it is much more likely that lymphocytes 'recognize' their appropriate microenvironment when they happen by chance and random locomotion to arrive there. They know when they arrive 'home' when they recognize their 'neighbours'. *de Sousa* [46] defined two sorts of neighbours for lymphocytes. The obvious neighbour which is another lymphocyte of the same or different population and an underlying neighbour is less obvious but include stromal cells, macrophages, dendritic and interdigitating cells.

Curtis and de Sousa [38] and *Curtis* [36] have also suggested that modulation of adhesive interactions between T and B lymphocytes within lymphoid organs plays an important role in the control of lymphocyte positioning and the speed of migration. T and B lymphocytes do differ in their adhesiveness in vitro; each type of lymphocyte releases a low molecular weight factor (IMF) capable of diminishing the adhesiveness of the other population of lymphocytes without altering the adhesiveness of its own population. There is some evidence [36, 38] that such factors do influence the rate of lymphocyte traffic, but the very fact that separate populations of T and B lymphocytes can position themselves in the 'right' micro-environment in the absence of the other lymphocyte population strengthens the concept that the major influences in environmental recognition are the 'underlying neighbours' [46]. In 1969 *de Sousa* [44] noted the differences in reticulum framework in the different compartments of the lymph node, and *Veldman* [252] identified a particular type of reticulum cell in the T areas of lymph nodes. It is now know that the interdigitating cell is characteristic of T areas in the spleen [249, 251] as well as the lymph nodes, and is also found in the thymus itself [117]. *Veldman* [252] also described the dendritic reticulum cell which is characteristic of B areas. It would now appear that the different compartments of spleen and lymph nodes have their own individual type of reticulum cell and their own mononuclear/macrophage type of cell [119], the reticulum cell being part of the fixed stroma of the tissue whilst the mononuclear/macrophage cell is a mobile component, and it seems likely that by interacting with these types of cells that lymphocytes 'discover' where they are in the lymphoid tissues. Morphological evidence of such interactions have been noticed in electron microscopy studies [249–253]. *Humphrey* [122], who has described the preparation of macrophages peculiar to the marginal zone of the spleen, noted that these macrophages when freshly isolated from the spleen often have numerous lymphocytes adhering to them.

Deprivation of lymph nodes of their supply of macrophages and interdigitating cells by severing of the afferent lymphatics results in the disappearance of nodular aggregation of (B) lymphocytes as well as the disappearance of germinal centres [119]. Peyer's patch tissue transplanted under the kidney capsule is also devoid of germinal centres [62]. More significantly it was noted that although injected labelled T and B lymphocytes entered the transplanted Peyer's patch tissue no defined ecotactic pattern could be discerned in autoradiographs [186]. It was not tested, however, whether the grafted tissue was monocyte- or macrophage-depleted.

E. The Selection of Lymphocytes which Enter Non-Lymphoid Tissues

In contrast to the organized lymphoid tissue the numbers of lymphocytes which traffic through most non-lymphoid tissues with the exception of the liver and mucosal surfaces is small. This is shown by the very small amount of radioactivity found in the tissues after the injection of labelled lymphocytes [9, 200, 207] and corresponds with the low number of lymphocytes found in afferent lymph draining all tissues except liver and gut mucosa [23, 58, 106, 163, 230]. For example, there are only 200–800 lymphocytes/mm^3 in afferent lymph draining skin or muscle, whilst the concentration of lymphocytes in efferent lymph or blood is 2,000–4,000/mm^3.

Moreover, not only are numbers of lymphocytes which normally enter the tissues restricted, but there appears to be some selection at the level of the capillary bed of the type of lymphocyte which enters most tissues, for the proportion of B:T lymphocytes in afferent lymph is very different from that in blood. T cells predominate in human afferent lymph [58] and other human extravascular fluids [155]. T cells also predominate in sheep afferent lymph for only 5–10% of lymphocytes are B lymphocytes [163, 218]. It seems unlikely that the paucity of B lymphocytes in skin and tissues is a result of some sort of exclusion mechanism or because B cells do not have the ability to enter tissues since they readily gain access to mucosal surfaces. It seems more likely that circumstances promote the extravasation of T lymphocytes along with other blood leucocytes but not B lymphocytes. Furthermore, those T lymphocytes which are present in tissues may not be representative of those in circulating blood, they have a higher proportion of cells with IgG Fc^+ receptors [155] and there is a greatly increased proportion of T cells bearing Ia antigen in human afferent lymph compared with T cells in the blood [83]. The capillaries of the skin and other tissues do not however exclude other leucocytes as do the HEV of lymph nodes. Monocytes and macrophages are always present in afferent lymph as well as lymphocytes [106, 230] and their proportion in sheep is 5–15% but can be as high as 50% in the rabbit [131].

Many of the large mononuclear cells present in afferent skin lymph do not usually phagocytose particulate matter. *Smith* et al. [230] observed lymphocytes attached to the plasma membrane (peripolesis) of macrophages as well as inside (emperipolesis) macrophages in sheep afferent lymph. More recently these frilly or veiled cells in afferent lymph have received extensive investigation in rabbits [131] and in pigs [54, 118]. In some species (rat, rabbit, man) they contain Birbeck granules; they have Fc and complement receptors and Ia antigen on their surfaces [54, 238].

1. Cellular Changes in Skin and Afferent Lymph following Antigenic Challenge

The very low baseline of lymphocytes in the tissues and skin, in particular, provides a good experimental system in which to examine the mechanisms which promote cell traffic and lymphocyte locomotion. The first and most regular change in cell population following almost all antigenic challenge is an increase in polymorphs in both skin and afferent lymph. This occurs within hours of: skin grafting [97] and kidney grafting [193]; induction of contact sensitivity to oxazolone in mice [48], to DNCB in sheep [102] and pigs [54]; injection of diphtheria toxoid in rabbits [131] and influenza virus in sheep [23]. As the numbers of polymorphs recede after 24 h the number of macrophages, veiled cells, increases [112, 163, 234]. Around 24 h after the application of the contact sensitivity with DNCB there is a large increase in cell aggregates or clusterings in afferent lymph composed of all the types in lymph although all clusters contain veiled cells [54, 234]. The clustering around and investigations of veiled cells by lymphocytes are very reminisent of the lymphocyte macrophage interactions which occur in vitro. It has been suggested that the Langerhans cell, the veiled and the interdigitating cells in the T areas of lymph nodes have a common lineage [54, 80, 135]. Certainly it is a reasonable supposition that the lymphocyte-Langerhans interactions in the skin [68] and the clusterings in afferent lymph could have a common purpose in the facilitation of macrophage-antigen-T cell interactions. In view of their size, it is surprising that so many large clusters manage to gain access to afferent lymphatics, and it may well be, in the presence of primed cells (i.e. at recall), that clusters become so large that they are restricted to the skin [54, 228], for there are fewer clusters in afferent lymph during secondary challenge. This could be the mechanism by which the traffic of specifically primed lymphocytes is delayed during passage through the skin (see below).

There may be an increase in lymphocytes and blast cells in afferent lymph from 2 days onwards but this is not an invariable occurrence [23, 97] after all antigens. There are, however, particular stimuli which always cause a large increase in lymphocyte extravasation. The largest and fastest primary stimulus is a renal allograft. *Pederson and Morris* [193] found a 20 to 30-fold increase in lymphocytes, particularly blast cells, in lymph draining a renal allograft within 48 h of transplantation. The dermal injection of allogeneic lymphocytes in sheep causes a comparable increase in lymphocytes and blast cells which takes 5–7 days to evolve [112]. Contact sensitizers also promote lymphocyte extravasation in pig [54] and sheep [98], but lymphocytes do not appear in the skin of mice until 4–5 days after primary sensitization with oxazolone [48] and even then they are a few in number. Large numbers of lymphocytes appear in lymph draining a chronic granuloma but the process takes from 20–40 days to build up to the maximum [231] and in so doing changes from immune inflammation to a typical delayed-type reaction.

The response to a second application of the same stimulus is very different. It has long been known that lymphocytes and the mononuclear cells appear in the skin in delayed-type reactions [254]. For example, lymphocytes are already present in the skin at 4 h after testing of oxazolone-sensitized mice although the numbers may not peak until 24 h [48] and *Drexhage* et al. [54] showed that the appearance and numbers of cells in the afferent lymph of pigs changes dramatically following a second application of DNCB to a sensitized pig. Large numbers of lymphocytes and other mononuclear cells appear and although many veiled cells are also present, which, as in primary sensitization, lymphocytes are 'interested' to touch, there are far fewer aggregates and clusters in lymph than following primary sensitization.

2. Immune Inflammation. An in vivo Test System of Lymphocyte Locomotion (table V)

The accessibility of tissues including skin to lymphocytes can be estimated by measuring the number of isotopically labelled cells which will enter the skin as well as by counting the numbers of cells in afferent lymph. It should be appreciated, however, that there need not be a direct correlation between the numbers and type of cell which will enter a tissue from the blood and those which are ready to leave and therefore appear in afferent lymph.

Comparatively, few lymphocytes enter normal skin and the amount of traffic is not only small but it is relatively rapid [200, 201]. Labelled cells peak within an hour of injection in the rat and the few that do enter have left by 24 h and this is consistant with the rapid transit time of labelled

Table V. Experimental Models of Lymphocyte Locomotion in vivo[1]

	Characteristics of lymphoid cells which move into tissue or mucosal sites	
	immune inflammation	mucosal sites
Cell type	activated T cell	activated T cell IgA and specific immunoglobulin-producing B cell
Mode of activation	contact sensitizers skin allografts intracellular parasites	normal intestinal contents and/or local antigen
Site of origin	peripheral or somatic lymph node (not mesenteric)	mesenteric (Peyer's patches or mesenteric lymph nodes) (not peripheral)
Site of accumulation	simple inflammation on skin or peritoneal cavity; sites of contact sensitizers bacterial antigen in peritoneal cavity; nematode infection	duodenum jejunum ileum large intestine caecum mammary glands vagina, lungs
Factors controlling accumulation	blood flow non-specific inflammation specific antigen lymphokines MHC-restricted (?)	blood flow occurs without antigen local antigen increases parasitic infection increases MHC-restricted (?)

[1] For references see text.

lymphocytes (blood – afferent lymph) of sheep [106]. By contrast after the primary application of the contact sensitizers oxazolone or picryl chloride to the skin not only are there 20–30 times more cells in the skin than normal [9, 10, 207] but the build-up is slower and does not reach a peak until 24 h or so [201, 208], and may be retained for 2–3 days [208].

Furthermore, experiments utilizing isotopically labelled cells have provided insight into the nature of the cell which has the facility to enter the skin (table V). In the lymph nodes draining the side of induction of delayed hypersensitivity reactions including skin allograft and contact sensitivity reactions there are generated many lymphoblast cells which are principally T cells. A series of studies have shown that these cells have a strong propensity to accumulate in sites of both immune and non-immune inflammation, and normal skin. For example, blast cells taken at 3 days from lymph nodes, draining the site of application of oxazolone and labelled with (^{125}I)UdR will, after transfer to syngeneic recipients, accumulate in large numbers in skin inflamed by the same contact sensitizers oxazolone, or picryl chloride or a non-immune inflammatory agent such as croton oil [9, 10, 207]. Such injected lymphoblasts will enter the skin much earlier in time and in much greater numbers than their natural sequence of appearance in the skin of the animal donating the lymphoblasts [48]. Thus, injected lymphoblasts will enter the skin within 2 h of the application of oxazolone at which time there would be only a few appearing naturally in the draining lymph nodes of the recipient and no lymphocytes in the skin [48]. But they indicate the receptive state of the skin and for the purposes of defining the factors involved in stimulated locomotion show that the normal time sequence of the naturally evolving lesion with the preliminary appearances of polymorphs and monocytes is not obligatory. Circumstances which favour extravasation of lymphoblasts occur within hours, whether or not they are available in circulation. Data concerning the cell type which shows the greatest propensity to enter inflammatory sites in the skin are summarized in table V. It is a T lymphoblast in S phase [10]; furthermore, the cell must be generated in peripheral lymph nodes and not mesenteric nodes [206–208]. Other cell types including monocytes and small circulating lymphocytes also enter but not with such enthusiasm as T lymphoblasts [10].

3. The Assembly of Lymphoblasts in the Peritoneal Cavity

Another model which has been used extensively in the study of in vivo lymphocyte locomotion is the assembly of lymphoblasts in the inflamed peritoneal cavity (table V); an experimental system very effectively used by *McGregor and Logie* [153]. In this system lymphoblasts are generated by subcutaneous injection of live intercellular parasites such as *Listeria monocytogenes* into the hind quarters of rats. These cells, although generated in peripheral lymph nodes, are routinely withdrawn by drainage from thoracic duct lymph, and after transfer to syngeneic recipients are persuaded to assemble in the peritoneal cavity after inflammation. Like the cells which migrate into inflamed skin they are activated T lymphocytes [175] and not small recirculating lymphocytes. There are many other reports which emphasize the differences between the migratory characteristics of cells which enter inflammatory exudates and normal recirculating small lymphocytes [27, 39, 81]. In man and guinea-pigs, peritoneal exudate lymphocytes have high E rosette-forming ability (i.e. they are activated) and a high proportion bear receptors for the Fc piece of IgG. Thus, they are similar to the T cells in skin afferent lymph. Functionally the cells which extravasate have many characteristics of short-lived effector cells [133, 175] and have the capacity to transfer immunity adoptively though whether they are directly cytotoxic or amplifier cells has not yet been demonstrated.

4. Lymphocyte Traffic through the Skin in Delayed-Type (CMI) Reactions.

There is now evidence that the nature of lymphocyte traffic through CMI reactions, is substantially different from that through non-immune and immune inflammation sites or through the sites of initial contact with antigen, for normal small recirculating lymphocytes localize (and migrate) through such areas [201; *Ottaway,* unpublished] as well as or possibly better than lymphoblasts. Initially, the flux of small lymphocytes through a CMI reaction is modest, only 4–5 times above background level, and does not approach that through a comparable volume of lymphoid tissue, but when the CMI reaction develops into a chronic granuloma in sheep [231] the number of small lymphocytes passing through becomes equivalent to that passing through a lymph node. Comparable increases in lymphocyte flux through granuloma have also been made in the rabbit [112] and the rat [202].

These observations emphasize the necessity of distinguishing between the cell types which enter the initial inflammatory or immune inflammatory reactions to an antigen, and later recall or delayed-type or CMI reactions. Associated with the increase in small lymphocyte flux through granuloma in sheep is a change in the appearance of the endothelium of small venules which come to resemble the morphologically distinct high-walled endothelium of lymph nodes. This is especially interesting because such structures although common in the lymph nodes of many species rarely occur in sheep lymph nodes [59]. Very similar changes in endothelium also occur in rats and rabbits during the development of chronic immunological and inflammatory lesions [11, 88] and these are probably also associated with CMI reactions. It is strange that vascular endothelium only excites comment when it changes into a pseudo-columnar form yet the lymphocyte selection processes take place in the absence of anything other than normal endothelium. As we have discussed subsets of activated T lymphoblasts are preferentially admitted together with monocytes and macrophages into the skin, for example, but not B lymphocytes. Yet the search for a 'T cell receptor' in skin capillaries does not attract much attention, and emphasis is placed on processes which will 'attract' lymphocytes and lymphoblasts.

F. The Factors which Promote Lymphoblast and Lymphocyte Extravasation

Various theories have been put forward to explain the increased accumulation of cells in sites of immune inflammation and CMI reactions, these include changes in vascular permeability, blood flow, antigen-directed locomotion and chemoattractants, and we now consider the relative importance of each of these factors.

1. The Role of Vascular Permeability

An increase in vascular permeability with consequent leakage of protein, occurs after antigen deposition [98, 102, 112, 205]. However, it has been known for some time that, even in a tissue such as the ovary where there is a high rate of leakage of blood protein [170], the cellular concentration in lymph from the ovary is nevertheless very low [230]. We [205] have shown that inducing vascular permeability by injection of serotinin does not promote lymphoblast extravasation, and although vascular permeability does occur during primary sensitization to oxazolone it is not directly related to lymphoblast migration [205]. In this respect, therefore, lymphoblasts behave like polymorphs whose migration from the blood is not dependent upon vascular permeability [123]. It should be borne in mind, however, that serum albumin does act as a chemokinetic agent [263] and could well accelerate the locomotion of lymphoblasts once they are within the tissues.

2. The Role of Blood Flow

Hay et al. [112] have measured blood flow to the skin in three types of immune reactions: The reaction to allogeneic lymphocytes, the evolution of a Freund's adjuvant-induced chronic granuloma and a delayed-type reaction to PPD. An increase in blood flow occurred in all three situations, up to 20–30 times after the injection of allogeneic lymphocytes down to a more modest 2–3 times after PPD. Recently the changes in blood flow and cell traffic following initial reaction to the contact sensitizer oxazolone [178] were measured simultaneously. The resultant increase in blood flow was directly related to the concomitant increase of migration of lymphoblasts to the skin. There was a much smaller increase in small lymphocyte migration which was not related to blood flow. At the same time, there was also an increase in blood flow to the draining lymph node (section C.5) but this was directly related to small lymphocyte and not blast cell traffic (fig. 11) [178]. Preliminary observations [*Ottaway,* unpublished] indicate, however, that at recall to the same sensitizer the relationship will change and that blood flow will be related to small lymphocytes.

3. Antigen-Promoted Migration

It is obviously of interest to determine whether or not the antigenic stimulus which promotes the activation of lymphocytes to lymphoblasts will subsequently modify their migration pathway or their net accumulation over time in the site of antigen deposition.

There is general agreement that S phase lymphoblasts have a ready capacity to leave the circulation at sites of non-immune inflammation, situations in which antigenic recognition can play little or no part. After injection lymphoblasts accumulate in skin sites inflamed by croton oil, turpentine [8, 10, 167] and in peritoneal exudates inflamed with various non-antigenic stimuli, e.g. glycogen, thiogycollate [145, 159, 208]. Even when antigens, such as contact sensitizers or preparations of bacterial antigens, are used to induce migration these often produce considerable non-immune inflammatory reactions as well as antigenic change, and the inflammatory reactions induced promote sufficient lymphoblast migration to mask any element of antigen-specific induced migration.

Nevertheless when peripheral lymph node lymphoblasts are presented with the alternative sites of extravasation they accumulate in sites where an element of antigenic specificity is present. Thus, oxazolone lymphoblasts will accumulate in the intestine of mice inflamed with *T. spiralis* but not if oxazolone is also applied in the ear skin, and picryl chloride was less effective in diverting oxazolone lymphoblasts than oxazolone itself [208]. These experiments, although suggestive of some specificity, nevertheless were not consistently so. More recently, *Ottaway and Parrott* [178], also using the contact sensitizers oxazolone and picryl chloride, did show consistent and specific increment of increased localization due to antigen which was comparable to that obtained by *Jungi and McGregor* [126, 128] when they measured the differential accumulation of specific antigen accumulation of specific antigen-activated S phase lymphoblasts in peritoneal exudates containing either *L. monocytogenes* mitogen or *F. tularensis* antigen. Others [3, 57, 242] have also detected some specific accumulation of lymphocytes but on challenge only in animals which had been primed previously and by using a careful control so as to be able to detect a minimal element of specificity over a major element of non-specific accumulation. An explanation for the comparatively large non-specific element of accumulation which is always present even when inflammation is small [3, 57] is afforded by the experiments of *Jungi and McGregor* [126] who showed that the presence of specifically sensitized lymphoblasts could substantially increase the assembly of lymphoblasts primed to another antigen. *Jungi and McGregor* [126] have postulated that the numbers of 'pioneer' lymphoblasts which make the initial contact with antigen can be quite small yet they attract substantial numbers of other lymphoblasts. The mechanism by which the lymphoblasts achieve this is not known but *Jungi and McGregor* [126, 128] put forward the idea that the 'pioneer' lymphoblasts release chemoattractive lymphokines after contact with the priming antigen and that the lymphokines non-specifically attract further lymphoblasts into the site. *Jungi and McGregor* [127] proffered a similar explanation when they showed that allogeneic lymphoblasts, although primed to the same antigen as syngeneic lymphoblasts, failed to enter inflammation sites when injected alone but did so if injected together with syngeneic lymphocytes. A limitation on these experiments was however that the peritoneal exudates were only sampled at 24 h. Recent experiments [*Shields and Parrott,* unpublished] have indicated that syngeneic lymphoblasts move more slowly through tissue than allogeneic lymphoblasts. It is tempting to speculate that the syngeneic lymphoblasts recognize the antigen on the Langerhans cells of the skin and are retained or even inhibited from leaving the skin but the allogeneic cells are not because they were unable to recognize antigen. Others have shown that there are MHC constraints on the transfer of contact sensitivity [248].

A similar explanation, i.e. increased retention, could account for the antigen-specific element in lymphoblast localization. When blood flow to the site of application of contact sensitizers was measured as well as the localization of lymphoblasts, it was found that the antigen-specific increase was achieved without a comparable increase in blood flow [178]. Thus, similar numbers of lymphoblasts would have been delivered to both antigen-specific and non-specific sites. Therefore, either more cells left the capillary bed when the specific antigen was present or the same number of cells entered regardless of antigenic specificity but more were retained when the specific antigen was present. Since sampling was only carried out at one time interval after injection, it is not possible to decide between these alternatives in these experiments though we favour the notion of antigen-specific retention rather than antigen-specific attraction.

On the basis of evidence presented here we propose that all lymphoblasts leave capillary beds because of their high random mobility and because of the presence in antigenically stimulated tissues of chemokinetic substances such as plasma proteins. This process will continue so long as inflammation and chemoattractants remain and may be exacerbated by the release of lymphokines as proposed by *Jungi and McGregor* [126]. At the same time there may be an element of specific retention of lymphoblasts because of antigen-specific interaction with macrophages and/or Langerhans cells but this is often masked by the high level of non-specific accumulation.

G. The Selection of Lymphocytes which Are Present in Mucosal Surfaces

1. The Composition of Mucosal Lymphocytes

The mucosal surfaces, including the gut, lungs and mammary glands, normally contain much larger numbers of lymphocytes than are present in other non-lymphoid sites. They also differ in type and effector functions. The most obvious distinguishing feature is the predominance in most species of IgA-producing cells [32]. An understanding of the mechanisms by which a predominance of IgA-producing cells in the gut mucosa is achieved is the problem whch has focused the attention of most workers in the field of lymphocyte traffic to the gut, although other immunoglobulin-producing cells are present and there are as many if not a slightly greater number of T cells present in the lamina propria [92], and T cells comprise the majority of lymphocytes between the epithelial cells covering the villi [62, 92, 185]. *Bull and Bockman* [21] found that 58% of mononuclear cells in human colon biopsies were T cells, 32% B cells and the remaining 10% were monocytes or macrophages.

The population of lymphoid cells in the lamina propria also differs from those in the circulating blood in the size and morphology of the cells therein. Most cells are larger than circulating lymphocytes, many have prominant nucleoli and basophilic cytoplasm and are actively motile when freshly separated [21]. Intra-epithelial lymphocytes are also medium to large in size and some are large and blast-like [159, 243]. They have a dense nuclei, granular cytoplasm, some rough endoplasmic reticulum and a well-developed Golgi apparatus [159]. Many epithelial lymphocytes contain tiny metachromatic granules [31, 93, 211].

The process of populating the gut tract with lymphoid cells commences after birth and exposure to luminal antigen. IgA-producing plasma cells appear for 10 days onwards [33] and epithelial T lymphocytes at 2–3 weeks [62]. There are few IgG- and no IgA-producing cells in intestinal lymph of the fetal lamb but during the first few weeks following birth the numbers of lymphocytes in the intestinal lymph in the new-born lamb increases 10-fold [169], and 20–35% are Ig^+ cells of which many are IgA-producing [59].

The main stimulus to the populating of the gut tract is therefore the presence of antigen in the lumen, and when that stimulus is reduced either by maintaining animals in a germ-free environment [32] or by transplanting gut so that it is free from intraluminal stimulation [62], then the numbers of lymphoid cells in the mucosa is very small. Administration of antigen locally, i.e. by feeding by introduction into the gut lumen by tube or injection, or placing within isolated ileal loops, will promote a local immune response, i.e. response in which the effector cells are primarily or solely restricted to the gut mucosa [17, 28] and directed to antigen in the lumen. Nevertheless, most of the lymphoid cells in the gut mucosa are not formed by cell division in situ but have entered by way of the circulating blood. Therefore, in the capillary beds of the mucosa some sort of selection process must take place which facilitates not only the predominance of IgA-forming cells but also the predominance of effector cells against antigen in the gut lumen. Once in the mucosa the cells do not survive for long.

Autoradiographic studies have shown that IgA-producing cells have a half-life of 4–7 days [161], and cell transfer studies [28, 121] indicate that specific antibody-forming cells die in situ after 1–2 weeks. Autoradiographic studies also show that T cells have a short life span [158, 191, 204].

It is not known whether T cells die in situ or leave the gut but cell traffic studies indicate that they do not reside there for longer than a few weeks [208, 236]. With such a short life span it is evident that the gut mucosa must be in receipt of large numbers of cells in order to sustain immune responsiveness. The origin of those cells, which subpopulation of lymphocytes they comprise, and how the appropriate cells are selected from the circulating blood are the questions which we now address.

2. The Source of Cells which Populate the Gut Tract

The gut-associated lymphoid tissues, namely tonsil, appendix and Peyer's patches, are all equipped with epithelium which has specialized micropinocytotic properties [181] which makes them especially suited to the sampling of the contents of the gut lumen. These tissues are the principal although not the only sites where priming of cells destined for the gut tract occurs.

In a whole series of elegant experiments *Cebra* et al. [28] and *Craig and Cebra* [34, 35] have demonstrated that Peyer's patches are the main source of IgA-producing cells. Studies of cell transfer into irradiated recipients showed that only Peyer's patch cells and not spleen or peripheral lymph node cells will repopulate, over a period of 10–14 days, the gut tract with IgA-producing cells. But such studies did not indicate the route the cells must take in order to reach the lamina propria, nor indeed whether other immunoglobulin-producing cells and T cells are also derived from Peyer's patches. The nature of the cells and the route which they take has been pieced together from a series of studies tracing the fate of specific antibody-producing cells and isotopically labelled preparations of separated mesenteric and thoracic duct cells [87, 92, 105, 186, 195].

3. The Type of Lymphocyte which Enters the Gut Mucosa

Numerous experiments have shown that only cells taken from tissues in the Peyer's patch, via the thoracic duct-lymph-blood stream cycle will enter the lamina propria. In practice, this means suspension of mesenteric lymph node cells and thoracic duct lymphocytes although the assumption is that most of the cells contained therein will have originated in Peyer's patches and will be carried through in the natural direction of lymph flow. *Griscelli* et al. [90] first emphasized the necessity of using lymphocytes of mesenteric origin when they showed that only lymphocytes obtained from mesenteric lymph nodes would migrate to the small intestine; cells taken from peripheral lymph nodes failed to migrate to the gut and were found in the spleen and peripheral lymph nodes. This pattern of tissue-directed migration [186] was confined to cells that took up ^{3}H-thymidine [90] in vitro and were large dividing cells or lymphoblasts or immunoblasts. This finding of a divergent migration pathway has been repeated with lymphoblasts from efferent peripheral lymph and intestinal lymph of sheep [103], and the selective preference of mesenteric lymphoblasts for the lamina propria is possessed both by B lymphoblasts [92, 154] and by purified mesenteric T blasts [92, 93, 206]. The only way in which peripheral lymphoblasts can be persuaded to enter the gut is if the animal is infected with the nematode *T. spiralis;* it then becomes transiently an inflammatory focus and some lymphoblasts of peripheral origin will migrate there as they would to the inflamed peritoneal cavity although their affinity for even an inflamed gut is not as great as for inflamed skin (section E. 2) and the affinity of lymphoblasts of peripheral origin for an inflamed gut is not as great as that of mesenteric lymphoblasts [107, 208].

The relative failure of small lymphocytes to localize in the gut mucosa whether taken from thoracic duct lymph and mesenteric lymph node or from peripheral lymph nodes or spleen has been demonstrated by means of autoradiography [87, 105, 186] and by measuring the radioactivity present in the intestine following injection of ^{51}Cr-labelled lymphocytes [75, 200, 206] from which blast cells have been removed by gradient separation [75] or passage through an intermediate recipient [75]. There is little increase in the localization of recirculating small lymphocytes in the intestinal mucosa even following primary oral antigenic challenge [206], but it is possible that during intensive or prolonged stimulus, e.g. inflammatory bowel disease and granuloma formation, that small lymphocytes could enter the mucosa.

4. The Efficiency of the Localization of Mesenteric Lymphoblasts in the Gut Mucosa

The efficiency with which lymphoblasts derived from Peyer's patches are transferred to the gut mucosa can be assessed by measuring the arrival of isotopically labelled mesenteric or thoracic duct lymph lymphoblasts in the small intestine. Significant numbers are found in the mucosa within an hour of injection [105, 180, 187, 206] and the numbers peak within 4–12 (fig. 12). Far fewer cells are found in the large intestine or caecum, Peyer's patches or mesenteric node [105, 180, 187, 206]. Immunoblasts are both fragile and short-lived cells and many die within a few hours as evidenced by the level of radioactive label recovered in the urine [236]. By 72 h after the injection of labelled cells, often only 5–6% of the injected radioactivity remains in the whole animal yet 70% of that is in the small intestine [208]. It would seem, therefore, that the small intestine is the preferred site of localization of lymphoblasts from the gut-associated tissues, and if the cells do not arrive then few survive beyond a few hours. What is the mechanism which ensures that only the 'right' cells are abstracted from the capillary beds of the gut mucosa? Many workers have attempted to relate the 'homing' potential of mesenteric lymphoblasts to the production of IgA.

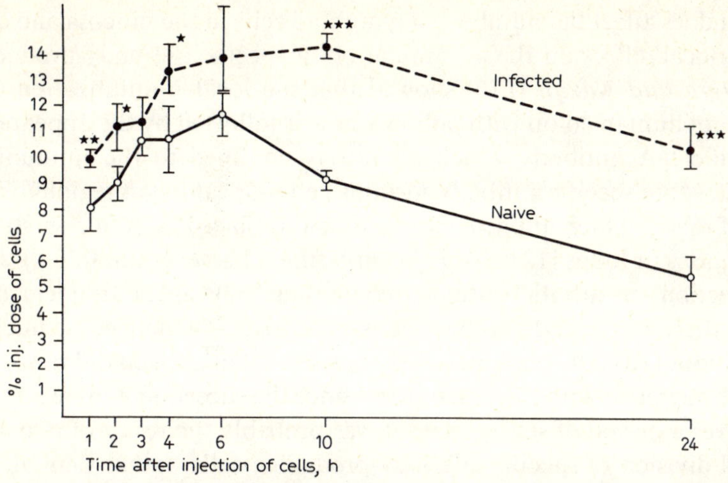

Fig. 12. Time course of the migration of (^{125}I) UdR-labelled mesenteric lymphoblasts to the intestine of naive recipient NIH mice (○) and NIH mice at 2 days after infection with *T. spiralis* (●) [data from ref. 179 and 180].

But coating of peripheral (T) lymphoblasts with IgA [207] does not facilitate their entry into the gut, and raising the level of serum IgA does not prevent mesenteric lymphoblasts from entering the gut [154]. Treatment of mice with anti-secretory piece does not support the postulate that IgA-producing cells are 'attracted' towards secretory piece as a tissue receptor [154]. And it would seem unlikely that any of these explanations could explain the 'homing' of T blasts.

5. The Role of Antigens in Mesenteric Lymphoblast Selection

The most simple hypothesis is that it is the presence of antigen in the gut lumen which attracts into the gut mucosa only those cells which have been primed in the gut-associated tissues [principally Peyer's patches] and which have had prior acquaintance with the appropriate antigenic stimulus. Indeed it is tempting to envisage mesenteric lymphoblasts as moving chemotactically down a concentration gradient towards antigen in the gut lumen. Lymphoblasts have high locomotor activity, and antigen specificity would seem to be an appropriate way of securing the arrival of mesenteric rather than peripheral blasts in the mucosa. Unfortunately examination of experimental data does not support this hypothesis.

Antigen does affect the numbers of lymphoid cells in the mucosa and does exert a 'local' effect on the accumulation of specific antibody-producing cells. *Ogra and Kazan* [176] showed that the local immunization of a segment of human colon with polio vaccine is followed by the appearance of specific IgA antibody which is entirely confined to the immunized segment. More specific antibody-forming cells appear in gut regions or gut loops of rats in which antigen cholera toxoid is placed than in non-stimulated regions or loops [121, 195]. At early times, however (up to 6 h), after the injection of specific antibody-containing cells from thoracic duct lymph, there were no differences between antigen-challenged or unchallenged loops [121]. It was concluded, therefore, that antigen did not influence the migration into the mucosa and since the subsequent increase took place over a period of several days it was probably the result of retention and cell division of specific antibody-producing cells rather than attraction of more cells into the mucosa [121]. Moreover, mesenteric blast cells can enter antigen-free gut grafts [186] and there secrete IgA into the lumen of the grafts [61]. Admittedly there are fewer such cells in grafts [62, 166] than in the normal intestine and it could well be that they are retained in the gut grafts because of inadequate lymphatic drainage [62, 186], but it does not explain why they entered in the first place although it could not have been the 'attraction' of antigen.

The distribution of IgA-producing cells along the gut tract to antigen in the gut lumen is uneven; there are far more IgA-producing cells in the duodenum than the ileum caecum or large intestine [32, 199] yet the major antigenic load of bacteria is in the distal ileum and colon. This problem has led to speculation that most of the IgA-producing cells must be directed against dietary rather than bacterial antigens [124]. However, even after oral immunization with bacteria or cholera toxin, specific IgA-producing cells [136] appear first in the duodenum and jejunum and only after multiple immunizations do the numbers of antibody-producing cells increase in the ileum and colon, although the majority of Peyer's patches (the source of the cells) are in the ileum. An answer to this problem has emerged from studies on both blast cell migration and measurement of blood flow to the gut.

6. The Role of Blood Flow in the Distribution of Lymphoblasts to the Gut

Like IgA-producing cells labelled lymphoblasts are not evenly distributed along the small intestine after intravenous injection. Far more

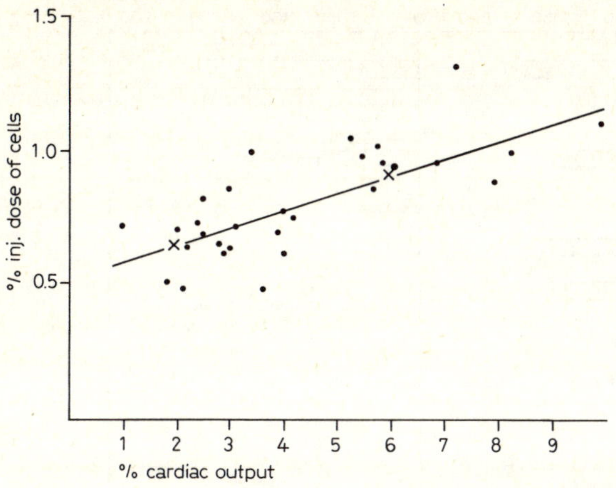

Fig. 13. The relationship between the localization of (^{125}I) UdR mesenteric lymphoblasts and the blood from (% cardiac output, CO) received by segments of the small intestine [data from ref. 187].

localize in the duodenum and jejunum than ileum, caecum or large intestine [156, 188] and this pattern is shown by both B and T blasts [156]. Moreover, this distribution pattern is apparent within an hour of injection of labelled cells and is present in mice fed elemental non-immunogenic diet [187]; it cannot, therefore, be the result of cell division per se or be the response to dietary antigen. The blood flow to the gut is also greater to the duodenum and jejunum than the ileum and this in its turn is greater than the blood flow to the caecum and large intestine [179, 180, 187] and, when blood flow and blast cell migration to gut segments are measured simultaneously, it is found that the distribution of blast cells is directly related to blood flow in the normal gut (fig. 13). From this it follows that the normal distribution of effector cells is determined by the pattern of blood flow (fig. 13) and not by antigen.

7. The Effect of Parasitic Infection in the Localization of Lymphoblasts

It has been shown that the localization of lymphoblasts is increased during with parasitic infections in *T spiralis* [206] and *N. brasiliensis* [145]. This increased localization is influenced not only by the presence of the parasite, but also by its location in the gut tract and by the stage of infection. For example, 2 days after the injection of NIH mice with *T. spiralis*

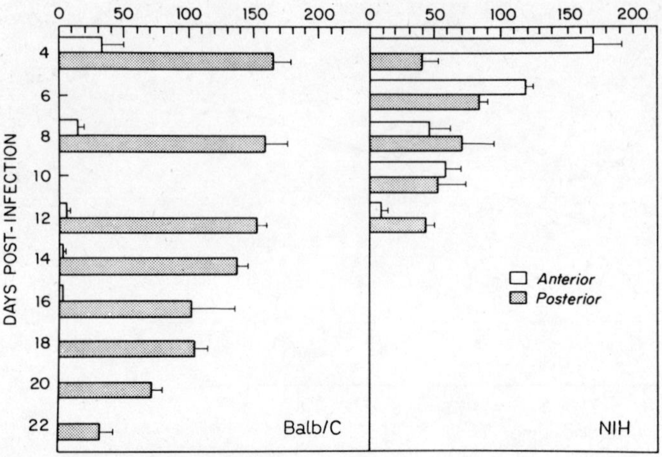

Fig. 14. The numbers of worms (*T. spiralis,* mean ±SE) recovered from the anterior and posterior halves of the intestine of NIH and Balb/C mice and various days of infection [data from ref. 156].

ova by mouth adult worms are established in the intestine. Initially, more worms are present in the anterior position of the intestine than the posterior but as the process of expulsion commences so there is a shift in location of the worm burden (fig. 14) further along the intestine. When lymphoblasts are injected into infected mice [156, 206] early after infection (2–4 days) the greatest increase in lymphoblast localization is in the anterior portion of the intestine, but later on as the worm burden shifts so does the localization of lymphoblasts. In Balb/c mice the worm burden is different from that in NIH mice in both time and location but lymphoblast localization nevertheless keeps in step with the worms (fig. 14, 15) [156].

However, the changes in localization of lymphoblasts in these experimental systems are not specific. For the change depends upon the stage of the infection of the recipient rather than on the stage of infection of the cell donors [206] or whether the cell donors are infected or not [145, 206]. Moreover, lymphoblasts from peripheral nodes [207] will also show enhanced localization in the small intestine of infected mice or rats, and will also show similar localization pattern changes during the course of infection as mesenteric lymphoblasts. The increase in blast cells in these parasitic infections cannot be explained by increased local blood flow in the intestine [179, 180] or by a local increase in vascular permeability [206]. Infection of mice with *T. spiralis* does increase blood flow [179] and

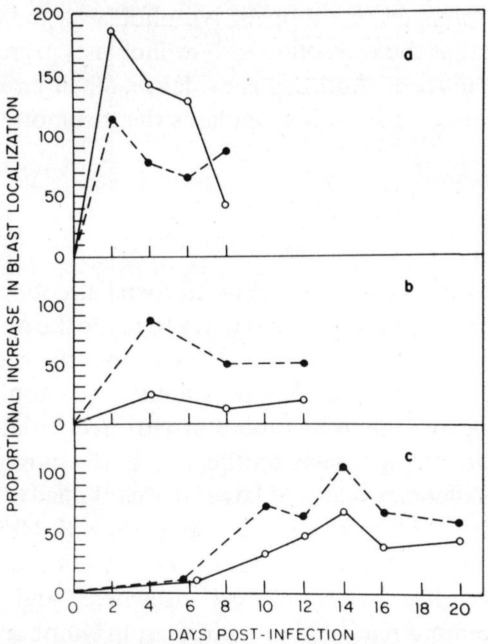

Fig. 15. Proportional increases in the 24 hour localization of (^{125}I) UdR-labelled mesenteric lymphoblasts in the anterior (○) and posterior (●) halves of the small intestine during *T. spiralis* infection. *a* NIH strain recipients of cells from donors infected 4 days previously. *b* Balb/C strain recipients of cells from donors infected 4 days previously. *c* Balb/C strain recipients of cells from donors infected 12 days previously. Proportional increase =

$$\frac{\text{\% injected radioactivity recovered in infected recipients} - \text{\% injected radioactivity recovered in uninfected recipients}}{\text{\% injected radioactivity recovered in uninfected recipients}} \times 100.$$

Data from *Manson-Smith et al.* [156].

vascular permeability [206], but does neither until 5–6 days after infection and that is after the peak increase in blast cell migration. At the time when the difference between naive and infected donors is maximum, i.e. 2–4 days after infection, the blood flow and therefore the delivery of cells to each segment of the gut is normal.

A recent time course study (fig. 12) [180] following the progress of mesenteric lymphoblasts injected into naive and *T. spiralis*-infected mice showed that the normal transit time for lymphoblasts through the mouse small intestine is around 6–8 h but in the infected gut it is much longer. Although more than half of the difference between infected and naive mice can be accounted for by prolonged retention of lymphoblasts, there is, nevertheless, clear evidence that the extraction of lymphoblasts is greater in infected than normal gut mucosa. And this is evidence, albeit circumstantial, that one form of antigenic stimulus increases the locomotion of lymphoblasts in vivo.

IV. Conclusions

One of the chief aims of this review has been to focus attention on lymphocyte locomotion as an active process, and to try to relate the migratory behaviour of lymphocytes in vivo to what is known about their locomotion in vitro. Obviously recirculation demands active locomotion; nevertheless, not all lymphocytes which locomote in vitro recirculate in vivo. Thus, lymphoblasts, though vigorously motile, are for the most part not recirculating cells. Such differences can and have been explained by the selectivity of interaction of recirculating small lymphocytes with HEV, a selectivity that is not shared by lymphoblasts. However, attention to the lymphocyte-HEV interaction has led to neglect elsewhere, and one purpose of this review is to remind readers that recirculating lymphocytes pass through many capillary beds and come into contact with endothelia in non-lymphoid organs, such as the lung, liver, etc. Different lymphocyte types have different transit times through these organs and the non-random distribution of lymphocytes may be determined as much by their behaviour in these non-lymphoid tissues as by their behaviour in lymphoid organs. Another simple mechanism for controlling lymphocyte distribution is the control of blood flow allowing or preventing access to given tissues. For example, the enhancing effect of antigen on blood flow to lymphoid tissues suggests an immunopharmacological regulation which is likely to be of considerable importance in the accumulation of lymphocytes in lymph nodes draining sites of antigenic deposition and the accumulation of effector cells in non-lymphoid sites, such as inflamed skin or the gut mucosa.

One important point that seems to emerge from all the work both in vitro and in vivo is that lymphocytes which have been stimulated or activated by antigens or mitogens show different locomotory properties from unstimulated cells. The latter cells are poorly motile in vitro but become motile on culture. Many of the motile cells in cultured or activated populations are lymphoblasts, but it is probable that post-blast small lymphocytes retain the high motility characteristic of blasts. Lymphoblasts in vivo not only have a higher affinity than small lymphocytes for other types of endothelia (as well as HEV) but they have a selective affinity for endothelia in different tissues. Lymphoblasts activated in peripheral lymph nodes do have an affinity for foci of inflammation and show a locomotor behaviour that is more like that of neutrophils than that of small lymphocytes in their predilection for inflammatory sites and their tendency to become sequestrated in pulmonary and other non-lymphoid capillary beds. Lymphoblasts activated in the lymphoid tissues draining the gut tract have an affinity for endothelia in the gut mucosa.

It may be asked how important reactions such as chemokinesis and chemotaxis are in controlling lymphocyte distribution in vivo. Since chemokinetic locomotion is influenced by molecules as ubiquitous as serum albumin, it is highly likely that the rate of active locomotion of lymphocytes within tissues is very much determined by molecules in their environment, either in free solution or bound to the surfaces across which they crawl. It is not so obvious that chemotaxis plays an important role in determining the direction in which lymphocytes move except in specialized circumstances. Other explanations are probably more plausible for explaining sorting out of lymphocytes within lymphoid tissue. The locomotor behaviour of lymphocytes in contact with other cell types, for example, their selective adhesion to lymphoid reticular cells [110] or to macrophages, and the possibility of contact guidance of lymphocyte locomotion in vivo, based on selective adhesiveness, seems to us worthy of exploration as a possible explanation of sorting out, for example, into T and B cell areas in lymph nodes. Unfortunately such studies are still quite difficult to put on a quantitative basis and in vitro models probably do not reflect the complexity of the environment in which the cell moves in living tissues.It is unlikely that lymphocytes are 'trapped' in sites of antigenic stimulation simply by chemotactic gradients. Lymphocytes do respond to such gradients in vitro; however, the response is a modest one, and it is our view at present that antigen-specific chemotaxis is probably a short-range phenomenon (such as could allow cell-cell cooperation) rather than a mechanism for calling lymphocytes out of the blood stream to foci of antigen deposition. Lymphoblasts do respond well to chemotactic gradients in vitro and it is probably no coincidence that blasts adhere to endothelia in inflammatory sites in vivo and migrate out of the blood stream at these sites whether the inflammatory sites are the site of application of contact sensitizer in the skin or infection of the gut tract with a nematode.

References

1 Alexander, E.; Henkart, P.: The adherence of human Fc receptor-bearing lymphocytes to antigen-antibody complexes. II. Morphologic alterations induced by the substrate. J. exp. Med. *143:* 329–347 (1976).

2 Allan, R.B.; Wilkinson, P.C.: A visual analysis of chemotactic and chemokinetic locomotion of human neutrophil leucocytes. Expl Cell Res. *111:* 191–203 (1978).

3 Allwood, G.G.: The migratory behaviour of T blasts to contact sensitivity reactions in actively and passively sensitized mice. Immunology *28:* 681–692 (1975).

4 Anderson, A.O.; Anderson, N.D.: Lymphocyte emigration from high endothelial venules in rat lymph nodes. Immunology 31: 731–748 (1976).

5 Anderson, N.D.; Anderson, A.O.; Wyllie, R.G.: Microvascular changes in lymph nodes draining skin allografts. Am. J. Path. *81:* 131–160 (1975).

6 Anderson, N.D.; Anderson, A.O.; Wyllie, R.G.: Specialized structure and metabolic activities of high endothelial venules in rat lymphatic tissues. Immunology *31:* 455–473 (1976).

7 Anderson, A.O.; Warren, J.T.: Lymphocyte chemotaxis under agarose (abstr.) Fed. Proc. *38:* 1429 (1979).

8 Andrews, P.; Ford, W.L.; Stoddart, R.W.: Metabolic studies of high-walled endothelium of post-capillary venules in rat lymph nodes; in O'Connor, Knight, Blood cells and vessel walls: functional interactions. Ciba Fdn Symp. *71:* 211–230 (1980).

9 Asherson, G.L.; Allwood, G.G.: Inflammatory lymphoid cells; cells in immunized lymph nodes that move to sites of inflammation. Immunology *22:* 493–502 (1972).

10 Asherson, G.L.; Allwood, G.G.; Mayhew, B.: Contact sensitivity in the mouse. XI. Movement of T blasts in the draining lymph nodes to sites of inflammation. Immunology *25:* 485–494 (1973).

11 Aström, K.E.; Webster, H.F.; Arnason, B.G.: The initial lesion in experimental allergic neuritis. J. exp. Med. *129:* 469–494 (1968).

12 Aswanikumar, S.; Corcoran, B.; Schiffmann, E.; Day, A.R.; Freer, R.J.; Showell, H.J.; Becker, E.L.; Pert, C.B.: Demonstration of a receptor on rabbit neutrophils for chemotactic peptides. Biochem. biophys. Res. Commun. *74:* 810–817 (1977).

13 Berman, L.: Lymphocytes and macrophages in vitro. Their activities in relation to functions of small lymphocytes. Lab. Invest. *15:* 1084–1099 (1966).

14 Beverley, P.C.L.: Lymphocyte heterogeneity; in Lever, Roelants, B and T cells in immune recognition, pp. 35–53 (Wiley & Sons, London 1977).

15 Biberfeld, P.: Uropod formation in phytohaemagglutinin (PHA) stimulated lymphocytes. Expl Cell Res. *66:* 433–445 (1971).

16 Biberfeld, P.; Biberfeld, G.; Perlmann, P.; Holm, G.: Cytological observations on the cytotoxic interaction between lymphocyte and antibody-coated monolayer cells. Cell. Immunol. *7:* 60–72 (1973).

17 Bienenstock, J.: The physiology of the local immune response and the gastrointestinal tract; in Brent, Holborow, Progress in immunology II, vol. 4, pp. 197–204 (North-Holland, Amsterdam 1974).

18 Bradfield, J.W.B.; Born, G.V.R.: Migration of rat thoracic duct lymphocytes through the spleen in vivo. Br. J. exp. Path. *54:* 509–517 (1973).

19 Braun, J.; Fujiwara, K.; Pollard, T.D.; Unanue, E.: Two distinct mechanisms for redis-

tribution of lymphocyte surface macromolecules. I. Relationship to cytoplasmic myosin. J. Cell Biol. *79:* 409–418 (1978).

20 Braun, J.; Sha'afi, R.I.; Unanue, E.R.: Cross-linking by ligands to surface immunoglobulin triggers mobilization of intracellular $^{45}Ca^{2+}$ in B lymphocytes. J. Cell Biol. *82:* 755–766 (1979).

21 Bull, D.M.; Bockman, M.A.: Isolation and functional characterization of human intestinal mucosal lymphoid cells. J. clin. Invest. *59:* 966–974 (1977).

22 Burwell, R.G.: Studies on primary and secondary immune responses of lymph nodes draining homografts of fresh cancellous bone (with particular reference to lymph node reactivity). Ann. N.Y. Acad. Sci. *99:* 821–860 (1962).

23 Cahill, R.N.P.; Frost, H.; Trnka, Z.: The effects of antigen on the migration of recirculating lymphocytes through single lymph nodes. J. exp. Med. *143:* 870–888 (1976).

24 Cahill, R.N.P.; Heron, I.; Poskitt, D.C.; Trnka, Z.: Lymphocyte recirculation in the sheep fetus; in O'Connor, Knight, Blood cells and vessel walls: functional interactions. Ciba Fdn Symp. *71:* 145–166 (1980).

25 Cahill, R.N.P.; Poskitt, D.C.; Frost, H.; Julius, M.H.; Trnka, Z.: Behaviour of sheep immunoglobulin-bearing and non-immunoglobulin-bearing lymphocytes isolated by nylon wool columns. Int. Archs Allergy appl. Immun. *57:* 90–96 (1978).

26 Cahill, R.N.P.; Poskitt, D.C.; Frost, H.; Trnka, Z.: Two distinct pools of recirculating T lymphocytes: migrating characteristics of modal and intestinal lymphocytes. J. exp. Med. *145:* 420–428 (1977).

27 Catanzaro, P.J.; Agniel, L.D.; Hogrefe, W.R.; Phillips, S.M.: Evidence that peritoneal lymphocytes are not related to the recirculating pool of lymphocytes. J. reticuloendoth. Soc. *23:* 459–468 (1978).

28 Cebra, J.J.; Kamat, R.; Gearhart, P.; Robertson, S.M.; Tseng, J.: The secretory IgA system of the gut. Ciba Fdn Symp. *46:* 5–28 (1977).

29 Center, D.M.; Wasserman, S.I.; Austen, K.F.: Alteration of rat splenic lymphocyte migration in vitro by the state of microtubule integrity. Cell. Immunol. *39:* 325–335 (1978).

30 Chang, T.W.; Celis, E.; Eisen, H.N.; Solomon, F.: Crawling movements of lymphocytes on and beneath fibroblasts in culture. Proc. natn. Acad. Sci. USA *76:* 2917–2921 (1979).

31 Collan, Y.: Characteristics of non-epithelial cells in the epithelium of normal rat ileum. Scand. J. Gastroent. *7:* suppl. 18, pp. 1–66 (1972).

32 Crabbé, P.A.; Heremans, J.F.: The distribution of immunoglobulin containing cells along the human gastrointestinal tract. Gastroenterology *51:* 305–316 (1966).

33 Crabbé, P.A.; Nash, D.R.; Bazin, H.; Eyssen, H.; Heremans, J.F.: Immunohistochemical observations on lymphoid tissues from conventional and germ-free mice. Lab. Invest. *22:* 448–457 (1970).

34 Craig, S.W.; Cebra, J.J.: Peyer's patches: an enriched source of precursors for IgA-producing immunocytes in the rabbit. J. exp. Med. *134:* 188–200 (1971).

35 Craig, S.W.; Cebra, J.J.: Rabbit Peyer's patches, appendix and popliteal lymph node B lymphocytes. A comparative analysis of their membrane immunoglobulin components and plasma cell precursor potential. J. Immun. *114:* 492–502 (1975).

36 Curtis, A.S.G.: Cell-cell recognition: position and patterning systems; in Curtis, Cell-cell recognition, pp. 31–82 (Cambridge University Press, London (1978).

37 Curtis, A.S.G.; de Sousa, M.A.B.: Factors influencing adhesion of lymphoid cells. Nature new Biol. *244:* 45–47 (1973).

38 Curtis, A.S.G.; de Sousa, M.A.B.: Lymphocyte interactions and positioning. I. Adhesive interactions. Cell. Immunol. *19:* 282–297 (1975).
39 Daniele, R.P.; Beacham, C.H.; Gorenberg, D.J.: The bronchoalveolar lymphocyte. Studies on the life history and lymphocyte traffic from blood to the lung. Cell. Immunol. *31:* 48–54 (1977).
40 de Bono, D.: Endothelial-lymphocyte interactions in vitro. 1. Adherence of nonallergised lymphocytes. Cell. Immunol. *26:* 78–88 (1976).
41 de Bruyn, P.P.H.: The amoeboid movement of the mammalian leukocyte in tissue culture. Anat. Rec. *95:* 177–192 (1946).
42 de Bruyn, P.P.H.; Michelson, S.; Thomas, T.B.: The migration of blood cells of the bone marrow through the sinusoidal wall. J. Morph. *133:* 417–438 (1971).
43 de Petris, S.: Preferential distribution of surface immunoglobulins on microvilli. Nature, Lond. *272:* 66–68 (1978).
44 de Sousa, M.A.B.: Reticulum arrangement related to the behaviour of cell populations in the mouse lymph node. Adv. exp. Med. Biol. *5:* 49–56 (1969).
45 de Sousa, M.: in Davies, Carter, The ecology of thymus-dependency. Contemp. Top. Immunobiol. *2:* 119–136 (1973).
46 de Sousa, M.: Microenvironment to a lymphoid cell is nothing more than interaction with its neighbours. Adv. exp. med. Biol. *66:* 165–167 (1976).
47 de Sousa, M.: Cell traffic; in Cuatrecasas, Greaves, Receptors and recognition, vol. 2, pp. 105–152 (Chapman & Hall, London 1976).
48 de Sousa, M.A.B.; Parrott, D.M.V.: Induction and recall in contact sensitivity changes in skin and draining lymph nodes of intact and thymectomized mice. J. exp. Med. *130:* 671–690 (1969).
49 de Sousa, M.; Pritchard, H.: The cellular basis of immunological recovery in nude mice after thymus grafting. Immunology *26:* 769–776 (1974).
50 de Sousa, M.A.B.; Parrott, D.M.V.; Pantelouris, E.M.: The lymphoid tissues in mice with congenital aplasia of the thymus. Clin. exp. Immunol. *4:* 637–644 (1969).
51 de Sousa, M.; Freitas, A.; Huber, B.; Cantor, H.; Boyse, E.A.: Migratory patterns of Ly subsets of T lymphocytes in the mouse. Adv. Exp. Biol. Med., pp. 51–54 (Plenum Press, New York 1979).
52 Dragsten, P.; Henkart, P.; Blumenthal, R.; Weinstein, J.; Schlessinger, J.: Lateral diffusion of surface immunoglobulin, Thy-1 antigen and a lipid probe in lymphocyte plasma membranes. Proc. natn. Acad. Sci. USA *76:* 5163–5167 (1979).
53 Dresser, D.W.A.; Taub, R.N.; Krantz, A.R.: The effect of localized injection of adjuvant material on the draining lymph node. Immunology *18:* 663–670 (1970).
54 Drexhage, H.A.; Lens, J.W.; Cvetanov, J.; Kamperdijk, E.W.A.; Mullink, H.; Balfour, B.M.: Veiled cells resembling Langerhans cells, present in lymph.; in van Furth Mononuclear phagocytes, pp. 235–272 (Nijhoff, Den Haag 1980).
55 Durkin, H.G.; Carboni, J.M.; Waksman, B.H.: Antigen-induced increase in migration of large cortical thymocytes (regulatory cells?) to the marginal zone red pulp of the spleen. J. Immun. *121:* 1075–1081 (1978).
56 Dvorak, H.F.; Mihm, M.C.; Dvorak, A.M.; Barnes, B.A.; Manseau, E.J.; Galli, S.J.: Rejection of first-set skin allografts in man. J. exp. Med. *150:* 322–337 (1979).
57 Emeson, E.: Migratory behaviour of lymphocytes with specific reactivity to alloantigens. II. Selective recruitment to lymphoid cell allografts and their draining lymph nodes. J. exp. Med. *147:* 13–24 (1978).

58 Engeset, A.; Frøland, S.S.; Bremer, K.: Studies of human peripheral lymph. II. Low lymphocyte count and few B-lymphocytes in peripheral lymph of patients with chronic lymphatic leukaemia. Scand. J. Haemat. *13:* 93–100 (1974).
59 Fahy, V.A.; Gerber, H.A.; Morris, B.; Trevella, W.; Zukoski, C.F.: The function of lymph nodes in the formulation of lymph; in Trnka, Cahill, Essays on the anatomy and physiology of lymphoid tissues. Monogr. Allergy, vol. 16, pp. 82–99 (Karger, Basel 1979).
60 Farr, A.G.; de Bruyn, P.P.H.: The mode of lymphocyte migration through post capillary venule endothelium in lymph nodes. Am. J. Anat. *143:* 59–92 (1975).
61 Ferguson, A.: Intraepithelial lymphocytes of the small intestine. Gut *18:* 921–937 (1977).
62 Ferguson, A.; Parrott, D.M.V.: The effect of antigen deprivation on thymus-dependent and thymus-independent lymphocytes in the small intestine of the mouse. Clin. exp. Immunol. *12:* 477–488 (1972).
63 Flanagan, J.; Koch, G.L.E.: Cross-linked surface Ig attaches to actin. Nature, Lond. *273:* 278–281 (1978).
64 Ford, W.L.: The kinetics of lymphocyte recirculation within the rat spleen. Cell Tiss. Kinet. *2:* 171–191 (1969).
65 Ford, W.L.; Simmonds, S.J.: The tempo of lymphocyte recirculation from blood to lymph in the rat. Cell Tiss. Kinet. *5:* 175–189 (1972).
66 Ford, W.L.; Sedgley, M.; Sparshott, S.M.; Smith, M.E.: The migration of lymphocytes across specialised endothelium. 2. The contrasting consequences of treating lymphocytes with trypsin or neuraminidase. Cell Tiss. Kinet. *9:* 351–361 (1976).
67 Ford, W.L.; Smith, M.E.; Andrews, P.: Possible clues to the mechanism underlying the selective migration of lymphocytes from the blood; in Curtis, Cell-cell recognition, pp. 359–392 (Cambridge University Press, London 1978).
68 Forsum, U.; Klareskog, L.; Tjernlund, U.M.; Peterson, P.A.: Significance of the expression of HLA-DR antigens on epidermal Langerhans' cells. Acta derm.-vener., Stockh. *58:* suppl. 79, pp. 37–43 (1978).
69 Freitas, A.A.: PhD thesis, Glasgow (1976).
70 Freitas, A.A.; de Sousa, M.: Control mechanisms of lymphocyte traffic. Modification of the traffic of ^{51}Cr-labelled mouse lymph node cells by treatment with plant lectins. Eur. J. Immunol. *5:* 831–838 (1975).
71 Freitas, A.A.; de Sousa, M.: Control mechanism of lymphocyte traffic. Altered migration of ^{51}Cr-labelled mouse lymph node cells pretreated in vitro with phospholipases. Eur. J. Immunol. *10:* 703–711 (1976).
72 Freitas, A.A.; de Sousa, M.: The role of cell interactions in the control of lymphocyte traffic. Cell. Immunol. *22:* 345–350 (1976).
73 Freitas, A.A.; de Sousa, M.: Control mechanisms of lymphocyte traffic. A study of the action of two sulphated polysaccharides on the distribution of ^{51}Cr and (^{3}H) adenosine-labelled mouse lymph node cells. Cell. Immunol. *31:* 62–76 (1977).
74 Freitas, A.A.; Rocha, B.; Chiao, J.W.; de Sousa, M.: Divergent actions of monomeric, dimeric and tetrameric concanavalin A on lymphocyte traffic. Cell. Immunol. *35:* 59–77 (1978).
75 Freitas, A.A.; Rose, M.L.; Parrott, D.M.V.: Mesenteric and peripheral lymph nodes in the mouse; a common pool of small T lymphocytes. Nature, Lond. *230:* 731–733 (1977).

76 Freitas, A.A.; Rose, M.L.; Rocha, B.: Random migration of thoracic duct T cells in the mouse. Cell. Immunol. (in press, 1980).
77 Frost, H.: The effect of antigen on the output of recirculating T and B lymphocytes from single nodes. Cell. Immunol. *37:* 390–396 (1978).
78 Frost, H.; Cahill, R.N.P.; Trnka, Z.: The migration of recirculating autologous and allogeneic lymphocytes through single lymph nodes. Eur. J. Immunol. *5:* 839–843 (1975).
79 Gabbiani, G.; Chapponier, C.; Zumbe, A.; Vassalli, P.: Actin and tubulin co-cap with surface immunoglobulins in mouse B lymphocytes. Nature, Lond. *269:* 697–698 (1977).
80 Gesner, B.M.; Ginsburg, V.: Effect of glycosidases on the fate of transfused lymphocytes. Proc. natn. Acad. Sci. USA *52:* 750–755 (1964).
81 Gillette, R.W.; Boone, C.W.; Swanson, M.H.: Unique homing properties of non-adherent peritoneal cells. Cell. Immunol. *9:* 445–452 (1973).
82 Glauert, A.M.; Sanderson, C.J.: The mechanism of K cell-(antibody-dependent) mediated cytotoxicity. III. An ultrastructural study of K cell projections and their possible role in cytotoxicity. J. Cell Sci. *35:* 355–366 (1979).
83 Godal, T.; Engeset, A.: A preliminary note on the composition of lymphocytes in human peripheral lymph. Lymphology *11:* 208–210 (1978).
84 Goldschneider, M.D.; McGregor, D.D.: Migration of lymphocytes and thymocytes in the rat. 11. Circulation of lymphocytes and thymocytes from blood to lymph. Lab. Invest. *18:* 397–406 (1968).
85 Gowans, J.L.: The effect of the continuous re-infusion of lymph and lymphocytes on the output of lymphocytes from the thoracic duct of unanaesthetized rats. Br. J. exp. Path. *38:* 67–78 (1957).
86 Gowans, J.L.: The recirculation of small lymphocytes from blood to lymph in the rat. J. Physiol., Lond. *146:* 54–69 (1959).
87 Gowans, J.L.; Knight, E.J.: The route of recirculation of lymphocytes in the rat. Proc. R. Soc. B *159:* 257–282 (1964).
88 Graham, R.C.; Shannon, S.L.: Peroxidase arthritis. II. Lymphoid cell-endothelial interactions during a developing immunologic inflammatory response. Am. J. Path. *69:* 7–24 (1972).
89 Grinnell, F.: Cellular adhesiveness and extracellular substrata. Int. Rev. Cytol. *53:* 65–144 (1978).
90 Griscelli, C.; Vassalli, P.; McCluskey, R.T.: The distribution of large dividing lymph node cells in syngeneic recipient rats after intravenous injection. J. exp. Med. *130:* 1427–1451 (1969).
91 Gupta, S.; Good, R.A.: Subpopulations of human T lymphocytes. III. Distribution and quantitation in peripheral blood, cord blood, tonsils, bone marrow, thymus, lymph nodes and spleen. Cell. Immunol. *36:* 263–270 (1978).
92 Guy-Grand, D.; Griscelli, C.; Vassalli, P.: The gut-associated lymphoid system: nature and properties of the large dividing cells. Eur. J. Immunol. *4:* 435–443 (1974).
93 Guy-Grand, D.; Griscelli, C.; Vassalli, P.: The mouse gut T lymphocyte, a novel type of T cell: nature, origin and traffic in mice, in normal and GVH conditions. J. exp. Med. *148:* 1661–1677 (1978).
94 Haemmerli, G.; Ploem, J.S.: Adhesion patterns of cell interactions revealed by reflection contrast microscopy. Expl Cell Res. *118:* 438–442 (1979).

95 Haemmerli, G.; Felix, H.; Sträuli, P.: Motility of L5222 rat leukemia cells in the flattened state. Evidence against emperopolesis. Virchows Arch. Abt. B Zellpath. *24:* 165–178 (1977).
96 Hall, B.M.; Dorsh, S.; Roser, B.: The cellular basis of allograft rejection in vivo. II. The nature of memory cells mediating second set graft rejection. J. exp. Med. *148:* 890–902 (1978).
97 Hall, J.G.: Studies of the cells in the afferent and efferent lymph of lymph nodes draining the site of skin homografts. J. exp. Med. *125:* 737–754 (1967).
98 Hall, J.G.: The effect of skin painting with oxazolone on the local extravasation of mononuclear cells in sheep; in O'Connor, Knight, Blood cells and vessel walls. Functional interactions. Ciba Fdn Symp. *71:* 197–204 (1980).
99 Hall, J.G., Morris, B.: The output of cells in lymph from the popliteal node of the sheep. Q. Jl exp. Physiol. *47:* 360-369 (1962).
100 Hall, J.G.; Morris, B.: The origin of the cells in the afferent lymph from a single lymph node. J. exp. Med. *121:* 901–910 (1965).
101 Hall, J.G.; Morris, B.: The immediate effect of antigens on the cell output of a lymph node. Br. J. exp. Path. *46:* 450–454 (1965).
102 Hall, J.G.; Smith, M.G.: Studies on the afferent and efferent lymph of lymph nodes draining the site of fluorodinitro-benzene. Immunology *21:* 69–79 (1971).
103 Hall, J.G.; Hopkins, J.; Orlans, W.: Studies on the lymphocytes of sheep. III. Destination of lymph-borne immunoblasts in relation to their tissue of origin. Eur. J. Immunol. *7:* 30–37 (1977).
104 Hall, J.G.; Morris, B.; Moreno, C.D.; Bessis, M.C.: The ultrastructure and function of the cells in lymph following antigenic stimulation. J. exp. Med. *125:* 91–110 (1967).
105 Hall, J.G.; Parry, D.M.; Smith, M.E.: The distribution and differentiation of lymph borne immunoblasts after intravenous injection into syngeneic recipients. Cell Tiss. Kinet. *5:* 269–287 (1972).
106 Hall, J.G.; Scollay, R.; Smith, M.: Studies on the lymphocytes of sheep. I. Re-circulation of lymphocytes through peripheral lymph nodes and tissues. Eur. J. Immunol. *6:* 117–120 (1976).
107 Harris, A.K.: Cell surface movements related to cell locomotion. (Ciba Fdn Symp. *14:* 3–26 (1973).
108 Harris, H.: The movement of lymphocytes. Br. J. exp. Path. *34:* 599–602 (1953).
109 Harris, H.: Role of chemotaxis in inflammation. Physiol. Rev. *34:* 529–562 (1954).
110 Haston, W.S.: A study of lymphocyte behaviour in cultures of fibroblast-like reticular cells. Cell. Immunol. *45:* 74–84 (1979).
111 Hay, J.B.; Hobbs, B.B.: The flow of blood to lymph nodes and its relation to lymphocyte traffic and the immune response. J. exp. Med. *145:* 31–44 (1977).
112 Hay, J.B.; Hobbs, B.B.; Johnston, M.G.; Movat, H.Z.: The role of hyperemia in cellular hypersensitivity reactions. Int. Archs Allergy Appl. Immun. *55:* 324–331 (1977).
113 Hay, J.B.; Johnston, M.G.; Vadas, P.; Chia, W.; Issekutz, T.; Movat, H.Z.: Relationships between changes in blood flow and lymphocyte migration induced by antigen; in Trnka, Cahill, Essays on the anatomy and physiology of lymphoid tissues. Monogr. Allergy, vol. 16, pp. 100–111 (Karger, Basel 1979).
114 Herman, P.G.: Microcirculation of organized lymphoid tissues; in Trnka, Cahill, Essays on the anatomy and physiology of lymphoid tissues. Monogr. Allergy, vol. 16, pp. 126–142 (Karger, Basel 1979).

115 Herman, P.G.; Lyonnet, D.; Fingerhut, R.; Tuttle, R.N.: Regional blood flow to the lymph node during the immune response. Lymphology *9:* 101–104 (1976).
116 Higuchi, Y.; Honda, M.; Hayashi, H.: Production of chemotactic factor for lymphocytes by neutral SH-dependent protease of PMN rabbit leukocytes from immunoglobulins especially IgM. Cell. Immunol. *15:* 100–108 (1975).
117 Hoefsmit, E.C.M.; Gerver, J.A.M.: Epithelial cells and macrophages in the normal thymus; in van Bekkum, Biological activity of thymic hormones, pp. 63–68 (Kooyker, Rotterdam 1975).
118 Hoefsmit, E.C.M.; Balfour, B.M.; Kamperdijk, E.W.A.; Cvetanov, J.: Cells containing Birbeck granules in the lymph and the lymph nodes. Functions and structure of the immune system. Adv. exp. med. Biol. *114:* 381–388 (1979).
119 Hoefsmit, E.C.M.; Kamperdijk, E.W.A.; Hendriks, H.R.; Beelen, E.W.A.; Balfour, B.M.: Lymph node macrophages; in Carr, Daems, The reticuloendothelial system, vol. I (in press, 1980).
120 Howard, J.: The life span and recirculation of marrow derived small lymphocytes from rat thoracic duct. J. exp. Med. *135:* 185–199 (1972).
121 Howard, J.G.; Hunt, S.V.; Gowans, J.L.: Identification of marrow-derived and thymus-derived small lymphocytes in the lymphoid tissues and thoracic duct lymph of normal rats. J. exp. Med. *135:* 200–219 (1972).
122 Humphrey, J.H.: Marginal zone and marginal sinus macrophages in the mouse are distinct populations. Adv. exp. med. Biol. *114:* 381–388 (1979).
123 Hurley, J.: An electron microscopic study of leucocytic emigration and vascular permeability in rat skin. Aust. J. exp. Biol. med. Sci. *41:* 171–186 (1963).
124 Husband, A.J.; Gowans, J.L.: The origin and antigen-dependent distribution of IgA containing cells in the intestine. J. exp. Med. *148:* 1146–1160 (1978).
125 Johnston, M.G.; Hay, J.B.; Movat, H.Z.: The modulation of enhanced vascular permeability by prostaglandins through alterations in blood flow (hyperaemia). Agents Actions *6:* 705–711 (1976).
126 Jungi, T.W.; McGregor, D.D.: Activated lymphocytes trigger lymphoblast extravasation. Cell. Immunol. *38:* 76–83 (1978).
127 Jungi, T.W.; McGregor, D.D.: Allogeneic restriction of the delayed inflammatory reaction in the rat. J. Immun. *121:* 456–463 (1978).
128 Jungi, T.W.; McGregor, D.D.: Role of complement in the expression of delayed-type hypersensitivity in rats: studies with cobra venom factor. Infect. Immunity *23:* 633–643 (1979).
129 Keller, H.U.; Wilkinson, P.C.; Abercrombie, M.; Becker, E.L.; Hirsch, J.G.; Miller, M.E.; Ramsey, W.S.; Zigmond, S.H.: A proposal for the definition of terms related to locomotion of leucocytes and other cells. Clin. exp. Immunol. *27:* 377–380 (1977).
130 Keller, H.U.; Wissler, J.H.; Hess, M.W.; Cottier, H.: Distinct chemokinetic and chemotactic responses in neutrophil granulocytes. Eur. J. Immunol. *8:* 1–7 (1978).
131 Kelly, R.H.; Balfour, B.M.; Armstrong, J.A .; Griffiths, S.: Functional anatomy of lymph nodes: peripheral lymph borne mononuclear cells. Anat. Rec. *190:* 5–22 (1978).
132 Koch, G.L.E.; Smith, M.J.: An association between actin and the major histocompatibility antigen H-2. Nature, Lond. *273:* 274–278 (1978).
133 Koster, F.T.; McGregor, D.D.: The mediator of cellular immunity. III. Lymphocyte traffic from the blood into the inflamed peritoneal cavity. J. exp. Med. *133:* 864–876 (1971).

134 Lackie, J.M.; Smith, R.P.C.: Interactions of leukocytes and endothelium; in Curtis, Pitts, Cell adhesion and Motility, pp. 235–272 (Cambridge University Press, London 1980).
135 Lampert, I.A.; Pizzolo, G.; Thomas, A.; Janossy, G.: Immunohistochemical characterization of cells involved in dermatopathic lymphadopathy. J. Pathol. *131:* 145–156 (1980).
136 Lange, S.; Hansson, H.; Molin, S.; Nygren, H.: Local cholera immunity in mice: intestinal antitoxin containing cells and their correlation with protective immunity. Infect. Immunity *23:* 743–750 (1979).
137 Lefford, M.J.; McGregor, D.D.: The lymphocyte mediators of delayed hypersensitivity: the early phase cells. Immunology *34:* 581–590 (1978).
138 Lewis, W.H.: Locomotion of lymphocytes. Bull. Johns Hopkins Hosp. *49:* 29–36 (1931).
139 Lewis, W.H.: The role of a superficial plasmagel layer in changes of form, locomotion and division of cells in tissue cultures. Arch. exp. Zellforsch. *23:* 1–7 (1939).
140 Lewis, W.H.; Webster, L.T.: Migration of lymphocytes in plasma cultures of human lymph nodes. J. exp. Med. *33:* 261–269 (1921).
141 Lipsky, P.E.; Rosenthal, A.S.: Macrophage lymphocyte interaction. I. Characteristics of the antigen-independent binding of guinea pig thymocytes and lymphocytes to syngeneic macrophages. J. exp. Med. *138:* 900–924 (1973).
142 Lipsky, P.E.; Rosenthal, A.S.: Macrophage-lymphocyte interaction. II. Antigen-mediated physical interactions between immune guinea pig lymph node lymphocytes and syngeneic macrophages. J. exp. Med. *141:* 138–154 (1975).
143 Loor, F.: Structure and dynamics of the lymphocyte surface in relation to differentiation recognition and activation. Prog. Allergy, vol. 23, pp. 1–153 (Karger, Basel 1977).
144 Lopez, L.R.; Vatter, A.E.; Talmage, D.W.: The requirement of viable thymocytes for species-specific attachment to and release from macrophages. J. Immun. *119:* 1668–1673 (1977).
145 Love, R.J.; Ogilvie, B.M.: *Nippostrongylus brasiliensis* and *Trichinella spiralis:* localization of lymphoblasts in the small intestine of parasitized rats. Expl Parasit. *41:* 124–132 (1977).
146 McCarty, J.; Goetzl, E.J.: Stimulation of human T lymphocyte chemokinesis by arachidonic acid. Cell. Immunol. *43:* 103–112 (1979).
147 McConnell, I.; Hopkins, J.; Lachmann, P.: Lymphocyte traffic through lymph nodes during cell shutdown; in O'Connor, Knight, Blood cells and vessel walls: functional interactions. Ciba Fdn Symp. *71:* 167–189 (1980).
148 McCutcheon, M.: Studies on the locomotion of leucocytes. III. The normal rate of locomotion of human lymphocytes in vitro. Am. J. Physiol. *69:* 279–282 (1924).
149 McCutcheon, M.: Chemotaxis in leukocytes. Physiol. Rev. *26:* 319–336 (1946).
150 McFarland, W.; Heilman, D.H.: Lymphocyte foot appendage: its role in lymphocyte function and in immunological reactions. Nature, Lond. *205:* 887–888 (1965).
151 McFarland, W.; Schechter, G.P.: The lymphocyte in immunological reactions in vitro: ultrastructural studies. Blood *35:* 683–688 (1970).
152 McFarland, W.; Heilmann, D.H.; Moorhead, J.F.: Functional anatomy of the lymphocyte in immunological reactions in vitro. J. exp. Med. *124:* 851–858 (1966).
153 McGregor, D.D.; Logie, P.S.: The mediator of cellular immunity. VII. Localization of sensitized lymphocytes in inflammatory exudates. J. exp. Med. *139:* 1415–1430 (1974).

154 McWilliams, M.; Phillips-Quagliata, J.M.; Lamm, M.E.: Characteristics of mesenteric lymph node cells homing to gut-associated lymphoid tissue in syngeneic mice. J. Immun. *115:* 54–58 (1975).

155 Manconi, P.E.; Fadda, M.F.; Cadoni, A.; Cornaglia, P.; Zaccheo, D.; Grifoni, V.: Subpopulations of T lymphocytes in human extravascular fluids. Int. Archs. Allergy appl. Immun. *56:* 385–390 (1978).

156 Manson-Smith, D.F.; Bruce, R.G.; Rose, M.L.; Parrott, D.M.V.: Migration of lymphoblasts to the small intestine. III. Strain differences and relationships to distribution and duration of *Trichinella spiralis* infection. Clin. exp. Immunol. *38:* 475–482 (1979).

157 Marchesi, V.T.; Gowans, J.L.: The migration of lymphocytes through the endothelium of venules in lymph nodes: an electron microscopy study. Proc. R. Soc. B *159:* 283–290 (1964).

158 Marsh, M.D.: Studies of intestinal lymphoid tissue. II. Aspects of proliferation and migration of epithelial lymphocytes in the small intestine of mice. Gut *16:* 674–682 (1975).

159 Marsh, M.D.: Studies of intestinal lymphoid tissue. I. Electron microscopic evidence of 'blast transformation' in epithelial lymphocytes of mouse small intestinal mucosa. Gut *16:* 665–674 (1975).

160 Marshall, W.H.; Roberts, K.B.: Continuous cinematography of human lymphocytes cultured with a phytohaemagglutinin including observations on cell division and interphase. Q. Jl exp. Physiol. *50:* 361–374 (1965).

161 Mattioli, C.A.; Tomasi, T.B.: The life span of IgA plasma cells from the mouse intestine. J. exp. Med. *138:* 452–460 (1973).

162 Meader, R.D.; Landers, D.F.: Electron and light microscopic observations on relationships between lymphocytes and intestinal epithelium. Am. J. Anat. *121:* 763–774 (1967).

163 Miller, H.R.P.; Adams, E.P.: Reassortment of lymphocytes in lymph from normal and allografted sheep. Am. J. Path. *87:* 59–80 (1977).

164 Miller, J.J.: Studies of the phylogeny and ontogeny of the specialized lymphatic tissue venules. Lab. Invest. *21:* 484–490 (1969).

165 Mitchell, J.: Lymphocyte circulation in the spleen: marginal zone bridging channels and their possible role in cell traffic. Immunology *24:* 93–107 (1973).

166 Moore, A.R.; Hall, J.: Evidence for a primary association between immunoblasts and small gut. Nature, Lond. *239:* 161–162 (1972).

167 Moore, A.R.; Hall, J.G.: Non-specific entry of thoracic duct immunoblasts into intradermal foci of antigens. Cell. Immunol. *8:* 112–119 (1973).

168 Mori, Y.; Akedo, H.; Tanigaki, Y.: Changes induced by concanavalin A in morphological and adhesive properties of suspension cells. Expl Cell. Res. *78:* 360–366 (1973).

169 Morris, B.: Effect of thymectomy on immunological responses in the sheep; Davies, Carter, Contemp. Top. Immunobiol. *2:* 39–62 (1973).

170 Morris, B.; Sass, M.D.: The formation of lymph in the ovary. Proc. R. Soc. B *164:* 577–591 (1966).

171 Nielsen, M.H.; Jensen, H.; Braendstrup, O.; Werdelin, O.: Macrophage-lymphocyte clusters in the immune response to soluble protein antigen in vitro. II. Ultrastructure of clusters formed during the early response. J. exp. Med. *140:* 1260–1272 (1974).

172 Nieuwenhuis, P.; Ford, W.L.: Comparative migration of B and T lymphocytes in the rat spleen and lymph nodes. Cell. Immunol. *23:* 254–267 (1976).

173 Nilsson, K.; Ponten, J.: Classification and biological nature of established human haemopoietic cell lines. Int. J. Cancer *15:* 321–341 (1975).

174 Norberg, B.; Rydgren, L.; Söderström, N.: Lymphocyte locomotion. Morphological criteria of the direction of lymphocyte movement. Lymphology *6:* 187–191 (1973).

175 North, R.J.: Cellular mediators of anti-Listeria immunity as an enlarged population of short-lived replicating T cells. J. exp. Med. *138:* 342–355 (1973).

176 Ogra, P.H.; Kazan, D.T.: Distribution of poliovirus antibody in the serum, nasopharynx and alimentary tract following segmental immunization of the alimentary tract with polio vaccine. J. Immun. *102:* 1423–1430 (1969).

177 O'Neill, G.J.; Parrott, D.M.V.: Locomotion of human lymphoid cells. I. Effect of culture and con A on T and non-T lymphocytes. Cell. Immunol. *33:* 257–267 (1977).

178 Ottaway, C.A.; Parrott, D.M.V.: Regional blood flow and its relationship to lymphocyte and lymphoblast traffic during a primary immune reaction. J. exp. Med. *150:* 218–230 (1979).

179 Ottaway, C.A.; Parrott, D.M.V.: The influence of regional blood flow and inflammation on the localization of lymphoblasts in the small intestine of the mouse; in Willoughby, Giroud, Inflammation: mechanisms and treatment (MTP Press, Lancaster 1980).

180 Ottaway, C.A.; Parrott, D.M.V.: Regional blood flow and the localization of lymphoblasts in the small intestine of the mouse. I. Examination of the small intestine. Immunology, in press (1980).

181 Owen, R.L.; Jones, A.L.: Epithelial cell specialization within human Peyer's patches: an ultrastructural study of intestinal lymphoid follicles. Gastroenterology *66:* 189–203 (1974).

182 Parrott, D.M.V.: The gut as a lymphoid organ. Clin. Gastroenterol. *5:* 211–228 (1976).

183 Parrott, D.M.V., de Sousa, M.A.B.: Changes in the thymus-dependent areas of lymph nodes after immunological stimulation. Nature, Lond. *212:* 1316–1317 (1966).

184 Parrott, D.M.V.; de Sousa, M.A.B.: Thymus-dependent and thymus-independent populations: origin, migratory patterns and lifespan. Clin. exp. Immunol. *8:* 663–684 (1971).

185 Parrott, D.M.V.; de Sousa, M.A.B.: B cell stimulation in nude (nunu) mice; in Rygaard, Povlsen, Proc. 1st Int. Workshop on Nude Mice, pp. 61–70 (Fischer, Stuttgart 1974).

186 Parrott, D.M.V.; Ferguson, A.: Selective migration of lymphoblasts within the small intestine. Immunology *26:* 571–588 (1974).

187 Parrott, D.M.V.; Ottaway, C.A.: The control of lymphoblast migration to the small intestine; in Ogra, Bienenstock, Mucosal immune system in health and disease. 81st Ross Conf. on Pediatric Research (in press, 1980).

188 Parrott, D.M.V.; Rose, M.L.: Migration pathways of T lymphocytes in the gut mucosa; in McGhee, Mestecky, The secretory immune system and caries Immunity, pp. 67–74 (Plenum Publishing, New York 1978).

189 Parrott, D.M.V.; de Sousa, M.A.B.; East, J.: Thymus-dependent areas in the lymphoid organs of neonatally thymectomized mice. J. exp. Med. *123:* 191–204 (1966).

190 Parrott, D.M.V.; Good, R.A.; O'Neill, G.J.; Gupta, S.: Heterogeneity of locomotion in human T cell subsets. Proc. natn. Acad. Sci. USA *75:* 2392–2395 (1978).

191 Parrott, D.M.V.; Tilney, N.L.; Sless, F.: The different migratory characteristics of lymphocyte populations from a whole spleen transplant. Clin. exp. Immunol. *19:* 459–474 (1975).

192 Pearson, L.D.; Simpson-Morgan, M.W.; Morris, B.: Lymphopoiesis and lymphocyte recirculation in the sheep foetus. J. exp. Med. *143:* 167–186 (1976).
193 Pederson, N.C.; Morris, B.: The role of the lymphatic system in the rejection of homografts: a study of lymph from renal transplants. J. exp. Med. *131:* 939–969 (1970).
194 Petroski, R.J.; Naccache, P.H.; Becker, E.L.; Sha'afi, R.I.: Effect of chemotactic factors on calcium levels of rabbit neutrophils. Am. J. Physiol. *237:* C43–C49 (1979).
195 Pierce, N.F.; Gowans, J.L.: Cellular kinetics of the intestinal immune response to cholera toxoid in rats. J. exp. Med. *142:* 1550–1563 (1975).
196 Pinkston, J.A.; Finch, S.C.; Iida, S.; Caplan, R.: An agarose plate method for the study of human T and B lymphocyte migration. Jap. J. exp. Med. *48:* 279–282 (1978).
197 Ponten, J.: Contact inhibition; in Becker, Cancer, vol. 4, pp. 55–100 (Plenum Publishing, New York 1975).
198 Porter, K.A.; Calne, R.Y.; Zukoski, C.F.: Vascular and other changes in 200 canine renal homotransplants treated with immunosuppressive drugs. Lab. Invest. *13:* 809–824 (1964).
199 Porter, P.; Noakes, D.E.; Allen, W.D.: Intestinal secretion of immunoglobulins in the pre-ruminant calf. Immunology *23:* 299–312 (1972).
200 Rannie, G.H.; Donald, K.J.: Estimation of the migration of thoracic duct lymphocytes to non-lymphoid tissues. Cell Tiss. Kinet. *10:* 523–541 (1977).
201 Rannie, G.H.; Ford, W.L.: Recirculation of lymphocytes: its role in implementing immune responses in the skin. Lymphology *11:* 193–201 (1978).
202 Rannie, G.H.; Smith, M.E.; Ford, W.L.: Lymphocyte migration into cell-mediated immune lesions is inhibited by trypsin. Nature, Lond. *267:* 520–522 (1977).
203 Reaven, E.P.; Axline, S.G.: Subplasmalemmal microfilaments and microtubules in resting and phagocytozing cultivated macrophages. J. Cell Biol. *59:* 12–27 (1973).
204 Röpke, L.; Everett, N.B.: Kinetics of intraepithelial lymphocytes in the small intestine of thymus-deprived and antigen-deprived mice. Anat. Rec. *185:* 101–108 (1976).
205 Rose, M.L.; Parrott, D.M.V.: Vascular permeability and lymphoblast extravasation into inflamed skin are not related. Cell. Immunol. *33:* 62–71 (1977).
206 Rose, M.L.; Parrott, D.M.V.; Bruce, R.G.: Migration of lymphoblasts to the small intestine. I. Effect of *Trichinella spiralis* infection on the migration of mesenteric lymphoblasts and mesenteric T lymphoblasts in syngeneic mice. Immunology *31:* 723–730 (1976).
207 Rose, M.L.; Parrott, D.M.V.; Bruce, R.G.: Migration of lymphoblasts to the small intestine. II. Divergent migration of mesenteric and peripheral immunoblasts to sites of inflammation in the mouse. Cell. Immunol. *27:* 36–46 (1976).
208 Rose, M.L.; Parrott, D.M.V.; Bruce, R.G.: The accumulation of immunoblasts in extravascular tissues including mammary gland, peritoneal cavity, gut and skin. Immunology *35:* 415–423 (1978).
209 Rosenthal, A.S.; Rosenstreich, D.L.: The lymphocyte uropod: a specialized surface suite for immunologic recognition. Biomembranes *5:* 1–24 (1974).
210 Rouse, B.T.; Wagner, H.: The in vivo activity of in vitro immunized mouse thymocytes. Transplantation *16:* 161–170 (1973).
211 Rudzik, O.; Bienenstock, J.: Isolation and characteristics of gut mucosal lymphocytes. Lab. Invest. *30:* 260–266 (1974).
212 Russell, R.J.; Wilkinson, P.C.; Sless, F.; Parrott, D.M.V.: Chemotaxis of lymphoblasts. Nature, Lond. *256:* 646–648 (1975).

213 Rydgren, L.; Norberg, B.; Håkansson, C.H.; von Mecklenburg, C.; Söderström, N.: Lymphocyte locomotion. 1. The initiation, velocity, pattern and path of locomotion in vitro. Lymphology *9:* 89–96 (1976).
214 Salvin, S.B.; Sell, S.; Nishio, J.: Activity in vitro of lymphocytes and macrophages in delayed hypersensitivity. J. Immun. *107:* 655–662 (1971).
215 Sanderson, C.J.; Glauert, A.M.: The mechanism of T cell-mediated cytotoxicity. II. Morphological studies of cell death by time-lapse microcinematography. Proc. R. Soc. Lond. B *192:* 241–255 (1977).
216 Sanderson, C.J.; Glauert, A.M.: The mechanism of T cell-mediated cytotoxicity. VI. T cell projections and their role in target cell killing. Immunology *36:* 119–129 (1979).
217 Schoefl, G.I.: The migration of lymphocytes across the vascular endothelium in lymphoid tissue. A re-examination. J. exp. Med. *136:* 568-588 (1972).
218 Schmalsteig, F.C.; Rudloff, H.B.; Goldblum, R.M.; Goldman, A.S.: The motility of human T lymphocytes. J. reticuloendoth. Soc. *20:* 331–339 (1976).
219 Scolley, R.; Hall, J.; Orlans, E.: Studies on the lymphocytes of sheep. II. Some properties of cells in various compartments of the recirculating lymphocyte pool. Eur. J. Immunol. *6:* 121–125 (1976).
220 Schreiner, G.F.; Unanue, E.R.: Anti-Ig-triggered movements of lymphocytes-specificity and lack of evidence for directional migration. J. Immun. *114:* 809–814 (1975).
221 Schreiner, G.F.; Unanue, E.R.: Membrane and cytoplasmic changes in B lymphocytes induced by ligand-surface immunoglobulin interaction. Adv. Immunol. *24:* 38–165 (1976).
222 Schreiner, G.F.; Braun, J.; Unanue, E.R.: Spontaneous redistribution of surface immunoglobulin in the motile B lymphocyte. J. exp. Med. *144:* 1683–1688 (1976).
223 Schreiner, G.F.; Fujiwara, K.; Pollard, T.D.; Unanue, E.R.: Redistribution of myosin accompanying capping of surface Ig. J. exp. Med. *145:* 1393–1398 (1977).
224 Seelig, L.L., Jr.; Holt, R.G.; Beer, A.E.: Cell kinetics and transepithelial migration of leukocytes in rat mammary epithelium during pregnancy and lactation; in Ogra, Dayton, Immunology of breast milk, pp. 159–172 (Raven Press, New York 1979).
225 Sharp, J.A.; Burwell, R.G.: Interaction 'peripolesis' of macrophages and lymphocytes after skin homografting or challenge with soluble antigens. Nature, Lond. *188:* 474–475 (1960).
226 Shields, J.M.; Wilkinson, P.C.: Relations between Fc receptor function and locomotion in human lymphocytes. Clin. exp. Immunol. *38:* 598–608 (1979).
227 Siegel, I.: Natural and antibody-induced adherence of guinea pig phagocytic cells to autologous and heterologous thymocytes. J. Immun. *105:* 879–885 (1970).
228 Silberberg-Sinakin, I.; Baer, R.L.; Rosenthal, S.A. The role of Langerhans cells in contact hypersensitivity. A review of findings in man and guinea-pig. J. invest. Derm. *66:* 210–217 (1976).
229 Smith, C.W.; Goldman, A.S.: The cells of human colostrum. I. In vitro studies of morphology and functions. Pediat. Res. *2:* 103–109 (1968).
230 Smith, J.B.; McIntosh, G.H.; Morris, B.: The traffic of cells through tissues: a study of peripheral lymph in sheep. J. Anat. *107:* 87–100 (1970).
231 Smith, J.B.; McIntosh, G.H.; Morris, B.: The migration of cells through chronically inflamed tissues. J. Path. *100:* 21–29 (1970).
232 Smith, M.E.; Martin, A.F.; Ford, W.L.: Migration of lymphoblasts in the rat. Preferential localization of DNA-synthesizing lymphocytes in particular lymph nodes and other

sites; in Trnka, Cahill, Essays on the anatomy and physiology of lymphoid tissues. Monogr. Allergy, vol. 16, pp. 203–232 (Karger, Basel 1979).

233 Smith, M.E.; Sparshot, S.M.; Ford, W.L.: The migration of lymphocytes across specialized vascular endothelium. III. Concanavalin A delays lymphocytes in normal traffic areas. Expl Cell Biol. *45:* 9–23 (1977).

234 Søeberg, B.; Sumerska, T.; Balfour, B.M.: The role of afferent lymph in the induction of contact sensitivities. Am. exp. med. Biol. *66:* 191–196 (1976).

235 Sprent, J.: Circulating T and B lymphocytes of the mouse. 1. Migratory properties. Cell. Immunol. *7:* 10–39 (1973).

236 Sprent, J.: Fate of H2-activated T lymphocytes in syngeneic hosts. 1. Fate in lymphoid tissues and intestines traced with ^{3}H-thymidine, ^{125}I-deoxyuridine and ^{51}chromium. Cell. Immunol. *21:* 278–302 (1976).

237 Sprent, J.: Migration and lifespan of lymphocytes; in Loor, Roelants, B and T cells in immune recognition, pp. 59–82 (Wiley & Sons, London 1977).

238 Spry, C.J.F.; Pflug, A.J.; Janossy, G.; Humphrey, J.H.: Large mononuclear (veiled) cells with 'Ia-like' membrane antigens in human afferent lymph. Clin. exp. Immunol. *39:* 750–755 (1980).

239 Stamper, H.B.; Woodruff, J.J.: Lymphocyte homing into lymph nodes: in vitro demonstration of the selective affinity of recirculating lymphocytes for high endothelial venules. J. exp. Med. *144:* 828–833 (1976).

240a Stamper, H.B.; Woodruff, J.J.: An in vitro model of lymphocyte homing. I. Characterization of the interaction between thoracic duct lymphocytes and specialized high-endothelial venules of lymph nodes. J. Immun. *119:* 772–780 (1977).

240b Stenman, S.; Vaheri, A.: Distribution of a major connective tissue protein, fibronectin, in normal human tissues. J. exp. Med. *147:* 1054–1064 (1978).

241 Strober, S.: Initiation of antibody responses by different classes of lymphocytes. V. Fundamental changes in the physiological characterisitics of virgin thymus-independent (B) lymphocytes and 'B' memory cells. J. exp. Med. *136:* 851–871 (1972).

242 Tilney, N.L.; Ford, W.F.: The migration of rat lymphoid cells into skin grafts. Some sensitized cells localize preferentially in specific allographs. Transplantation *17:* 12–21 (1974).

243 Toner, P.G.; Ferguson, A.: Intraepithelial cells in the human intestinal mucosa. J. Ultrastruct. Res. *34: 329–344 (1971).*

244 Trnka, Z.; Cahill, R.N.P.: Essays on the anatomy and physiology of lymphoid tissues. Monogr. Allergy, vol. 16. (Karger, Basel 1979).

245 Trnka, Z.; Cahill, R.N.P.: Aspects of the immune response in single lymph nodes; in Trnka, Cahill, Essays on the anatomy and physiology of lymphoid tissues. Monogr. Allergy, vol. 16, pp. 245–259 (Karger, Basel 1979).

246 Unanue, E.R.; Karnovsky, M.J.: Ligand-induced movement of lymphocyte membrane macromolecules. V. Capping, cell movement and microtubular function in normal and lectin-treated lymphocytes. J. exp. Med. *140:* 1207–1220 (1974).

247 Unanue, E.; Ault, K.A.; Karnovsky, M.J.: Ligand-induced movement of lymphocyte-surface macromolecules. IV. Stimulation of cell motility by anti-Ig and lack of relationship to capping. J. exp. Med. *139:* 295–312 (1973).

248 Vadas, M.A.; Miller, J.F.A.P.; Whitelaw, A.M.; Gamble, J.R.: Regulation by the H-2 gene complex of delayed type hypersensitivity. Immunogenetics *4:* 137–153 (1977).

249 van Ewijk, W.; Verzijden, J.H.M.; der Kwast, T.H. van; Luijex-Meijer, S.W.M.: Recon-

stitution of the thymus dependent area in the spleen of lethally irradiated mice. A light and electron microscopical study of the T-cell microenvironment. Cell Tiss. Res. *149:* 43–60 (1974).
250 van Ewijk, U.; Brons, N.H.; Rozing, J.: Scanning electron microscopy of homing and recirculating lymphocyte populations. Cell. Immunol. *19:* 245–261 (1975).
251 Veerman, A.J.P.: On the interdigitating cells in the thymus-dependent area of the rat spleen: a relation between the mononuclear phagocyte system and T lymphocytes. Cell Tiss. Res. *148:* 247–257 (1974).
252 Veldman, J.E.: Histophysiology and electron microscopy of the immune response; Ph D thesis, Groningen (1970).
253 Veldman, J.E.; Kenning, F.J.: Histophysiology of cellular immunity reactions in B-cell deprived rabbits. Virchows Arch. Abt B Zellpath. *28:* 203–216 (1978).
254 Waksman, B.H.: A comparitive histopathological study of delayed hypersensitivity reactions; in Wolstenholme, O'Connor, Cellular aspects of immunity. Ciba Fdn Symp. 280–329 (Churchill, London 1960).
255 Waksman, B.H.: The homing pattern of thymus-derived lymphocytes in calf and neonatal mouse Peyer's patches. J. Immun. *111:* 878–884 (1973).
256 Ward, P.A.; Offen, C.D.; Montgomery, J.R.: Chemoattractants of leukocytes, with special reference to lymphocytes. Fed. Proc. *30:* 1721–1724 (1971).
257 Ward, P.A.; Unanue, E.R.; Goralnick, S.J.; Schreiner, G.F.: Chemotaxis of rat lymphocytes. J. Immun. *119:* 416–421 (1977).
258 Weisberger, A.S.; Guyton, R.A.; Heinle, R.W.; Storaali, J.P.: The role of the lungs in removal of transfused lymphocytes. Blood *6:* 916–925 (1952).
259 Werdelin, O.; Braendstrup, O.; Pedersen, E.: Macrophage-lymphocyte clusters in the immune response to soluble protein antigen in vitro. I. Role of lymphocytes and macrophages in cluster formation. J. exp. Med. *140:* 1245–1259 (1974).
260 Wilkinson, P.C.; Allan, R.B.: Chemotaxis of neutrophil leucocytes towards substratum-bound protein attractants. Expl. Cell Res. *117:* 403–412 (1978).
261 Wilkinson, P.C.; Allan, R.B.: Assay systems for measuring leukocyte locomotion: an Overview; in Gallin, Quie, Leukocyte chemotaxis, pp. 1-24 (Raven Press, New York 1978).
262 Wilkinson, P.C.; Allan, R.B.: The locomotor behaviour of human blood monocytes in chemotactic and chemokinetic environments and the role of the substratum in monocyte locomotion; in van Furth, Mononuclear phagocytes. Functional aspects, pp. 475-500 (Nijhoff, Den Haag 1980).
263 Wilkinson, P.C.; Parrott, D.M.V.; Russell, R.J.; Sless, F.: Antigen-induced locomotor responses in lymphocytes. J. exp. Med. *145:* 1158–1168 (1975).
264 Wilkinson, P.C.; Roberts, J.A.; Russell, R.J.; McLoughlin, M.: Chemotaxis of mitogen-activated human lymphocytes and the effects of membrane-active enzymes. Clin. exp. Immunol. *25:* 280–287 (1976).
265 Wong, S.Y.; Longenecker, B.M.; Pazderka, F.; Ruth, R.F.: Immobilisation of lymphocytes at surfaces by lectins. Expl. Cell Res. *92:* 428–434 (1975).
266 Woodruff, J.J.: Role of lymphocyte surface determinants in lymph node homing. Cell. Immunol. *13:* 378–384 (1974).
267 Woodruff, J.J.; Kuttner, B.J.: In vitro adherence of lymphocytes to the high endothelium of lymph nodes. Ciba Fdn Symp. *71:* 243–256 (1980).
268 Woodruff, J.F.; Woodruff, J.J.: Virus-induced alterations of lymphoid tissues. IV. The

effect of Newcastle disease virus on the fate of radiolabelled thoracic duct lymphocytes. Cell. Immunol. *10:* 78–85 (1974).
269 Woodruff, J.J.; Woodruff, J.F.: Influenza A virus interactions with murine lymphocytes. I. The influence of influenza virus A/Japan 305 (H2N2) on the pattern of migration of recirculating lymphocytes. J. Immun. *117:* 852–858 (1976).
270 Woodruff, J.J.; Katz, I.M.; Lucas L.E.; Stamper, H.B.: An in vitro model of lymphocyte homing. II. Membrane and cytoplasmic events involved in lymphocyte adherence to specialized high endothelial venules of lymph nodes. J. Immun. *119:* 1603–1610 (1977).
271 Zatz, M.M.; Lance, E.M.: The distribution of ^{51}Cr-labelled lymphocytes in antigen stimulated mice. Lymphocyte trapping. J. exp. Med. *134:* 224–241 (1971).
272 Zigmond, S.H.: Mechanisms of sensing chemical gradients by polymorphonuclear leukocytes. Nature, Lond. *249:* 450–452 (1974).
273 Zigmond, S.H.: Ability of polymorphonuclear leukocytes to orient in gradients of chemotactic factors. J. Cell Biol. *75:* 606–616 (1977).
274 Zigmond, S.H.: Gradients of chemotactic factors in various assay systems; in van Furth, Mononuclear phagocytes, functional aspects, pp. 461–473 (Nijhoff, The Hague 1980).
275 Zigmond, S.H.; Hirsch, J.G.: Leukocyte locomotion and chemotaxis. New methods for evaluation, and demonstration of a cell-derived chemotactic factor. J. exp. Med. *137:* 387–410 (1973).
276 Zigmond, S.H.; Sullivan, S.J.: Sensory adaptation of leukocytes to chemotactic peptides. J. Cell Biol. *82:* 517–527 (1979).

Delphine M.V. Parrott, Department of Bacteriology and Immunology,
Western Infirmary, Glasgow G11 6NT (Scotland)